Notes on Numerical Fluid Mechanics and Multidisciplinary Design

Volume 141

Notes on Numerical Fluid Mechanics and Multidisciplinary Design publishes state-of-art methods (including high performance methods) for numerical fluid mechanics, numerical simulation and multidisciplinary design optimization. The series includes proceedings of specialized conferences and workshops, as well as relevant project reports and monographs.

More information about this series at http://www.springer.com/series/4629

Rudibert King
Editor

Active Flow and Combustion Control 2018

Papers Contributed to the Conference "Active Flow and Combustion Control 2018", September 19–21, 2018, Berlin, Germany

Springer

Editor
Rudibert King
Chair of Measurement and Control
Institute of Process and Plant Technology
Berlin, Germany

ISSN 1612-2909 ISSN 1860-0824 (electronic)
Notes on Numerical Fluid Mechanics and Multidisciplinary Design
ISBN 978-3-030-07464-7 ISBN 978-3-319-98177-2 (eBook)
https://doi.org/10.1007/978-3-319-98177-2

This Springer imprint is published by the registered company Springer Nature Switzerland AG
The registered company address is: Gewerbestrasse 11, 6330 Cham, Switzerland

Preface

The ability to manipulate flow fields started more than 100 years ago when Prandtl presented his concept of boundary layer control in the year 1904. Meanwhile, open-loop control and closed-loop control of flow instabilities let to what is known today as active flow control (AFC). AFC has the potential to save huge amounts of costs for land, air, and sea vehicles by reducing drag or increasing lift. Noise emitted by flow engines can be mitigated by AFC, mixing in reaction systems can be improved, or unsteady cooling concepts can be exploited to reduce costly cooling fluids to name just a few additional areas of application. AFC is inherently interdisciplinary needing expertise at least from experimental, theoretical and numerical fluid mechanics, acoustics, metrology, mathematics, and control theory. In the year 1998, this led to the creation of a collaborative research center CRC 557 Control of turbulent shear flows at Technische Universität Berlin, which was funded by the Deutsche Forschungsgemeinschaft (DFG) for 12 years. This CRC organized the first conference on Active flow control in the year 2006, followed by Active flow control II in 2010. In contrast to many other meetings, the interdisciplinary discussion was stimulated by the avoidance of parallel sessions. Invited lectures allowed the International Program Committee to set a clear focus and to guarantee high-quality contributions. Besides talks coming from internationally renowned experts, the local CRC presented the actual results.

Some of the projects of the CRC 557 were already devoted to the mitigation of thermo-acoustic instabilities that might occur in burners of turbomachines. Combing new ideas and local expertise of AFC and combustion control ended in the formulation of a new CRC proposal. This CRC 1029 Substantial efficiency increase in gas turbines through direct use of coupled unsteady combustion and flow dynamics was granted by the DFG in 2012 for a first 4-year period. It allowed the CRC 1029 to announce a follow-up conference on Active flow and Combustion control 2014. Resulting from the new challenges faced by the CRC, the scope of the conference was extended to combustion control as well, and the name was adopted accordingly. Meanwhile, the CRC is in its second funding period organizing Active flow and Combustion control 2018, for which the support by the DFG is gratefully acknowledged. The successful format of the conference was unchanged with invited lectures and single-track sessions only. Not all presenters could prepare a manuscript for this volume, but it still presents a well-balanced combination of theoretical, numerical, and experimental state-of-the-art results of active flow and combustion control.

As in the former conferences, experimental results of AFC applied to flight applications, theoretical investigations of AFC, actuators, and model reduction were the main topics of the meeting. These were complemented by contributions resulting from the scope of the CRC 1029, for which AFC and combustion control will have to be highly interlinked. The focus of CRC 1029's research is the increase of the efficiency of a gas turbine by more than 10% by the exploitation and control of innovative combustion concepts and unsteady characteristics of a machine. The major contribution to an efficiency increase is expected from a thermodynamically motivated change from a constant pressure to a constant volume combustion. Besides the more classical pulsed detonation combustion, a shockless explosion concept was proposed in the CRC. In the meantime, this portfolio of constant volume combustion schemes is extended by the rotating detonation combustion, for which new results are described in this volume as well. As all combustion concepts produce highly dynamic boundary conditions for the remaining parts of a turbomachine, an efficiency increase will only be possible if these unsteady effects can be controlled. It is the vision of the CRC 1029 that this is possible by AFC applied in the compressor, turbine, or in the cooling system. First results are presented here as well.

All papers in this volume have been subjected to an international review process. We would like to express our sincere gratitude to all involved reviewers, to the International Program Committee, and to DFG for supporting this conference.

Finally, the members of CRC 1029 are indebted to their respective hosting organizations, TU Berlin and FU Berlin, for the continuous support, and to Springer and the editor of the series Notes on Numerical Fluid Mechanics and Multidisciplinary Design, W. Schröder, for handling this volume.

Berlin, Germany
June 2018

Rudibert King
Chairman of AFCC 2018 and CRC 1029

Contents

Part I Active Flow Control

Sparse Model of the Lift Gains of a Circulation Control Wing with Unsteady Coanda Blowing . . . 3
Richard Semaan, M. Yosef El Sayed and Rolf Radespiel

Unsteady Roll Moment Control Using Active Flow Control on a Delta Wing . . . 19
Xiaowei He, Mathieu Le Provost, Xuanhong An and David R. Williams

Implementing Rotating Stall Control in a Radial Diffuser Using Microjet Arrays . . . 33
Jennifer Gavin, Erik Fernandez, Prabu Sellappan, Farrukh S. Alvi, William M Bilbow and Sun Lin Xiang

High Frequency Boundary Layer Actuation by Fluidic Oscillators at High Speed Test Conditions . . . 53
Valentin Bettrich, Martin Bitter and Reinhard Niehuis

Model Predictive Control of Ginzburg-Landau Equation . . . 75
Mojtaba Izadi, Charles R. Koch and Stevan S. Dubljevic

A Qualitative Comparison of Unsteady Operated Compressor Stator Cascades with Active Flow Control . . . 91
Marcel Staats, Jan Mihalyovics and Dieter Peitsch

Transitioning Plasma Actuators to Flight Applications . . . 105
David Greenblatt, David Keisar and David Hasin

Part II Combustion Control

Effect of the Switching Times on the Operating Behavior of a Shockless Explosion Combustor . . . 121
Fatma C. Yücel, Fabian Völzke and Christian O. Paschereit

Part Load Control for a Shockless Explosion Combustion Cycle 135
Florian Arnold, Giordana Tornow and Rudibert King

Knock Control in Shockless Explosion Combustion by Extension of Excitation Time ... 151
Lisa Zander, Giordana Tornow, Rupert Klein and Neda Djordjevic

Reduced Order Modeling for Multi-scale Control of Low Temperature Combustion Engines ... 167
Eugen Nuss, Dennis Ritter, Maximilian Wick, Jakob Andert, Dirk Abel and Thivaharan Albin

Part III Constant Volume Combustion

The Influence of the Initial Temperature on DDT Characteristics in a Valveless PDC ... 185
Fabian E. Völzke, Fatma C. Yücel, Joshua A. T. Gray, Niclas Hanraths, Christian O. Paschereit and Jonas P. Moeck

Types of Low Frequency Instabilities in Rotating Detonation Combustors ... 197
Vijay Anand and Ephraim Gutmark

Influence of Operating Conditions and Residual Burned Gas Properties on Cyclic Operation of Constant-Volume Combustion 215
Quentin Michalski, Bastien Boust and Marc Bellenoue

Part IV Data Assimilation and Model Reduction

Validation of Under-Resolved Numerical Simulations of the PDC Exhaust Flow Based on High Speed Schlieren ... 237
M. Nadolski, M. Rezay Haghdoost, J. A. T. Gray, D. Edgington-Mitchell, K. Oberleithner and R. Klein

On the Loewner Framework for Model Reduction of Burgers' Equation ... 255
Athanasios C. Antoulas, Ion Victor Gosea and Matthias Heinkenschloss

Model Reduction for a Pulsed Detonation Combuster via Shifted Proper Orthogonal Decomposition ... 271
Philipp Schulze, Julius Reiss and Volker Mehrmann

Part V Numerical Aspects in Combustion

Control of Condensed-Phase Explosive Behaviour by Means of Cavities and Solid Particles ... 289
Louisa Michael and Nikolaos Nikiforakis

An Open and Parallel Multiresolution Framework Using Block-Based Adaptive Grids 305
Mario Sroka, Thomas Engels, Philipp Krah, Sophie Mutzel, Kai Schneider and Julius Reiss

A 1D Multi-Tube Code for the Shockless Explosion Combustion 321
Giordana Tornow and Rupert Klein

Part VI Unsteady Cooling

Experimental Study on the Alteration of Cooling Effectivity Through Excitation-Frequency Variation Within an Impingement Jet Array with Side-Wall Induced Crossflow 339
Arne Berthold and Frank Haucke

Effects of Wall Curvature on the Dynamics of an Impinging Jet and Resulting Heat Transfer 355
G. Camerlengo, D. Borello, A. Salvagni and J. Sesterhenn

Map Estimation for Impingement Cooling with a Fast Extremum Seeking Algorithm 367
Benjamin Fietzke, Matthias Kiesner, Arne Berthold, Frank Haucke and Rudibert King

Index 379

Part I
Active Flow Control

Sparse Model of the Lift Gains of a Circulation Control Wing with Unsteady Coanda Blowing

Richard Semaan, M. Yosef El Sayed and Rolf Radespiel

Abstract The present study investigates and models the lift gains and losses generated by the superposition of a periodic actuation component onto a steady component on an airfoil with a highly deflected Coanda flap. The periodic actuation is provided by two synchronized specially-designed valves that deliver actuation frequencies up to 30 Hz and actuation amplitudes up to 20% of the mean blowing intensity. The lift gains/losses response surface is modeled using a data-driven sparse identification approach. The results clearly demonstrate the benefits of superimposing a periodic component onto the steady actuation component for a separated or partially-attached flow, where up to $\Delta C_l = 0.47$ lift increase is achieved. On the other hand, this same superimposition for an attached flow is detrimental to the lift, with up to $\Delta C_l = -0.3$ lift reduction compared to steady actuation with similar blowing intensity is observed.

Keywords High-lift · Sparse modeling · Coanda actuation

1 Introduction

Active flow control (AFC) with periodic actuation has been proven to bring aerodynamic benefits (e.g. Greenblatt and Wygnanski [1]; Barros et al. [2]; and Chabert et al. [3]). These benefits stem from the power reductions obtained compared to steady actuation. The success of periodic actuation is rooted in its exploitation of the flow

R. Semaan (✉) · M. Y. El Sayed · R. Radespiel
Technische Universität Carolo-Wilhelmina zu Braunschweig,
Hermann-Blenk-Str. 37, 38108 Braunschweig, Germany
e-mail: r.semaan@tu-bs.de
URL: https://www.tu-braunschweig.de/ism

R. King (ed.), *Active Flow and Combustion Control 2018*,
Notes on Numerical Fluid Mechanics and Multidisciplinary Design 141,
https://doi.org/10.1007/978-3-319-98177-2_1

instabilities, contrary to steady actuation which attempts to alter the flow topology by momentum injection.

For aircraft, circulation control in combination with high-lift devices offers several advantages over traditional high-lift configurations. The basic concept of circulation control involves the Coanda principle, where energy is introduced into the flow by means of a thin jet ejected tangentially from a slit near the trailing edge. The main advantage of circulation control is an increased lift output, which makes shorter take-offs and landings possible. This technology was first patented by Davidson [4] in 1960 and has since been repeatedly investigated (Lachmann [5]; Wood and Nielson [6]; and Englar [7]). A circulation control wing (CCW) with steady jets, even at very small mass flow rates, has been shown to yield lift coefficients that are comparable or superior to conventional high-lift systems (Sexstone et al. [8]; Smith [9]). A particular variant of circulation control is the Coanda flap, where the objective is to keep the flow attached over a highly deflected flap by blowing a jet tangentially over its specially designed upper surface. This concept has been previously investigated and geometrically optimized in several previous studies [10–12].

Efficiency requirements demand that the lift gained through the use of circulation control be as large as possible in comparison to the momentum coefficient of the blown jet, which is usually acquired by engine bleed. This ratio is referred to as the lift gain factor. An increase in the lift gain factor can be achieved through periodic blowing. Two studies during the mid-1970s investigated pulsed blowing associated with circulation control (Oyler and Palmer [13]; Walters et al. [14]). Results from these experiments indicated that pulsed blowing reduced the mass requirements for CCW. More recently, periodic blowing on a circulation control wing with circular trailing edge was examined (Jones et al. [15]), where a 50% reduction in the required mass flow for a required lift coefficient was achieved. It is worth to note that the benefits of periodic excitation targeting flow instabilities have also been demonstrated in other flow control applications, such as pulsed actuation over a flap (e.g. Petz and Nitsche [18]) and acoustic excitation (e.g. Greenblatt and Wygnanski [1]; Seifert et al. [19]).

In this study, we investigate and model the lift gains/losses from superimposed steady and periodic Coanda actuation. The investigation is based on experimental measurements in the large water tunnel facility at the Technische Universität Braunschweig using the configuration studied by Burnazzi and Radespiel [20] as part of the Coordinated Research Centre 880 (CRC 880). The periodic actuation is achieved using two specially-designed high-speed proportional valves. The results demonstrate the benefits of superimposing a periodic actuation component onto the steady one, where up to $\Delta C_l = 0.47$ lift increase compared to the corresponding steady blowing is observed.

2 Experimental Setup

2.1 Model

The experimental 2D model is a modified DLR-F15 airfoil with a $c = 300$ mm chord length and a 1 m span. It features a highly deflected Coanda-flap and a drooped leading edge, as seen in Fig. 1. The Coanda-flap has a length of $c_{fl} = 0.25 \cdot c$ (i.e. $c_{fl} =$ 75 mm) and is deflected by 65° for a landing configuration. The Coanda jet is blown over the flap shoulder through a $0.00067 \cdot c$ slit following design considerations presented in Radespiel et al. [12]. The jet is supplied through a plenum inside the model, which is connected on both sides of the model to a KSB Movitec VF 32-7 PD multi-stage inline-pump installed outside the tunnel that delivers flow rates up to 10 l/s at 8 bar pressure.

The droop nose shape was reached through a parametric study that maximized the lift [20]. The geometry is morphed from the clean nose by deflecting the leading edge down by 90° over a length of 0.2 c and increasing the leading edge thickness by 60%. The reference coordinate system, shown in Fig. 1, is that of the clean airfoil, where the leading edge coincides with the origin. Henceforth, all subsequent dimensions are with respect to this reference coordinate system.

2.2 Facilities

The experiment is carried out in the large water tunnel facility (Großer Wasserkanal Braunschweig (GWB)) at the Technische Universität Braunschweig [21] (Fig. 1b). The facility is a Göttingen-type closed return tunnel, with a 6 m long and 1 m ×1 m test section. The flow is driven by a 1.5 m diameter one-stage axial pump powered by a variable frequency drive 160 kW electric motor, yielding flow velocities up to 6 m/s in the test section. To inhibit cavitation, the tunnel can be pressurized up to 2 bar above ambient pressure.

The choice of a water tunnel is based on time scale considerations. Due to the much smaller kinematic viscosity of water compared to air (about 1/16th that of air

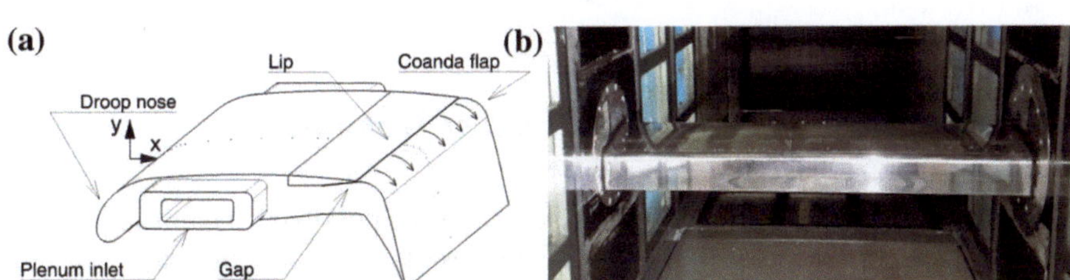

Fig. 1 **a** Schematic and **b** a picture of the experimental model installed in the water tunnel

at room temperature), the same Reynolds number for a given model size is reached at a fraction of the free stream velocity. Higher Reynolds numbers can be achieved by heating the water (up to 40 °C are possible in the GWB). Therefore, it is possible in a water tunnel to capture a considerably larger portion of the flow dynamics than in air at similar Reynolds number, since flow phenomena happen on larger time scales. The larger time scales also make water tunnels better suited for closed-loop flow control experiments, which would be the focus of our future efforts.

Although Reynolds numbers of up to $\mathrm{Re} = \frac{V_\infty c}{\nu_\infty} = 2.5 \times 10^6$ are possible, the Reynolds number in the current study is limited to 2.0×10^6 at 4.5 m/s due to concerns about exceeding the load limit of the model. The subscript ∞ denotes freestream conditions.

2.3 *High-Speed Proportional Valves*

Due to the lack of commercially-available high-frequency proportional valves for water, two custom-made valves are purposely constructed (see Fig. 2). The valves build on the body of Danfoss EV210B, where the electromagnetic coils are lengthened and the core tubes are extended and modified. To keep the valves from overheating, each electromagnetic coil is wrapped in a water cooling coil, which in turn is encased in thermally conductive epoxy. The valves allow high flow rates of up to 3 l/s and actuation frequencies of up to 30 Hz. The flow rate fluctuation levels are however dependent on the actuation frequency. The blowing intensity amplitude decreases with frequency as the core tube inertia becomes increasingly difficult to overcome. This amplitude drop can be compensated (to some extent) by increasing the power. The valves are controlled by a special electronic circuit, illustrated in Fig. 2. A signal produced by a signal generator is sent to a pulse width modulation (PWM) unit and then to a solid state relay. This electronic circuit is a cheaper alternative to using massive amplifiers, albeit at the cost of restricting the signal to a rectangular train type. An adjustable (0 to 60 V) voltage power source provides the necessary power. To minimize hydraulic dampening, the two valves are attached on either side of the airfoil model close to both plenum inlets (see Fig. 1). Two Prandtl probes installed between the valves and the model plenum are used to measure the flowrate and thus the actuation intensity in real time.

2.4 *Instrumentation and Measurement Technique*

Since the main objective of this work is to examine the possible lift gains brought by superimposing periodic actuation, all presented results are surface pressure measurements and corresponding lift coefficients. The pressure distributions are measured using 64 pressure taps along the mid-span section connected to high precision (0.1% FS error) Keller PD-X33 pressure transducers. Each ensemble is sampled at 100 Hz

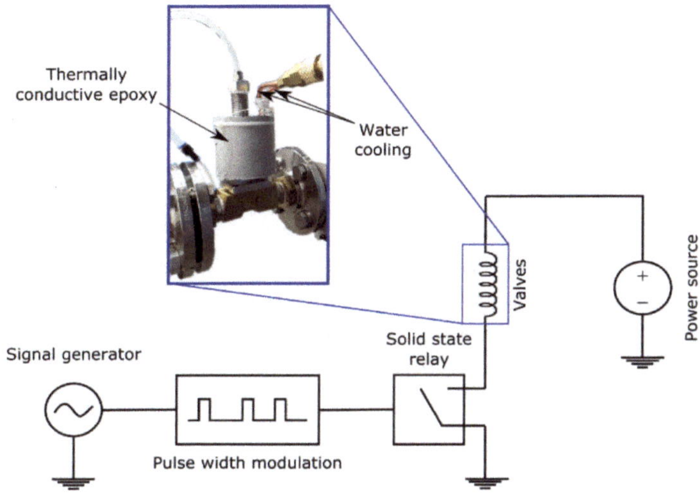

Fig. 2 Schematic of the electronic circuit controlling the two in-house constructed proportional valves, along with a picture of one of the valves

for 5 s. Besides the pressure taps, the model is instrumented with seven real-time miniature piezo-resistive pressure Keller sensors with 300 kHz possible eigenfrequency that are flush mounted onto the Coanda flap. These real-time pressure sensors will provide the input for the upcoming closed-loop control measurement campaign.

2.5 *Test Cases*

Periodic actuation is characterized by three parameters, the actuation frequency, the blowing intensity, and the blowing amplitude. The actuation frequency f^a is usually presented in its non-dimensional form F^+, defined as [1]

$$F^+ = \frac{f^a \, c_{fl}}{V_\infty}, \tag{1}$$

where c_{fl} is the flap chord length. In this study, the actuation frequency can reach a maximum of 1 at $\mathrm{Re} = 1 \times 10^6$, imposed by the valves mechanical limitations. The blowing intensity of a circulation control wing is usually characterized by the non-dimensional momentum coefficient c_μ [17]. The momentum coefficient was first introduced by Poisson-Quinton and Lepage [22] as the blown jet thrust normalized by the product of the free stream dynamic pressure and a reference area

$$c_\mu = \frac{\rho_{jet} \, V_{jet}^2 \, h}{\frac{1}{2}\rho_\infty \, V_\infty^2 \, S_{ref}}, \tag{2}$$

where V_{jet} is the jet nozzle average exit velocity, h is the slit height, and S_{ref} is the projected clean wing area. The momentum coefficient, in this case, contains both the mean and the periodic component. For sinusoidal actuation, the momentum coefficient can be expressed as

$$c_{\mu} = C_{\mu} + C_{\mu 1}\, \cos(2\pi f^{a}\, t) = C_{\mu} + C_{\mu 1}\, \cos(\frac{2\pi V_{\infty}}{c_{fl}} F^{+}\, t)\,, \tag{3}$$

where C_{μ} is the mean steady component, and $C_{\mu 1}$ is the oscillation amplitude.

The measurements are acquired for a range of momentum coefficients, starting from the natural unactuated case $C_{\mu} = 0$ and up to $C_{\mu} = 0.06$, where the flow is fully attached to the flap. Measurements are conducted for three Reynolds numbers $\mathrm{Re} = 1 \times 10^{6}$, 1.5×10^{6}, and 2.0×10^{6} and for angles of attack up to 22°.

3 Methodology

3.1 *Identification of Sinusoidal Actuation*

The type of actuation signal (e.g. sinusoidal, square, etc) can affect the flow's aerodynamic response. With a square wave voltage input from the PWM, the custom-built valves are designed to deliver sinusoidal forcing. This sinusoidal flow results from viscous dampening effects of the flow, and from the valve mechanism itself: the coil cannot react instantly to the square voltage input, as it requires a finite time to build a strong-enough magnetic field to push the rod. The valves operate in fully-open mode by default. When the magnetic coils are energized, the core rods are pushed downward to constrict the flow. The closing force is dependent on the actuation frequency (rod inertia) and on the pressure acting against the rod (blowing intensity). For certain actuation frequency/pressure settings, the rod/flow interactions yield non-sinusoidal flow rates (and hence momentum coefficients). Figures 3a and 4a show one sinusoidal and one non-sinusoidal time trace of the momentum coefficient, respectively. Non-sinusoidal forcing occurs in $\approx$13% of the total number of test cases, and yield a different aerodynamic response than that of the sinusoidal one. This motivated the identification and deletion of non-sinusoidal actuation signals. The procedure involves four steps:

1. Fourier transform the momentum coefficient trace, $\mathcal{F}(c_{\mu})$ (e.g. Figs. 3b and 4b).
2. Identify the peaks in the Fourier transform.
3. Compute the ratio $\mathfrak{R}$ between the highest peak and the second highest *non-harmonic* peak.
4. Enforce a $\mathfrak{R} > 5$ threshold for sinusoidal classification.

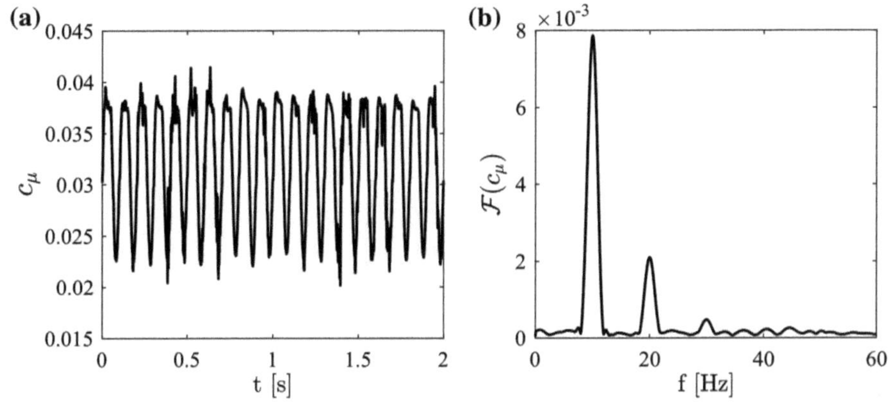

Fig. 3 **a** A sinusoidal time trace of the momentum coefficient, and **b** its corresponding Fourier transform

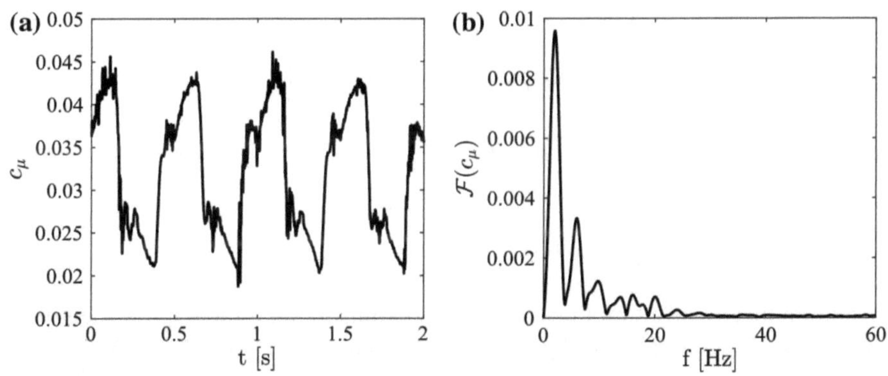

Fig. 4 **a** A non-sinusoidal time trace of the momentum coefficient, and **b** its corresponding Fourier transform

All momentum coefficient signals with $\mathfrak{R} < 5$ are thus considered non-sinusoidal and are removed. The threshold is simply reached by trial and error. The Fourier transformation peak also served to identify the actuation amplitude $C_{\mu 1}$ in Eq. (3).

3.2 *Sparse Modeling of the Lift Gains/Losses*

This section describes the methodology to identify a sparse nonlinear model of the lift gains/losses. The motivation behind promoting model sparsity is to reduce complexity and to avoid overfitting. The current approach is comparable to the sparse identification of nonlinear dynamical systems (SINDY) method [23], but for a response surface. Specifically, we are concerned with identifying a sparse nonlinear model for the lift gains/losses based on the experimentally-measured data for a range of actuation parameters C_μ, $C_{\mu 1}$ and F^+ (refer to Eq. 3). In other words, we seek to identify the following model

$$\Delta \mathbf{C_l} = \Delta \mathbf{C_l}(\mathbf{X}) \,, \tag{4}$$

where,

$$\Delta \mathbf{C_l} = \begin{bmatrix} \Delta C_{l,1} \\ \Delta C_{l,2} \\ \vdots \\ \Delta C_{l,m} \end{bmatrix} \tag{5}$$

are the lift gains/losses for m test cases, and

$$\mathbf{X} = \begin{bmatrix} x_1 \\ x_2 \\ \vdots \\ x_m \end{bmatrix} = \begin{bmatrix} \hat{C}_{\mu,1} & \hat{C}_{\mu 1,1} & \hat{F}_1^+ \\ \hat{C}_{\mu,2} & \hat{C}_{\mu 1,2} & \hat{F}_2^+ \\ \vdots & \vdots & \vdots \\ \hat{C}_{\mu,m} & \hat{C}_{\mu 1,m} & \hat{F}_m^+ \end{bmatrix} , \tag{6}$$

are the normalized actuation settings, $\hat{C}_\mu = \frac{C_\mu}{\max\{C_\mu\}}$, $\hat{C}_{\mu 1} = \frac{C_{\mu 1}}{\max\{C_{\mu 1}\}}$, $\hat{F}^+ = \frac{F^+}{\max\{F^+\}}$. The normalization ensures that

$$-1 \le \hat{C}_\mu, \hat{C}_{\mu 1}, \hat{F}^+ \le 1 \,,$$

a condition that greatly simplifies the sparse optimization problem in the identification process. Modeling begins by constructing an augmented library $\Theta(\mathbf{X})$ consisting of candidate nonlinear functions of the columns of $\mathbf{X}$. For example, $\Theta(\mathbf{X})$ may consist of constant and polynomial terms:

$$\Theta(\mathbf{X}) = \begin{bmatrix} | & | & | & | & \\ 1 & \mathbf{X} & \mathbf{X}^{P_2} & \mathbf{X}^{P_3} & \dots \\ | & | & | & | & \end{bmatrix} . \tag{7}$$

Here, $\mathbf{X}^{P_i}$ denote higher polynomials. For example, $\mathbf{X}^{P_2}$ denotes the quadratic nonlinearities in $\mathbf{X}$, given by:

$$\mathbf{X}^{P_2} = \begin{bmatrix} \hat{C}^2_{\mu,1} & \hat{C}_{\mu,1}\hat{C}_{\mu 1,1} & \hat{C}_{\mu,1}\hat{F}_1^+ & \hat{C}^2_{\mu 1,1} & \hat{C}_{\mu 1,1}\hat{F}_1^+ & \hat{F}_1^{+^2} \\ \hat{C}^2_{\mu,2} & \hat{C}_{\mu,2}\hat{C}_{\mu 1,2} & \hat{C}_{\mu,2}\hat{F}_2^+ & \hat{C}^2_{\mu 1,2} & \hat{C}_{\mu 1,2}\hat{F}_2^+ & \hat{F}_2^{+^2} \\ \vdots & \vdots & \vdots & \vdots & \vdots & \vdots \\ \hat{C}^2_{\mu,m} & \hat{C}_{\mu,m}\hat{C}_{\mu 1,m} & \hat{C}_{\mu,m}\hat{F}_m^+ & \hat{C}^2_{\mu 1,m} & \hat{C}_{\mu 1,m}\hat{F}_m^+ & \hat{F}_m^{+^2} \end{bmatrix} . \tag{8}$$

For the current study, the highest polynomial order is set to four. Hence,

$$\Theta(\mathbf{X}) = \begin{bmatrix} | & | & | & | & | \\ 1 & \mathbf{X} & \mathbf{X}^{P_2} & \mathbf{X}^{P_3} & \mathbf{X}^{P_4} \\ | & | & | & | & | \end{bmatrix} . \tag{9}$$

Each column of $\Theta(\mathbf{X})$ represents a candidate function for the model in Eq. (4). There is a large number of possible entries in this matrix of nonlinearities. Since we are seeking a parsimonious model, only a few of these nonlinearities are active. Moreover, the measured lift coefficient contains experimental uncertainties. The solution to this overdetermined system with noise is obtained by solving

$$\Delta \mathbf{C}_\mathbf{l} = \Theta(\mathbf{X})\Xi + \eta \mathbf{Z} , \tag{10}$$

which yields the sparse vectors of coefficients $\Xi = [\xi_1 \xi_2 \dots \xi_n]$. Here, $\mathbf{Z}$ represents white noise with vanishing mean and unit variance while η denotes noise gain. System (10) can be easily solved using the LASSO [24], which is an ℓ_1-regression that promotes sparsity. The LASSO solves the minimization problem,

$$\hat{\beta}^{\text{lasso}} = \text{argmin}_\beta \left\{ \frac{1}{2} \sum_{i=1}^{m} \left(y_i - \beta_0 - \sum_{j=1}^{p} x_{ij}\beta_j \right)^2 + \lambda \sum_{j=1}^{p} \beta_j \right\} \tag{11}$$

where y_i is the response ($\Delta C_{l i}$) at observation i, λ is a positive regularization parameter, and β_0 and β are the model coefficients. Hence, as λ increases, the number of nonzero components of β decreases. To select the proper regularization parameter, we examine the cross-validated mean-square error of the model for a range of λ's (Fig. 5). As expected, the mean square error reduces with smaller values of λ, which

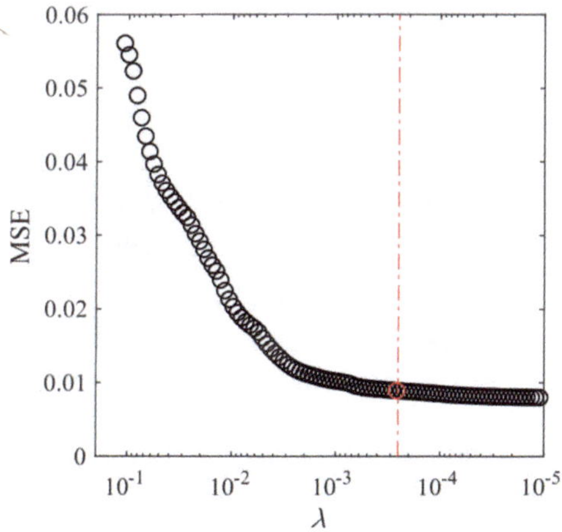

Fig. 5 The cross-validated mean square error for a range of λ's

also indicate higher model complexity. The red circle and dashed line indicate the selected regularization parameter λ, which yields fifteen nonactive coefficients from a total of thirty five, and a mean-square error of $\approx$0.008. This selection is a compromise between complexity and accuracy. For $\alpha = 0°$ and Re $= 1.5 \times 10^6$, 706 measured test cases covering four mean steady momentum coefficients ($C_\mu = 0.03 - 0.05$) are used to train the model, which reads

$$\begin{aligned}\Delta C_l = &\ 0.1874 + 2.8211\hat{C}_{\mu,1} - 0.0443\hat{F}^+ - 0.3891\hat{C}_\mu - 2.2557\hat{C}_{\mu,1}\hat{F}^+ \\ &- 0.0232\hat{C}_{\mu,1}\hat{C}_\mu - 0.5286\hat{C}_{\mu,1}^3 - 1.1201\hat{C}_{\mu,1}^2\hat{C}_\mu - 6.2910\hat{C}_{\mu,1}\hat{F}^{+2} \\ &- 2.0115\hat{C}_{\mu,1}\hat{C}_\mu^2 + 0.0972\hat{F}^+ + 0.0043\hat{F}^+\hat{C}_\mu^2 + 0.7102\hat{C}_{\mu,1}^3\hat{F}^+ \\ &+ 1.9343\hat{C}_{\mu,1}^2\hat{F}^+\hat{C}_\mu + 4.7541\hat{C}_{\mu,1}\hat{F}^{+3} - 0.6339\hat{C}_{\mu,1}\hat{F}^{+2}\hat{C}_\mu \\ &+ 4.8698\hat{C}_{\mu,1}\hat{F}^+\hat{C}_\mu^2 - 0.6847\hat{C}_{\mu,1}\hat{C}_\mu^3 - 0.0002\hat{F}^{+2}\hat{C}_\mu^2 + 0.1589\hat{C}_\mu^4\end{aligned}$$

4 Results

4.1 Steady Actuation

The pressure distributions over the airfoil with increasing blowing intensity are shown in Fig. 6 for zero angle of attack and Re $= 1.0 \times 10^6$ and Re $= 1.5 \times 10^6$, respectively. The benefit of the droop nose is clear. The low and rounded suction peak at the leading edge reduces the adverse pressure gradient and thereby enhances the boundary layer over the suction side and its receptivity to actuation. This in turn

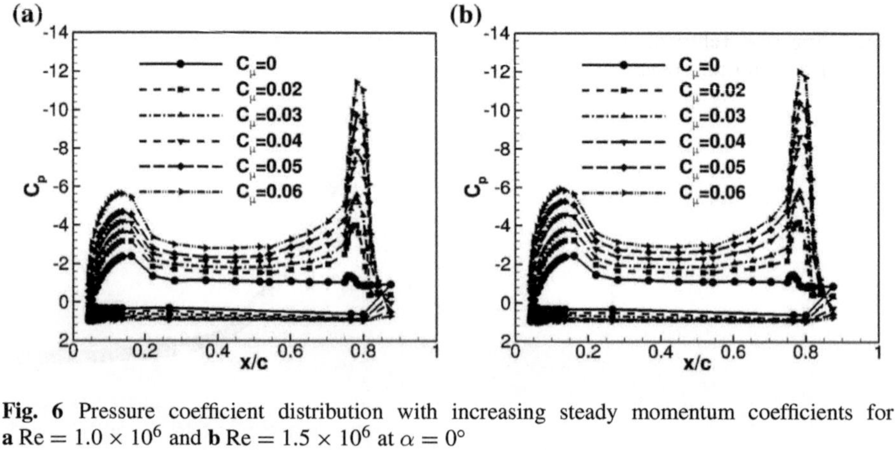

Fig. 6 Pressure coefficient distribution with increasing steady momentum coefficients for **a** Re $= 1.0 \times 10^6$ and **b** Re $= 1.5 \times 10^6$ at $\alpha = 0°$

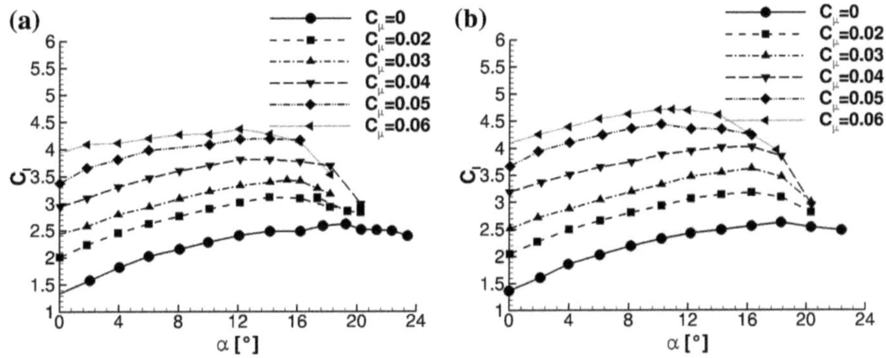

Fig. 7 The effect of the steady momentum coefficient on the lift coefficient over a range of angles of attack at **a** Re $= 1.0 \times 10^6$ and **b** Re $= 1.5 \times 10^6$, respectively

yields improved aerodynamic performance, such as increased lift and higher stall angles compared to a traditional droop nose [25].

The second suction peak near the trailing edge results from the locally high rate of flow turning, which is enabled the jet momentum injection that increases with higher blowing intensities. Stronger actuation also increases the circulation, shifting the stagnation point away from the leading edge over the pressure side. The increase in actuation causes the separation to gradually recede over the flap up to $C_\mu = 0.05$ where the flow becomes fully attached. This gradual mitigation of separation and the added circulation result in steady lift gains, as shown in Fig. 7 for six blowing intensities at two Reynolds numbers. The lower stall angles with higher C_μ can be also observed. Figure 7 also shows the Reynolds number influence on the aerodynamic performance, with higher Reynolds numbers yielding higher lift coefficients. Further increase in the Reynolds number yields smaller lift gains (not shown).

4.2 *Unsteady Actuation*

The lift gains/losses distributions from superimposing a periodic component onto the steady actuation are presented in Fig. 8 for two test cases at Re $= 1.5 \times 10^6$ and $\alpha = 0°$ with mean blowing intensities of $C_\mu = 0.03$ and $C_\mu = 0.05$, respectively. The test cases are selected for their different aerodynamic characteristics (see Figs. 6 and 7), where the flow is partially separated for $C_\mu = 0.03$ and is attached for $C_\mu = 0.05$ during steady actuation ($F^+ = 0$). The contour distributions are those of the modeled ΔC_l, and the dots designate the measured test cases. The addition of a periodic component to actuation has an opposite effect on the two cases. For the $C_\mu = 0.03$ case, unsteady forcing yields lift increases throughout the tested actuation frequency and amplitude range. The highest lift gains with respect to the reference steady case is $\Delta C_l = 0.47$ achieved at $F^+ = 0.17$ and $C_{\mu 1} = 0.016$. On the other hand, unsteady actuation is detrimental to the lift coefficient for the $C_\mu = 0.05$ case almost across

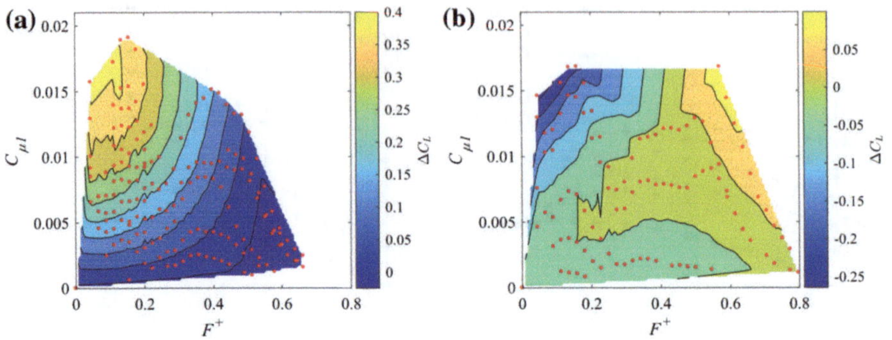

Fig. 8 Modeled ΔC_l distributions for a range of actuation amplitudes $C_{\mu 1}$ and actuation frequencies F^+ for **a** $C_\mu = 0.03$ and **b** $C_\mu = 0.05$ at Re $= 1.5 \times 10^6$

the entire actuation parameter range. In fact, the lift coefficient decreases by as much as $\Delta C_l = -0.3$ at $F^+ = 0.11$ and $C_{\mu 1} = 0.015$.

The evolution of the lift gain with increasing momentum coefficient can be examined using the previously-developed model. Figure 10 shows the lift gain distribution over a range of C_μ for a fixed actuation amplitude $C_{\mu 1} = 0.016$ and a fixed actuation frequency $F^+ = 0.11$. As the figure shows, the lift gains gradually decrease with increasing momentum coefficient until they become lift losses starting $C_\mu \approx 0.04$, when the flow starts to fully attach. The highest lift gains are achieved at the lowest evaluated mean steady momentum coefficient, where the flow is partially attached. Since there exist no measurements at lower C_μ, it is not clear if these lift gains continue their increase. It is worth to note that even if the lift gains were highest at lower C_μ, the overall lift is low and is of no interest to a STOL-capable aircraft.

Figure 9 presents the modeled ΔC_l distributions for a range of actuation amplitudes $C_{\mu 1}$ and actuation frequencies F^+ for (a) $\alpha = 0°$ and (b) $\alpha = 12°$ at $C_\mu = 0.035$ and Re $= 1.5 \times 10^6$. Compared to Fig. 8a, the two distributions exhibit a similar shape

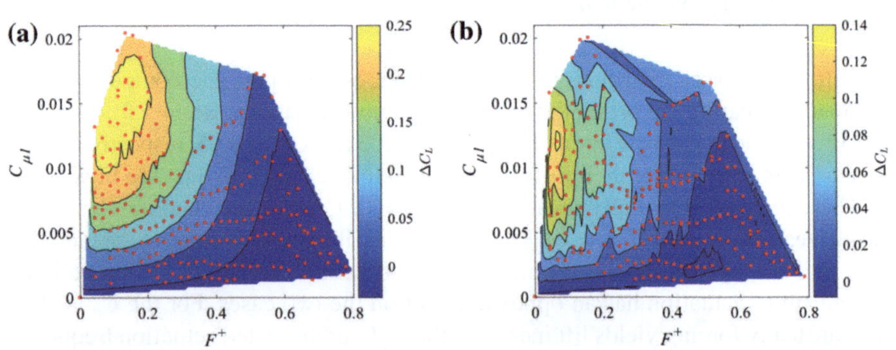

Fig. 9 Modeled ΔC_l distributions for a range of actuation amplitudes $C_{\mu 1}$ and actuation frequencies F^+ for **a** $\alpha = 0°$ and **b** $\alpha = 12°$ at $C_\mu = 0.035$ and Re $= 1.5 \times 10^6$

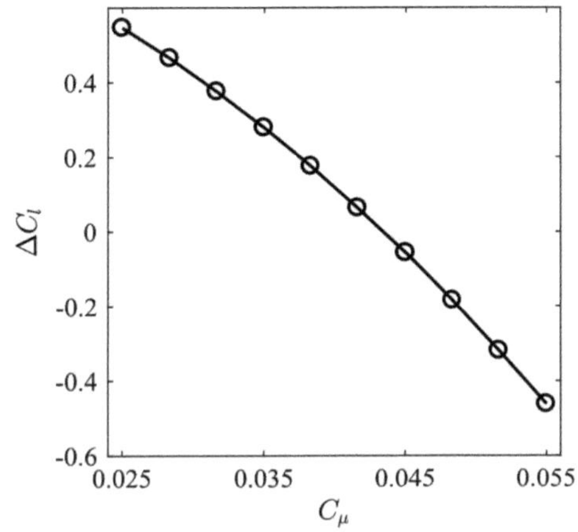

Fig. 10 The evolution of ΔC_l over a range of blowing intensities for $C_{\mu 1} = 0.016$ and $F^+ = 0.11$ at $\alpha = 0°$

with the highest lift gains at $F^+ \approx 0.1$. However, between $\alpha = 0°$ and $\alpha = 12°$ the maximum lift gains are halved from $\Delta C_l = 0.3$ to 0.14. This reduction in lift gains is attributed to uncertainty in measuring the lift coefficient for the steady-blown cases, as shown in Fig. 11, where a comparison between an actuated ($F^+ = 0.11$ and $C_{\mu 1} = 0.012$) and the non-dynamically actuated ($F^+ = 0$) lift coefficient over a range of angles of attack is shown. As can be seen, the lift gains (ΔC_l) remain constant up to $\alpha = 10°$, followed by a small drop starting at 12°. This small drop in lift gain is caused by an unexpected small increase in the lift coefficient for the steady blown case. We suspect that this small increase is an anomaly in the measurements, which will be addressed in our next campaign.

For all cases, the maximum lift gains occur at amplitudes smaller than the largest measured one and smaller than the corresponding mean momentum coefficient C_μ.

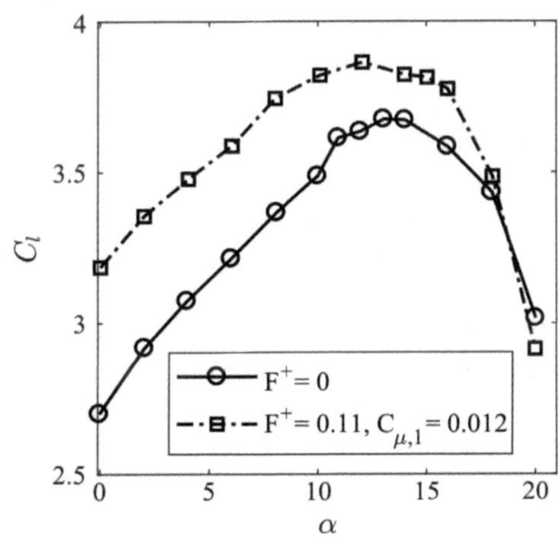

Fig. 11 C_l versus α for an actuated and a non-dynamically actuated case for $C_\mu = 0.035$ and $\mathrm{Re} = 1.5 \times 10^6$

Hence, the maximum aerodynamic gains are achieved at amplitudes that do not require complete closing/opening of the valves. This demonstrates the advantage of having actuation with independently-controlled, steady and unsteady components, something switching valves cannot perform. Moreover, the highest lift gains are achieved at an actuation frequency of $F^+ \approx 0.1$, which is not harmonically related to the natural shedding frequency of the flow, $St = \frac{f \cdot c_{fl}}{V_\infty} = 0.23$ [16]. It is not entirely clear what the significance of this frequency is. Without any field wake measurements and spectral analysis, it is difficult to accurately pinpoint the physical mechanisms yielding the relative lift increase. However, based on previous studies [1, 19] and current observations, it is possible to infer the underlying phenomena. Separated or partially-separated flows are more receptive to unsteady actuation targeting some flow instabilities. Here, the actuation frequency can excite or dampen certain mode(s) to delay separation and to increase circulation. On the other hand, periodic actuation on attached flows can only disturb the attachement without significant benefits. Moreover, oscillations associated with dynamic actuation generate additional vorticity in the flow field contributing to adverse changes in lift and drag. Attached flows can only attain higher lift through added circulation, which cannot be provided by periodic forcing. Time-resolved PIV measurements are planned in a future campaign to confirm these hypotheses.

5 Conclusion

This study investigates and models the lift gains and losses from superimposing a periodic actuation component on a steady component for a high-lift configuration with a highly deflected Coanda flap. The assessment relies on pressure measurements along the model centerline. Steady actuation is shown to increase the lift coefficient significantly up to $C_{l,max} = 4.72$ at $\mathrm{Re} = 1.5 \times 10^6$. This high lift coefficient enables short take-off and landing (STOL) capabilities. Further lift gains are achieved through the superposition of a periodic forcing component. For test cases where the flow is partially separated during steady actuation, superimposed periodic forcing yields lift increases through most of the actuation parameter range. The highest lift gain with respect to the reference steady case reaches $\Delta C_l = 0.47$. On the other hand, superimposed periodic actuation is detrimental to the lift coefficient for attached flows through most of the actuation parameter range. In fact, the lift coefficient decreases by as much as $\Delta C_l = -0.3$. The highest aerodynamic gains are achieved at amplitudes that do not require complete closing/opening of the valves, demonstrating the advantage of having actuation with independently-controlled steady and unsteady components. The highest gains are also achieved at an actuation frequency of $F^+ \approx 0.1$, which is not harmonically related to the natural shedding frequency of the flow. Further investigations and time-resolved PIV measurements are required to understand the underlying phenomena leading to the lift increase/decrease.

Acknowledgements We acknowledge the funding of the Collaborative Research Centre (CRC 880) "Fundamentals of High Lift of Future Civil Aircraft" supported by the Deutsche Forschungsgemeinschaft (DFG) and hosted at the Technische Universität Carolo-Wilhelmina zu Braunschweig.

References

1. Greenblatt, D., Wygnanski, I.: The control of flow separation by periodic excitation. Prog. Aerosp. Sci. **36**(7), 487–545 (2000)
2. Barros, D., Borée, J., Noack, B.R., Spohn, A., Ruiz, T.: Bluff body drag manipulation using pulsed jets and Coanda effect. J. Fluid Mech. **805**, 422–459 (2016)
3. Chabert, T., Dandois, J., Garnier, A.: Experimental closed-loop control of separated-flow over a plain flap using extremum seeking. Exp. Fluid **57**(37) (2016)
4. Davidson, I.M.: Aerofoil boundary layer control system. Brit. Pat. **913**, 754 (1960)
5. Lachmann, G.V.: Boundary Layer and Flow Control: Its Principles and Application. Pergamon Press, New York (1961)
6. Wood, N.J., Nielsen, J.N.: Circulation control airfoils—Past, present and future. In: AIAA Paper-85, p. 0204 (1985)
7. Englar, R.: Circulation control pneumatic aerodynamics: blown force and moment augmentation and modifications; past, present, and future. In: AIAA Paper-2000, p. 2541 (2000)
8. Sexstone, M.G., Huebner, L.D., Lamar, J.E., McKinley, R.E., Torres, A.O., Burley, C.L., Scott, R.C., Small, W.J.: Synergistic airframe-propulsion interactions and integrations. Technical Report No. NASA/TM-1998-207644. NASA Langley Research Center; Hampton, VA United States (1998)
9. Smith, A.M.O.: High-lift aerodynamics. J. Aircr. **12**(6), 501–530 (1975)
10. Jensch, C., Pfingsten, K.C., Radespiel, R., Schuermann, M., Haupt, M., Bauss, S.: Design aspects of a gapless high-lift system with active blowing. In: Deutscher Luft- und Raumfahrtkongress, vol. 58, Aachen, Germany (2009)
11. Radespiel, R., Pfingsten, K.-C., Jensch, C.: Flow analysis of augmented high-lift systems. In: Radespiel, R., Rossow, C.C., Brinkmann, B.W. (eds.) Hermann Schlichting-100 Years, pp. 168–189 (2009)
12. Radespiel, R., Burnazzi, M., Casper, M., Scholz, P.: Active flow control for high lift with steady blowing. Aeronaut. J. **120**(1223), 171–200 (2016)
13. Oyler, T.E., Palmer, W.E.: Exploratory Investigation of Pulse Blowing for Boundary Layer Control. Technical Report No.: NR72H-12, North American Rockwell, Columbus division (1972)
14. Walters, R.E., Myer, D.P., Holt, D.J.: Circulation Control by Steady and Pulsed Blowing for a Cambered Elliptical Airfoil. Technical Report No.:TR-32, West Virginia University, Aerospace Engineering (1972)
15. Jones, G.S., Viken, S.A., Washburn, A.E., Jenkins, L.N., Cagle, C.M.: An active flow circulation controlled flap concept for general aviation aircraft applications. In: AIAA paper-2002, p. 3157 (2002)
16. El Sayed, Y., Semaan, R., Sattler, S., Radespiel, R.: Wake characterization methods of a circulation control wing. Exp. Fluids **58**(10), 144 (2017)
17. El Sayed, M.Y., Beck, N., Kumar, P., Semaan, R., Radespiel, R.: Challenges in the experimental quantification of the momentum coefficient of circulation controlled wings. In: New Results in Numerical and Experimental Fluid Mechanics XI, pp. 533–543. Springer (2018)
18. Petz, R., Nitsche, W.: Active separation control on the flap of a two-dimensional generic high-lift configuration. J. Aircr. **44**(3), 865–874 (2007)
19. Seifert, A., Greenblatt, D., Wygnanski, I.J.: Active separation control: an overview of Reynolds and Mach number effects. Aerosp. Sci. Technol. **8**(7), 569–582 (2004)

20. Burnazzi, M., Radespiel, R.: Design and analysis of a droop nose for coanda flap applications. AIAA J. Aircr. **51**(5), 1567–1579 (2014)
21. Scholz, P., Sattler, S., Wulff, D.: Der Große Wasserkanal 'GWB'-Eine Versuchsanlage für zeitauflösende Messungen bei großen Reynoldszahlen. Deutscher Luft-und Raumfahrtkongress, Stuttgart, Germany (2013)
22. Poisson-Quinton P., Lepage L.: Survey of French research on the control of boundary layer and circulation. In: Lachmann, G.V. (eds.) Boundary layer and Flow Control: Its Principles and Application, vol. 1, pp. 21–73 (1961)
23. Brunton, S.L., Proctor, J.L., Nathan, J.K.: Discovering governing equations from data by sparse identification of nonlinear dynamical systems. Proc. Natl. Acad. Sci. **113**(15), 3932–3937 (2016)
24. Tibshirani, R.: Regression shrinkage and selection via the lasso. J. Roy. Stat. Soc. Ser. B (Methodol), 267–288 (1996)
25. Burnazzi, M., Radespiel, R.: Assessment of leading-edge devices for stall delay on an airfoil with active circulation control. CEAS Aeronaut. J. **5**(4), 359–385 (2014)

Unsteady Roll Moment Control Using Active Flow Control on a Delta Wing

Xiaowei He, Mathieu Le Provost, Xuanhong An and David R. Williams

Abstract A feedforward controller is designed to attenuate the roll moment coefficients produced by forced roll motion of a delta-wing type model. Active flow control effectors in the form of variable strength pneumatic slot-jets are located along the trailing edge of the model. The control effectors produce a roll moment coefficient proportional to the momentum coefficient. Direct measurements of the roll moment are made with the model in a wind tunnel. Black-box models for the plant and disturbance are identified, and used in the design of the feedforward controller. The effectiveness of the feedforward controller in attenuating disturbance roll moments produced by forced roll maneuvers is evaluated with periodic and pseudo-random maneuvers. Near the design point of the for the controller the root-mean-square value of the net roll moment is four times smaller than the roll moment without control.

Keywords Active flow control · Unsteady aerodynamics · Roll control

1 Introduction

During landing aircraft fly at low speeds (approximately 30 % above the stall speed) and at high angles of attack (around $\alpha = 10° - 15°$), conditions that make them susceptible to the adverse effects of turbulence, such as, partial wing stall. Landing on aircraft carriers can be particularly challenging due to the region of strong velocity gradients and turbulence at the approach end of the flight deck, which is known as the 'burble' [3]. To maintain roll control in this environment, conventional aircraft

X. He · X. An · D. R. Williams (✉)
Illinois Institute of Technology, Chicago, IL 60616, USA
e-mail: david.williams@iit.edu

M. L. Provost
University California Los Angeles, Los Angeles, CA 90095, USA

R. King (ed.), *Active Flow and Combustion Control 2018*,
Notes on Numerical Fluid Mechanics and Multidisciplinary Design 141,
https://doi.org/10.1007/978-3-319-98177-2_2

use mechanical ailerons (or elevons) that deflect in opposite directions, producing roll moments by differential lift on the left and right wings. However, elevons lose effectiveness when the flow separates, because the differential lift is reduced.

Control effectors based on active flow control (AFC) techniques offer the potential for increased control authority when the flow is separated. Examples of AFC for vehicle flight control include axisymmetric bluff body trajectory control using synthetic jet actuators, which is being investigated by Lambert et al. [11]. Unsteady lift and pitching moment control on nominally 2-D wings has been explored by Woo et al. [17, 18], An et al. [1], and Reißner [13]. AFC for unsteady lift control in a surging flow wind tunnel was demonstrated by Kerstens et al. [9] on a 3-D semi-circular planform wing. Tailless aircraft control with variable strength pneumatic jets is being investigate by Williams and Seidel [16].

The focus of this paper is the application of AFC to dynamic roll moment control. In particular, we attempt to design a feedforward control algorithm that will attenuate roll moment disturbances created by a forced roll motion of a tailless aircraft model. It is understood that the forced roll motion does not produce unsteady aerodynamics that are equivalent to an aircraft flying through a velocity gradient. However, it is our contention that a controller that effectively attenuates the forced roll moments would also be effective for naturally occurring external flow disturbances. The control approach is based on the assumption that superposition of a plant model response to actuation and a disturbance model will accurately predict the behavior of the aircraft. Identifying low-dimensional models that account for unsteady aerodynamic effects of the external disturbances is explored using the Sparse Identification of Nonlinear Dynamics (SINDy) algorithm introduced by Brunton et al. [2]. This approach allows the important inputs to the disturbance model to be identified. The feedforward controller is constructed from the product of an inverted nominal plant model, a filter, and a disturbance model. The effectiveness of the control approach is evaluated by using periodic and pseudo-random forced roll motion inputs. The measured roll moment coefficients are compared to predicted values.

The hardware used in the experiment is described in Sect. 2. The architecture and components of the feedforward controller are presented in Sect. 3. The model identification procedures used, including discussion of the SINDy approach are given in Sect. 4. The system performance with and without active flow control is presented in Sect. 5 for periodic and pseudo-random forced rolling motions. The conclusions are discussed in Sect. 6.

2 Experimental Setup

2.1 Model and Wind Tunnel

The model used for these measurements is the tailless UAS shown in Fig. 1. The design is a hybrid of the Lockheed-Martin ICE-101 planform [5] with profiles from the NATO SACCON design [8]. The centerline (root) chord of the model is $c_r =$ 330 mm with a span $b = 287$ mm. The leading edge sweep angle is $\Lambda = 65°$. The mean aerodynamic chord is $c_{mac} = 214$ mm with a corresponding Reynolds number $Re_{c_{mac}} = 9.6 \times 10^4$. The overall mass of the model including sensors is 0.13 kg, and the corresponding moment of inertia about the x-axis of the model is $I_{xx} =$ 3.12×10^{-4} kg m^2.

Measurements are done in the Fejer Unsteady Flow Wind Tunnel, whose test section length is 2.1m and square cross section has dimensions 0.61m $\times$ 0.61m. The freestream speed was 6.5 m/s for all measurements reported. A dSPACE microLab-Box system is used to acquire data and provide control at a rate of 1000 samples/sec.

The suction surface of the model is instrumented with 12 surface pressure sensors (AllSensor 1-inch H^2O) mounted internally in the upper surface skin of the model. The sensors have a response time of 0.3 ms. Because the distance between the pressure port opening and the sensor is less than 2 mm, the nominal bandwidth of the sensors is on the order of 1 kHz, which is more than sufficient to detect the experimental signals of interest that are below 10 Hz.

The roll motion of the model is controlled by a Hitec servo motor (HSB-9360TH). The servo motor is connected to the internal force transducer with a tube that carries instrumentation and active flow control lines. The roll motion amplitude is $\phi = \pm 20°$, and the maximum achievable roll frequency is 2.8 Hz. The roll rates used in this study are $f = 0.5$, 1.0, 1.5, and 2.0 Hz, which correspond to dimensionless frequencies $k = \frac{\pi f c}{U} = 0.052$, 0.103, 0.155, and 0.207. The angle of attack of the model is controlled with a pair of Copley servo tubes. The angles of attack are manually fixed at $\alpha = 0°$, 5°, 10°, 15°, and 20° for the current investigation.

The time-varying forces and moments are measured with an ATI Nano 25 force transducer. The transducer is located internally in the model at $0.54c_r$ from the nose of the model. The roll-moment is defined with the positive x-axis aligned with the model centerline and pointing toward the nose. A positive roll corresponds to the right wing moving downward.

2.2 AFC Actuators as Control Effectors

The active flow control actuators are a pair of downward blowing slot jets located at the trailing edge of the model as shown in Fig. 2. The flow rates to the left and right side actuators is controlled by the voltage to a pair of Clippard proportional valves (EV-P-10-4050). The slot jet exits are have a width of 0.2 mm and a length of

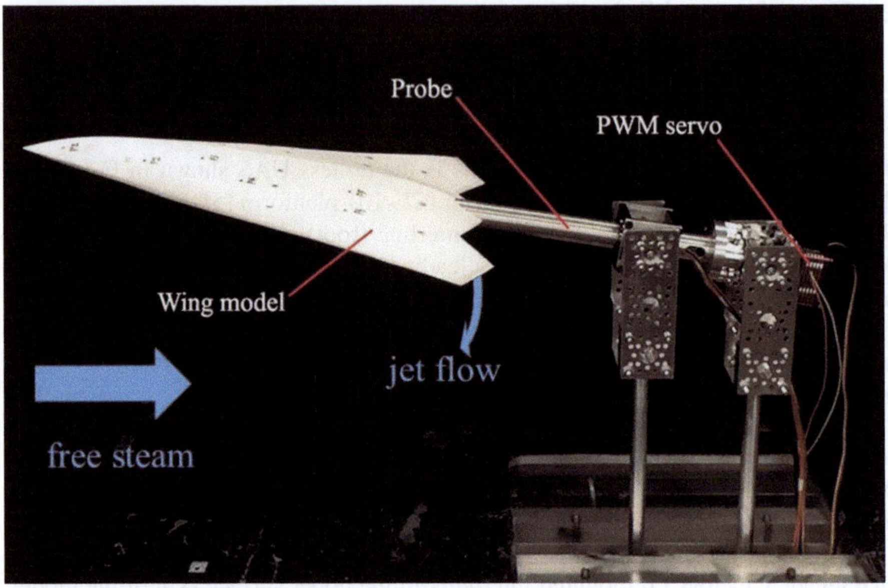

Fig. 1 ICE/SACCON model connected to the roll and pitch mechanism in wind tunnel test section

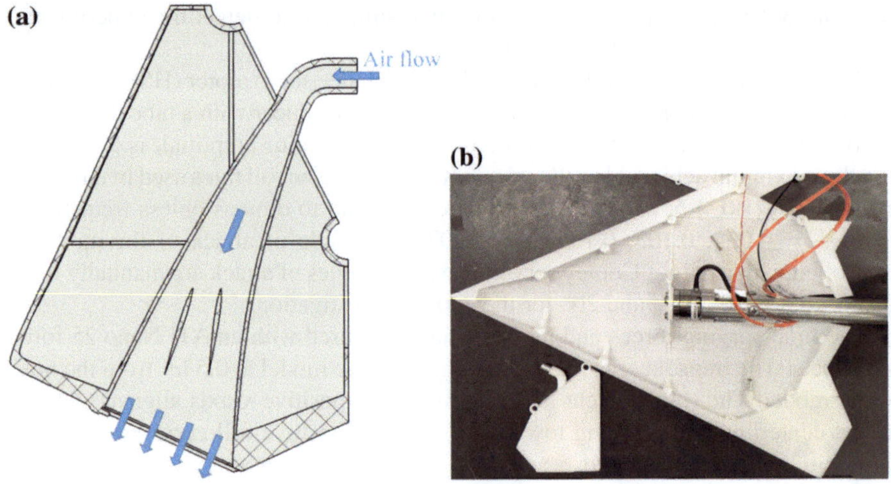

Fig. 2 Active flow control system. **a** Slot-jet actuator module, **b** Interior of model showing air supply tubing to actuator

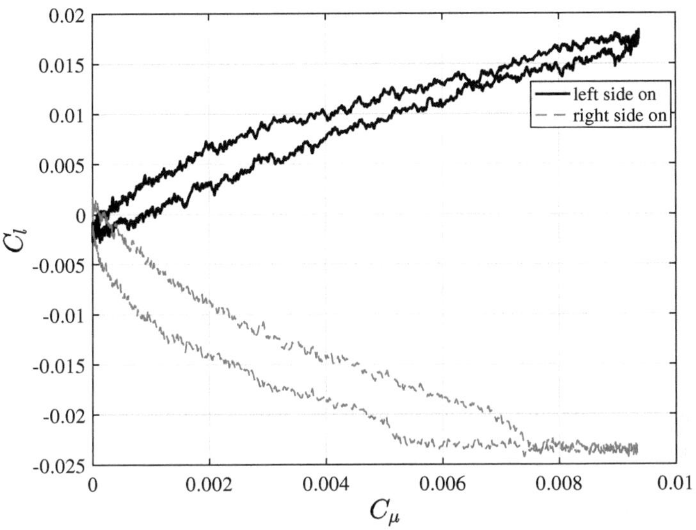

Fig. 3 Steady state calibration of left and right active flow control effectors at $\alpha = 20°$

30 mm. The jet exits over the top of a cylindrical surface with a 2 mm diameter, which produces a downward directed jet by the Coanda effect. The actuators are modular plug-in designs that allow different designs to be used. The strength of actuation is measured by the momentum coefficient, which is defined as $C_\mu = \frac{\varrho V_{jet}^2 A_{jet}}{q_\infty S}$. Here q_∞ is the dynamic pressure, S is the planform area, V_{jet} is the jet exit velocity, and A_{jet} is the slot exit area. The maximum achievable momentum coefficient is $C_\mu = 0.009$.

Figure 3 shows the roll moment coefficient produced by the left and right wing active flow control actuators. Under steady state conditions the actuators have enough control authority to produce roll moments above $C_l = 0.015$, which is sufficient for dimensionless roll rates up to $k = 0.02$ and roll amplitudes $\pm 30°$. The hysteresis is caused by the Clippard valves. The average between the two branches of the hysteresis loops is used for the controller, which ultimately leads to noise in the feedforward controller.

3 Feedforward Control

The feedforward control architecture shown in Fig. 4 assumes that the net roll moment ($C_{l-measure}$) of the wing results from the linear superposition of the roll motion disturbance (ΔC_{l-d}) and the actuator influence on the plant (C_{l-act}). Even though this approach ignores potential nonlinear interactions between the plant and the disturbance, the linear assumption has proven to be effective in several different

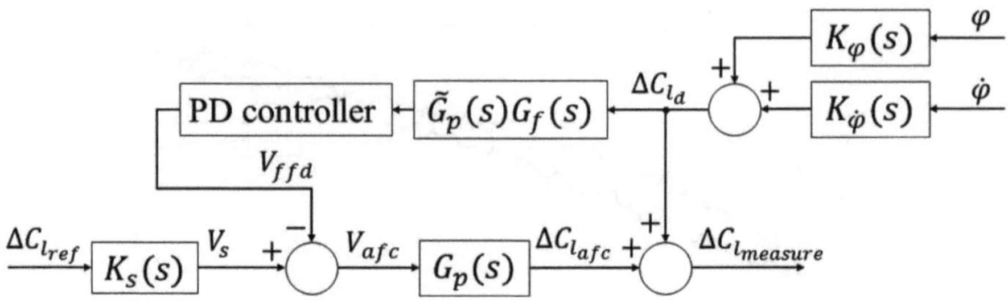

Fig. 4 Block diagram of the feedforward control architecture

active flow control experiments, such as, 2-D bluff body control [7], 3D bluff body gust control [12], 3D wing gust alleviation [9], and stator wake control [10].

The disturbance model (G_d) and plant model (G_p) are identified from experimental measurements. The disturbance model is obtained using an algorithm developed by Brunton et al. [2], which is discussed in Sect. 4.1. A third-order black box plant model is identified with the prediction error method. Details are provided in Sect. 4.2. The minimum phase component of the nominal plant model is inverted and combined with a low-pass filter, G_f to form the feedforward controller. A PD controller is included with the low-pass filter to act as a lead compensator.

4 Model Identification

4.1 Disturbance Model Identification for the Rolling Wing

The choice of kinematic inputs to use for the disturbance model is not obvious. The roll angle, roll rate, roll rate derivative, and possible nonlinear combinations of those inputs were all considered to be potentially important in determining the output disturbance roll moment. One way to deal with this uncertainty is to use the Sparse Identification of Nonlinear Dynamics (SINDy) algorithm developed by Brunton et al. [2]. This approach is to assume a large set of candidates exists, and then identify terms that are not active, which are then removed from the model. The procedure is applied at each angle of attack of interest to obtain models for each α.

The disturbance model is initially assumed to have the form

$$\frac{dC_l}{dt} = AC_l + Bf(\phi, \dot\phi, \ddot\phi) \tag{1}$$

or

$$\frac{dC_l}{dt} = [A \quad B]\begin{bmatrix} C_l \\ f \end{bmatrix} \tag{2}$$

where C_l is the roll moment coefficient, ϕ is the roll angle and f is a vector of the input variables. Here, we chose the input f to be

$$f = [\phi\dot{\phi}, \phi\ddot{\phi}, \dot{\phi}\ddot{\phi}, \phi, \dot{\phi}, \ddot{\phi}]^T \tag{3}$$

Before seeking the values of A and B, Eq. 2 is rewritten in the form of

$$Y = \Theta(X)\Xi \tag{4}$$

where Y is the training dataset (experimental dataset in this case), $\Theta(X)$ is a matrix whose elements are a pool of candidate terms, and initally contains all the time series data of C_l, ϕ, $\dot{\phi}$, $\ddot{\phi}$, etc. Thus,

$$Y = \frac{dC_l}{dt} \tag{5}$$

$$\Theta = \begin{bmatrix} C_{l\ roll\ exp} \\ \phi C_{l\ roll\ exp} \\ \phi\dot{\phi} \\ \phi\ddot{\phi} \\ \dot{\phi}\ddot{\phi} \\ \phi \\ \dot{\phi} \\ \ddot{\phi} \end{bmatrix} \tag{6}$$

Ξ is the coefficient vector which contains the A and B matrices. The goal is to find a vector Ξ by using sparse regression techniques. One option is to use the least absolute shrinkage and selection operator(LASSO) [15]:

$$\Xi = \text{argmin}\, ||\Theta\Xi' - Y||_2 + \lambda||\Xi'||_1. \tag{7}$$

As it is shown in the flow map, Fig. 5, L^2 minimization is performed on Eq. 4 first to generate a initial guess of Ξ. Next, if there is any Ξ_i smaller than λ, then we eliminate the corresponding Θ_i, update the library of Θ and repeat the process until all elements inside Ξ satisfy $\Xi_i > \lambda$.

The model is identified using a concatenated time series of periodic rolling cases at 0.5, 1.0, 1.5 and 2.0 Hz with ±20° of amplitude. The sparse coefficient vector computed by the SINDy regression is

$$\Xi = [-0.9114, 0, 0, 0, 0, -0.0208, -0.0028]^T \tag{8}$$

Thus, for $\alpha = 10°$ the dynamic model is obtained as

$$\frac{dC_l}{dt} = -0.9114C_l - 0.0208\dot{\phi} - 0.0028\ddot{\phi} \tag{9}$$

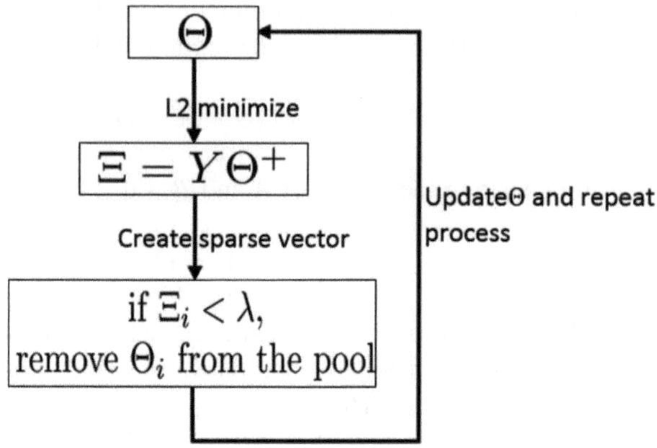

Fig. 5 SINDy process flow map

When the angle of attack was increased to $\alpha = 20°$ then a nonlinear term became significant.

$$\frac{dC_l}{dt} = -6.1038C_l + 0.0496\phi\dot{\phi} + 0.1762\phi - 0.976\dot{\phi} - 0.0018\ddot{\phi} \qquad (10)$$

The model is then validated by a random rolling maneuver. The correlation coefficient between the model estimation and the experimental data is 0.9926 for the 1.5 Hz periodic cases shown in Fig. 6c, and 0.9253 for the random case shown in Fig. 6d.

4.2 Plant Model Identification with Actuation

The plant model predicts the roll moment coefficient response to active flow control actuator input. Conventional system identification techniques that use the prediction error method are used to identify 3rd order models. At each angle of attack a family of 20 different models are identified using pseudo-random binary (prbs) input voltages to the active flow control valves. The binary signal voltage values ranged between 2 and 10 V. The 0–8 V prbs signal used for input is shown in the lower part of Fig. 7a. The upper part of the figure compares to the measured output roll moment coefficient to the model prediction. The family of plant models is shown in Fig. 7b. There is generally good agreement among the models, except for a few outliers that occurred at low voltage levels. The black dashed line shows the nominal plant model that was obtained by averaging the models not considered to be outliers. The transfer function for the nominal plant at $\alpha = 10°$ is $\tilde{G}_p(s) = \frac{-4.045\times10^{-3}s^2 - 0.2959s + 58.35}{s^3 + 56.13s^2 + 2922s + 44252}$, which has a right-half-plane (RHP) zero at 89 s^{-1}. The RHP zero is analogous to a time delay of approximately 0.022 s, which is short compared to the typical periods of the disturbance motion.

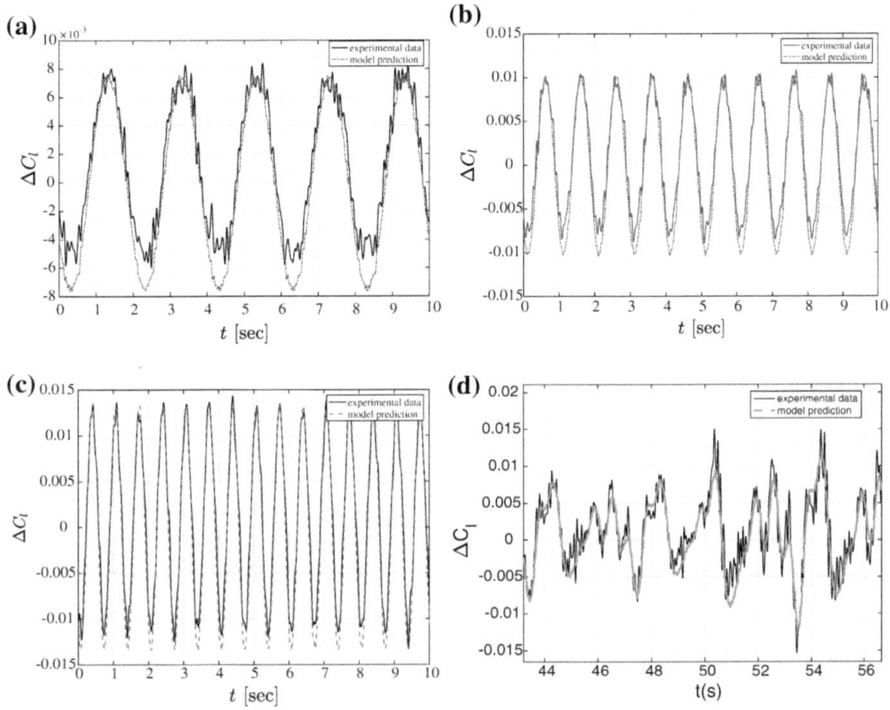

Fig. 6 Comparison of the disturbance model prediction and the experimental data for the periodic rolling at $\alpha = 10°$, **a** 0.5 Hz, **b** 1.0 Hz, **c** 1.5 Hz, **d** pseudo-random pitching

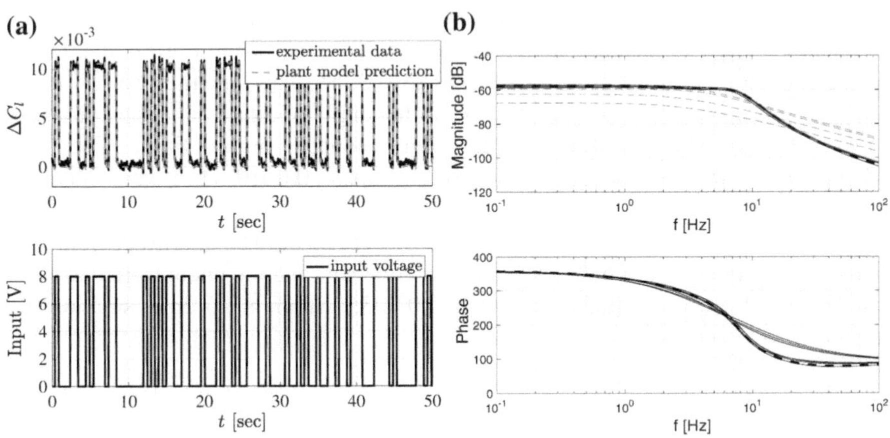

Fig. 7 Plant model identification. **a** Comparison of measured and modeled prbs signals. **b** Bode plot of family of plant models and nominal model (heavy dashed)

5 Feedforward Control Results

The minimal phase component of the plant model is inverted and combined with the disturbance model, a low-pass filter, and a PD controller to form the feedforward controller. The objective is to attenuate the effects of periodic and pseudo-random roll motion on the roll moment coefficient. The results are shown in the following two subsections.

5.1 Roll Moment Attenuation During Periodic Roll Maneuvers

The feedforward control results for forced periodic roll motion at $1Hz(k = 0.0102)$ are shown in Fig. 8 at three angles of attack, $\alpha = 5°$, $10°$, and $15°$. The PD compensator was tuned for the conditions at $\alpha = 10°$. The uncontrolled roll moment is shown by the dashed line. The solid line shows the roll moment with feedforward control, and the linear model prediction of the roll moment with control is shown by the dash-dot line. At the lowest angle of attack, $\alpha = 5°$ shown in Fig. 8a, the amplitude of the roll moment oscillations is small (root mean square, r.m.s. = 0.0047). It can be seen that the controller is not very effective in attenuating the roll moment, because the reduced r.m.s. value is only 0.0037. The simulation predicted the controlled r.m.s. = 0.0016, which is significantly lower than the value achieved in experiment. The simulation does not account for measurement noise, which is relatively strong at low roll rates. Thus, the inputs to the feedforward controller are low signal-to-noise ratio signals, and it appears that the higher frequency noise is being amplified.

When the angle of attack is increased to $\alpha = 10°$ as shown in Fig. 8b, the uncontrolled roll moment r.m.s. value increases to 0.0066. The signal-to-noise ratio is stronger. The controller is more effective at this angle of attack, and the controlled r.m.s. = 0.0016, which is more than a 4 times reduction in amplitude. The r.m.s. value achieved in simulation is again lower (r.m.s. = 0.0010) than experimentally measured.

The case for angle of attack $\alpha = 15°$ is shown in Fig. 8c. The amplitude of the roll moment coefficient is twice the value seen at $\alpha = 10°$, and the uncontrolled r.m.s. is 0.0122. With control the r.m.s. is reduced to 0.0035, which is 3.5 times smaller than the uncontrolled case.

5.2 Roll Moment Attenuation During Pseudo-Random Roll Maneuvers

A sequence of four sine waves with different amplitudes and phases was superposed to create the pseudo-random roll maneuver. Comparisons between the uncontrolled

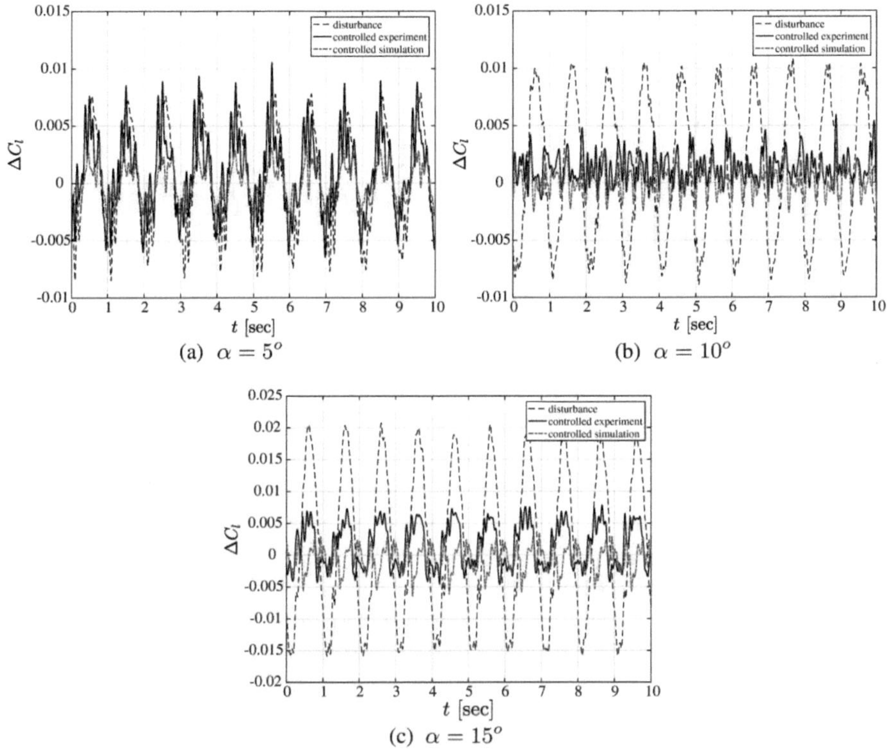

Fig. 8 Periodic roll motion comparison of the measured uncontrolled C_l (dashed line), the measured C_l with control (solid line), and the predicted response(dash-dot) at **a** $\alpha = 5°$, **b** $\alpha = 10°$, **c** $\alpha = 15°$

roll moments, the experimental controlled cases, and the simulation prediction of the controlled cases are shown for three angles of attack in Fig. 9.

The low angle of attack $\alpha = 5°$ result is shown in Fig. 9a. The r.m.s. level of the disturbance roll moment is $\Delta C_{l_{rms}} = 0.0026$. The experimental control has little effect, and only attenuates the disturbance to an r.m.s. level of $\Delta C_{l_{rms}} = 0.0022$. The simulation predicted a significantly lower r.m.s. level of $\Delta C_{l_{rms}} = 0.0012$. From the time series signals it appears that the controller is actually exciting high frequencies in the roll moment.

The data in Fig. 9b shows for the $\alpha = 10°$ angle of attack that the controller becomes more effective. The black-box models used in the controller design were identified at this angle of attack, so it is not surprising that the best level of performance is achieved during the 'on-design' conditions. The uncontrolled r.m.s. level is $\Delta C_{l_{rms}} = 0.0035$. The experimental control case reduces the r.m.s. to $\Delta C_{l_{rms}} = 0.0015$, which is close to the simulated value of $\Delta C_{l_{rms}} = 0.0014$.

In Fig. 9c the angle of attack is increased to $\alpha = 15°$. The uncontrolled r.m.s. level more than doubles relative to the $\alpha = 10°$ case to $\Delta C_{l_{rms}} = 0.0083$. Although the controller is able to reduce the r.m.s. value to $\Delta C_{l_{rms}} = 0.0052$, it is also clear

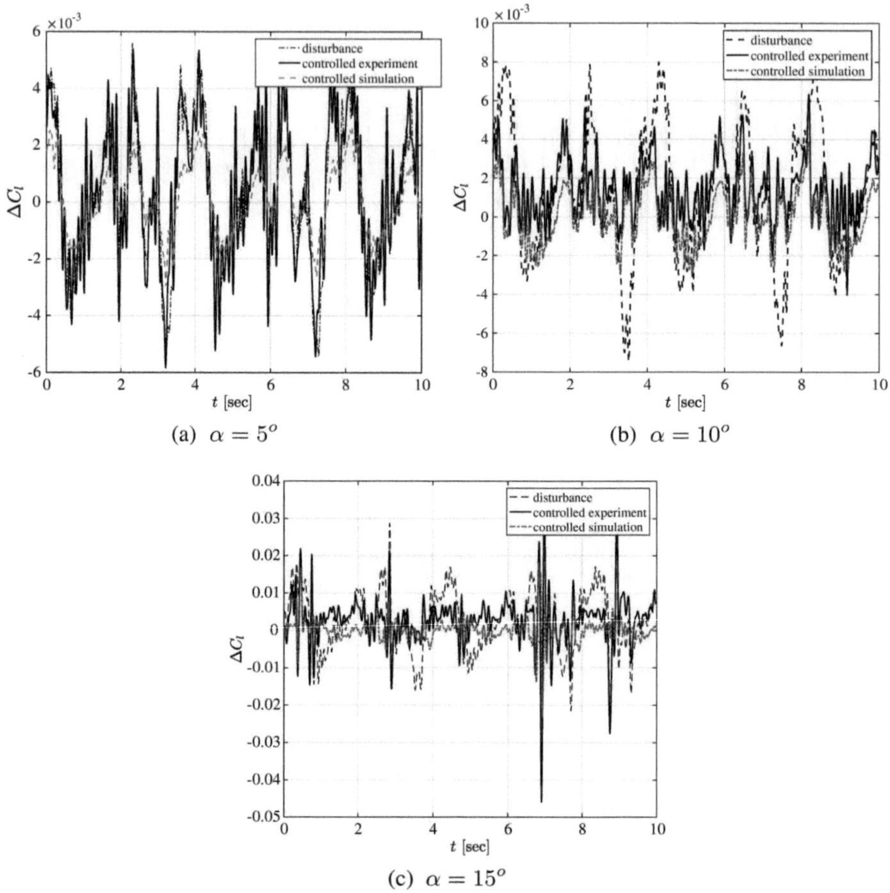

Fig. 9 Pseudo-random motion comparison of the measured uncontrolled C_l (dashed line), the measured C_l with control (solid line), and the predicted controlled response(dash-dot) at **a** $\alpha = 5°$, **b** $\alpha = 10°$, **c** $\alpha = 15°$

from the time series data that the controller is exciting higher frequencies. Large amplitude spikes in ΔC_l are seen near $t = 3$, 7, and 9 s, which come from the feedforward controller. In some cases the spike amplitudes exceed the amplitude of the uncontrolled roll moment, so the controller is making things worse.

The next step is to include feedback in the control architecture. So far that has not been attempted, because a suitable surrogate signal for the instantaneous roll moment has not been identified. The current configuration of surface pressure sensors is capable of identifying the disturbance roll moment, but they are not able to identify the roll moment associated with the actuator.

6 Conclusion

Active flow control is used to attenuate the roll moment coefficient produced by forced rolling maneuvers with a Δ-wing (ICE/SACCON) aircraft model. A feedforward controller is designed based on low-dimensional models for the roll moment response to actuation and roll motion. The disturbance model is identified using the SINDy optimization algorithm. The plant model for the response to actuation is identified using the prediction error method. The control is applied during forced periodic and quasi-random oscillatory maneuvers of the model. The controller is able to reduce the r.m.s. level of the roll moment coefficient by a factor of four in the periodic (1 Hz, $k = 0.102$) case at $\alpha = 10°$, which is the 'on-design' condition for the controller. Controller performance is degraded at angles of attack both above and below the design point, but is still effective in reducing the r.m.s. value of the roll moment coefficient for both periodic and pseudo-random forced disturbances. In the case of pseudo-random forcing the controller acted to excite higher frequency disturbances.

Acknowledgements The authors gratefully acknowledge support of the Office of Naval Research (ONR) under Grant N001416-1-2622 with Dr. Ken Iwanski and Dr. Brian Holm-Hansen as program managers.

References

1. An, X., Grimaud, L., Williams, D.R.: Feedforward control of lift hysteresis during periodic and random pitching maneuvers. In: King, R. (ed.) Active Flow and Combustion Control 2014. NNFM, vol. 127, pp. 55–69. Springer, Heidelberg (2015). https://doi.org/10.1007/978-3-319-11967-0_4
2. Brunton, S.L., Proctor, J.L., Kutz, J.N.: Discovering governing equations from data by sparse identification of nonlinear dynamical systems. Proc. Nat. Acad. Sci. **113**(15), 3932–3937 (2016). Nat. Acad. Sci
3. Cherry, B.E., Constantino, M.M.: The burble effect: Superstructure and flight deck effects on carrier air wake. In: Proceedings of the American Society of Naval Engineers Launch and Recovery Symposium 2010, Launch, Recovery and Operations of Manned and Unmanned Vehicles from Marine Platforms, Arlington, VA (2010), http://www.dtic.mil/dtic/tr/fulltext/us/a527798.pdf
4. Cook, M.V.: Flight Dynamics Principles: A Linear Systems Approach to Aircraft Stability and Control, pp. 20–21. Butterworth-Heinemann (2012)
5. Gillard, W.J.: Innovative control effectors (Configuration 101). Technical Report AFRL-VA-WP-TR-1998-3043. Air Force Research Laboratory Wright Patterson (1998)
6. Gursul, I.: Review of unsteady vortex flows over slender delta wings. J. Aircr. **42**(2), 299–319 (2005). https://doi.org/10.2514/1.5269
7. Henning, L., Pastoor, M., King, R., Noack, B.R., Tadmor, G.: Feedback control applied to the bluff body wake. In: King, R. (ed.) Active Flow Control. NNFM, vol. 95. Springer, pp. 369–390 (2007)
8. Huber, K.C., Vicroy, D., Schütte, A., Hübner, A.: UCAV model design and static experimental investigations to estimate control device effectiveness and stability and control capabilities. In: AIAA Paper 2014–2002 (2014)

9. Kerstens, W., Pfeiffer, J., Williams, D.R., King, R., Colonius, T.: Closed-loop control of lift for longitudinal gust suppression at low Reynolds numbers. AIAA J **49**(8), 1721–1728 (2011). https://doi.org/10.2514/1.J050954
10. Kiesner, M., King, R.: Multivariable closed-loop active flow control of a compressor stator cascade. AIAA J. (2017). https://doi.org/10.2514/1.J055728
11. Lambert, T.J., Vukasinovic, B., Glezer, A.: Aerodynamic flow control of wake dynamics coupled to a moving bluff body. In: AIAA Paper 2016–4081 (2016). https://doi.org/10.2514/6.2016-4081
12. Pfeiffer, J.: Closed-loop active flow control for road vehicles under unsteady cross-wind conditions. Ph.D. Thesis, Technische Universität Berlin (2015)
13. Reißner, F.: Hysteresis modeling and comparison of controller effectiveness on a pitching airfoil. Master's thesis, Technische Universität Berlin (2015)
14. Taylor, G., Wang, Z., Vardaki, E., Gursul, I.: Lift enhancement over flexible nonslender delta wings. AIAA J. **45**(12), 2979–2993 (2007)
15. Tibshirani, R.: Regression shrinkage and selection via the lasso. J. R. Stat. Soc. Ser. B (Method) JSTOR, 267–288 (1996)
16. Williams, D.R., Seidel, J.: Crossed-actuation AFC for lateral-directional control of an ICE-101/Saccon UCAV. In: AIAA Paper 2016–3167 (2016)
17. Woo, G., Crittenden, T., Glezer, A.: Transitory control of a pitching airfoil using pulse combustion actuation. In: AIAA Paper 2008–4324 (2008)
18. Woo, G., Crittendon, T., Glezer, A.: Transitory separation control of dynamic stall on a pitching airfoil. In: King, R. (ed.), Active Flow Control II. NNFM, vol. 108, pp. 3–18. Springer, Heidelberg (2010)

Implementing Rotating Stall Control in a Radial Diffuser Using Microjet Arrays

Jennifer Gavin, Erik Fernandez, Prabu Sellappan, Farrukh S. Alvi, William M. Bilbow and Sun Lin Xiang

Abstract This study is part of our effort to implement and refine microjet-based flow control in realistic and challenging applications. Our goal is to reduce/eliminate rotating stall in the radial diffuser of a production compressor used in commercial heating, ventilation, and air conditioning (HVAC) systems, using microjet arrays. We systematically characterize the flow using pressure and velocity field measurements. At low load conditions, the flow is clearly stalled over a range of RPM where the presence of two rotating stall cells was documented. Circular microjet arrays were integrated in the diffuser and the flow response to actuation was examined. The array closest to the initiation of stall cells was most effective in reattaching the flow. Control led to a very significant increase in the stall margin, reducing the minimum operational mass flow rate to 14% of the design flow rate, half of the original 28% flow rate before microjet control was implemented. The results will show that the parameters found be most effective in the simple configurations proved to be near-optimal for the present surge control application in a much more complex geometry. This provides us confidence that the lessons learned from prior studies can be extended to more complex configurations.

Keywords Active flow control · Microjets · Stall · Radial diffuser · PIV

J. Gavin · E. Fernandez · P. Sellappan (✉) · F. S. Alvi
Department of Mechanical Engineering, Florida Center for Advanced Aero-Propulsion (FCAAP), FAMU-FSU College of Engineering, Tallahassee, FL 32310, USA
e-mail: psellappan@fsu.edu

F. S. Alvi
e-mail: falvi@fsu.edu

W. M. Bilbow · S. L. Xiang
Danfoss Turbocor, Tallahassee, FL 32310, USA

R. King (ed.), *Active Flow and Combustion Control 2018*,
Notes on Numerical Fluid Mechanics and Multidisciplinary Design 141,
https://doi.org/10.1007/978-3-319-98177-2_3

1 Introduction

Flow separation is ubiquitous in a wide range of internal and external flows, ranging from the simplest, e.g. on the leeward side of a cylinder or a backward facing ramp to the very complex such as in turbomachinery components of propulsion, power generation and other systems. Separating flows are nearly always undesirable as they lead to unsteadiness, increased aerodynamics loads (steady and unsteady), reduced pressure, reduced operational envelope (for aircraft, propulsion and power systems) resulting in a loss in performance and efficiency. As such, control of separated flows has been the focus of research for a number of decades where passive and active methods have been explored in the academic and applied fluid dynamics community.

Some of the more pervasive passive methods that have been explored for separation control, include vortex generators such as vanes, bumps, dimples and ramps among others [1, 2]. While passive methods have sometimes shown a favorable influence on separated flows, their impact is often limited to a narrow range of flow conditions as passive devices cannot be adjusted to changes in operational conditions. Active methods, ones that requires energy input, have shown more promise, in large part due to their potential ability to adapt to changing conditions. Hence a range of actuators for active flow control have been explored including acoustic excitation through speakers [3, 4]; synthetic jets (Zero Net Masss Flux, ZNMF actuators [5–7] that have been examined experimentally and computationally [8, 9]; dielectric barrier discharge plasma actuators [10], plasma arc actuators [11, 12], plasma driven jets [13–15], and many more. A recent review by Cattafesta and Sheplak [16] treats this topic in a comprehensive manner. One actuation technique that has shown considerable promise is the use of strategically located, very small-relative to the relevant length scale, high momentum, jets commonly referred to as 'microjets'.

Microjet-Based Flow Control—This control approach has been implemented in a wide array of fundamental-canonical and application-driven flows, including by the present authors. The applications include the use of steady and unsteady microjet arrays for the control of flow oscillations in cavities [17–19], aeroacoustics of free jets [20–22] and the control of the highly unsteady supersonic impinging jets [23, 24]. In almost all these applications significant improvements were achieved through this active flow/noise control approach. As discussed in these references, the primary mechanism for their effectiveness is the ability of the microjets arrays, often injected in a wall-normal direction, to generate coherent streamwise vortices which increase longitudinal momentum near the wall and streamwise vorticity through the jets in crossflow (JICF) mechanism. The enhanced mixing so achieved can be leveraged for the control of different flow—applications through a careful choice of the location and operational conditions of actuation in the context of the base flow. An abbreviated discussion regarding the physical mechanisms is provided in Sect. 2.2. More details can be found in Refs. [24–26].

AFC of Separated Flows—Control of flow separation is an flow application where microjet-based control has been particularly effective. The primary reason for this is the JICF induced, enhanced mixing which energizes the near-wall, low momentum fluid by efficiently mixing it with the higher momentum fluid in the upper region of the boundary layer [25–27]. Separation control has been examined extensively, starting with very fundamental flows and advancing to more complex, realistic, application-driven configurations—a very brief review of a selected number of these efforts follows next. The initial studies examining the use of microjet arrays for the control of boundary layer separation were conducted on a modified Stratford ramp that was nominally two-dimensional, although the separated flowfield was highly three-dimensional. The ramp was equipped with multiple linear arrays of steady microjets that allowed for individual arrays to be activated as needed. The angle of attack of the ramp was also variable allowing the adverse pressure gradient and hence the size of separation to be controlled. The effect of steady microjets in controlling increasingly larger separated flowfields was systemically examined where microjets were able to completely eliminate the strongest separation generated in this study. Further, the cost of this control in terms of the requisite mass flow through the actuator array was very low, fraction of a percent, once the optimal actuation conditions were identified. Details of this study may be found in Kumar and Alvi [25].

This control strategy was next explored for controlling separation induced stall on the wing of an RC aircraft during a high α maneuver. Using tuft based visualization and GPS data the ability to extend the stall angle was demonstrated [28]. Steady microjet actuation for separation control was also demonstrated on a *Low Pressure Turbine* (LPT) blade in a simplified linear cascade. The efficacy of control on an L1A blade (L1A is the arbitrarily designated model name), which is an Air Force Research Laboratory (AFRL) designed configuration with a higher loading than conventional low-pressure turbine profiles, was demonstrated over a range of conditions. Separation was completely eliminated on the blade with AFC resulting in a very significant reduction in the integrated wake loss coefficient [29].

Present Study—The study described in this paper is part of our continuing effort to implement, demonstrate and refine microjet-based active flow control in increasingly realistic and challenging applications. The results of the studies in canonical configurations were used to guide the design of the actuators for the present. Our goal is to reduce rotating stall in the radial diffuser of a production compressor used for commercial HVAC systems. These compressors use R-134a refrigerant where energy added to the fluid by the impeller is recovered as high pressure when the refrigerant is decelerated through a radial diffuser. For 'low-load' conditions, where the mass flow rate through the compressor is reduced, the reduced velocity and hence momentum of the fluid as it moves radially outwards through the diffuser makes it susceptible to separation which leads to rotating stall [30]. This is obviously problematic as it limits the compressor efficiency, increases vibration and noise and limits the viable operational range of the machine.

Passive methods such as vortex generators, vanes, and bumps, as well as active methods such as acoustic excitation have been utilized with limited success. They require high degrees of customization in order to remain effective, and even when optimally designed are only effective for specific flow ranges. Kurokawa et al. [31] have shown that customized radial grooves completely suppressed rotating stall in a vaneless diffuser, however, the implementation resulted in an overall loss of pressure in the diffuser. Alternatively, low-speed blowing has been found to be effective in controlling flow separation in a vaneless diffuser, but resulted in a loss of impeller-diffuser combination performance [32].

The goal of this study was to attempt to eliminate or significantly reduce stall at the low load conditions, corresponding to low mass flow rate, thus improving efficiency and extending the operational range. Implementing AFC in a production turbomachinery system is non-trivial due to the system complexity, limited space and access, and limited measurement capability. Furthermore, limited information is available in open literature regarding the rotating stall flowfield, even for simple configurations and even less so for production systems- these are challenges that had to be addressed in this project.

2 Experimental Hardware and Methods

2.1 Model Details—Radial Diffuser and Compressor

The experiments were conducted at the Florida Center for Advanced Aero-Propulsion at the Florida State University. The TT300 compressor was supplied by Danfoss Turbocor Compressors Inc. The original design (Fig. 1a) features a diffuser built into a large shroud, creating a high degree of difficulty in reaching the diffuser. In order to improve access, while realistically modeling the problems faced by the production line of compressors, the compressor was modified in collaboration with the manufacturer to provide simplified access to the unbladed radial diffuser during testing. The multistage compressor was modified for testing with a single stage. The shroud was removed and the impeller shaft was elongated to bring the diffuser to the edge of the compressor body. The flow path was slightly modified, creating a diffuser that was purely radial, but retained the size (radius, $R_{Diffuser} = 127$ mm) and basic shape of the production model diffuser. Air replaced R-134a as the working fluid and the impeller was also modified for the alternative working fluid. Schematic of the modified diffuser-compressor configuration is shown in Fig. 1b and the general flow direction of the working fluid in the diffuser is indicated in Fig. 4.

2.2 Physical Mechanism and Implementation of Microjets in Diffuser

Steady microjets injected into the diffuser crossflow (see Fig. 2) alter the flowfield both locally and globally by promoting mixing in the boundary layer. This is due to the creation of counter rotating vortex pairs (CVP), which are streamwise oriented vortices, that transfer high momentum fluid from the freestream into the boundary

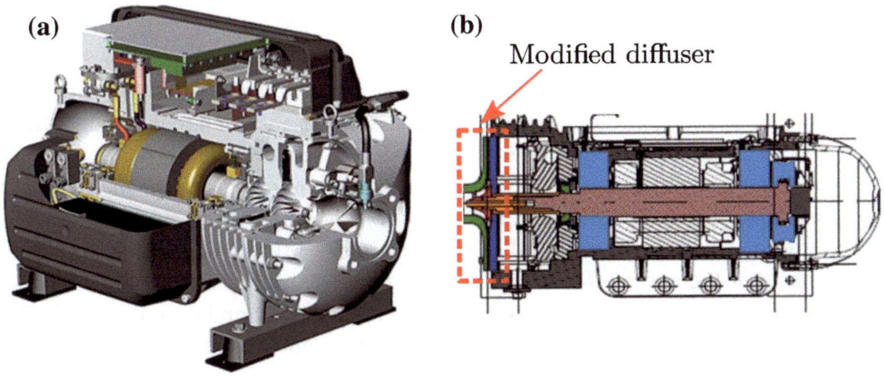

Fig. 1 **a** Three-dimensional cut-away of production configuration TT300 compressor, and **b** cross-section of modified diffuser configuration used for testing

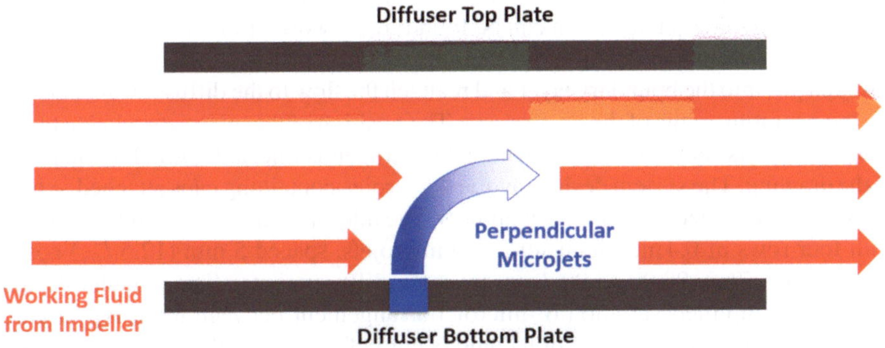

Fig. 2 Schematic of microjet injection into diffuser cross-flow

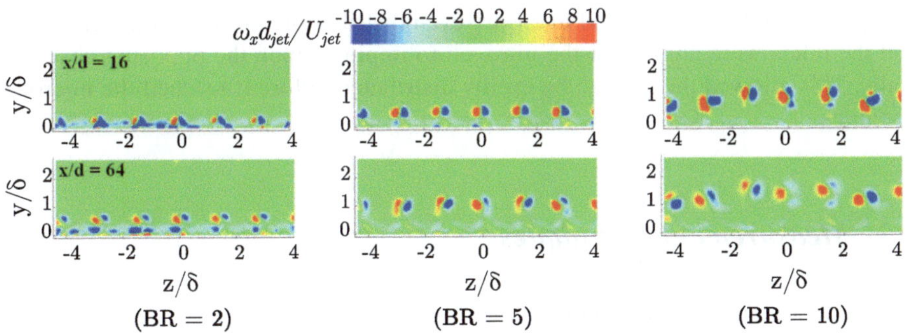

Fig. 3 Flowfield induced by microjets injected into a crossflow, with inter-microjet spacing of $25d_{jet}$, for different microjet blowing ratios. Streamwise vorticity contours along flow normal planes obtained from stereo-PIV measurements at two different streamwise locations ($x/D = 16$—top row, $x/D = 64$—bottom row). Figure has been redrawn to improve print quality and is similar to the original figure from Fernandez and Alvi [27]

layer. A representative example from Fernandez et al. [27] showing the growth and evolution of multiple CVPs at different blowing ratios can be seen in Fig. 3, which indicates the flowfield in planes normal to the flow and downstream of steady microjet injection. The vorticity contours clearly indicate the presence of streamwise vortices created by multiple microjets. The microjet spacing, location and blowing ratio for the present surge control application were informed by prior studies in canonical configurations that were aimed at exploring the fundamental physical mechanism behind microjet-based control. Studies by Fernandez [33] and Fernandez et al. [27] discovered that the most effective blowing ratio, BR, ranged between 1.5 and 3, as this is where the CVP remain within the boundary layer and efficiently mix the low momentum and high momentum fluid within the boundary layer. Similarly, these studies also demonstrated that while inter-jet spacing of $12.5d_{jet}$ and $25d_{jet}$ were both effective, in terms of benefit versus cost, $25d_{jet}$ was the most efficient. This spacing maximized the interactions between neighboring CVPs in enhancing inter-boundary layer mixing. A more detailed analysis of the fundamental physical mechanism underpinning microjet actuation can be found in the work of Fernandez [33].

Microjet assemblies, consisting of multiple microjets, were designed to inject nitrogen gas into the boundary layer and reattach the flow to the diffuser wall, extending the operating range of the compressor. The position of the microjets relative to the working fluid is shown in Fig. 4a, where the nitrogen is injected normal to the plane of the drawing. The system design involved microjet-containing caps attached to four individual pressurized stagnation chambers. The microjet caps were aluminum plates with four rows of 0.4 mm diameter (d_{jet}) microjets, spaced 5 mm ($12.5d_{jet}$) apart, located 63.5, 76.2, 88.9 and 99.7 mm from the diffuser center. The modular design allows for simplified variation of microjet arrangement between tests. By varying the selected microjet row, injection could occur before, on, or after the approximate boundary of stall cells. Blocking alternate microjet openings allowed for tests with $25d_{jet}$ inter-microjet spacing. The stagnation chambers were individually filled with nitrogen. Two inlets were used to ensure even distribution throughout the chamber, and the inlets were angled for the nitrogen to impinge upon the opposing chamber walls. This allowed the fluid to be evenly distributed before it reached the microjet openings.

2.3 Measurement Techniques

Steady Pressure Measurements Sixteen static pressure taps were placed at incremental distances from the center of the diffuser, as shown in Fig. 5. Surface pressures were scanned by a Scanivalve™ pressure transducer. The steady pressure data were used to determine the approximate radial location of stall, shown by rapid rise in the static pressure to atmospheric conditions (As seen in Fig. 7, to be discussed later).

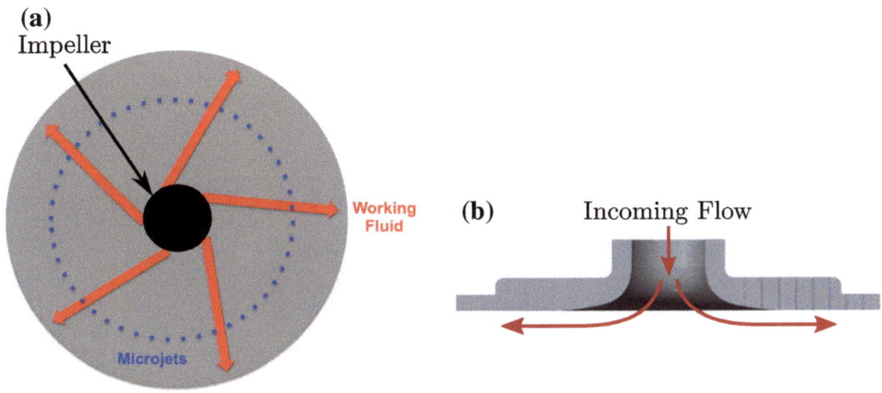

Fig. 4 **a** Diffuser back plate schematic. Microjets injected normal to the plane of the drawing, and **b** diffuser upper plate showing airflow inlet, tapered to accommodate smooth flow from the impeller

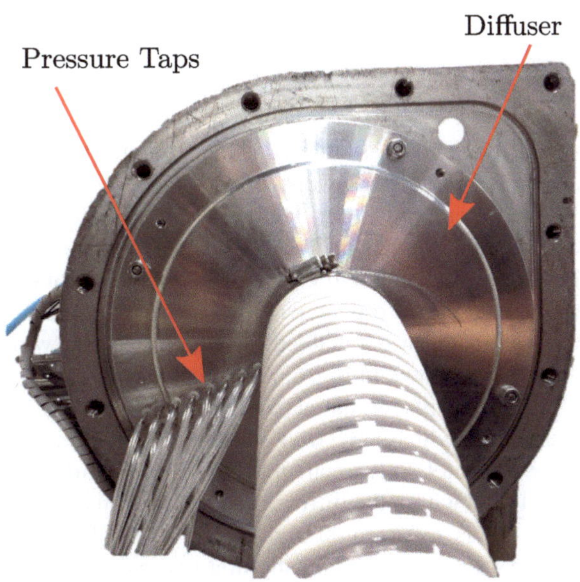

Fig. 5 Steady pressure taps are shown placed at incremental distances from the diffuser center, with tubing attached

Unsteady Pressure Measurements Using the radial location of stall determined by the steady pressure measurements, Endevco® piezoresistive pressure transducers with a 1 psig range were installed at various intervals along the stall radius. Two transducers were located 77 mm from the center of the diffuser, and five were located at a radial distance of 111 mm. These radial distances were chosen both to encompass stalled flow region, as well as to accommodate the geometric constraints of the diffuser. The two located 77 mm from the center were spaced 12.7 mm apart, and the other five were located with 12.7, 25.4, 38.1, and 50.8 mm spacing between each tap. The machined holes for Endevco transducers may be seen in Fig. 6. Transducer

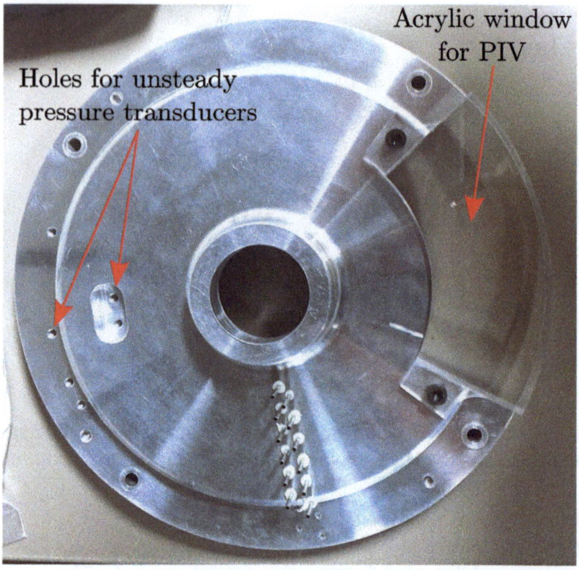

Fig. 6 Diffuser top plate shown with holes for unsteady pressure transducers and an acrylic window added for optical access

signals were low-pass filtered using a Stanford Research Systems filter and sampled at a rate of 5120 Hz. Power spectral density (PSD) estimates were calculated through Welch's method using a Hanning window and 75% overlap. This data was used to determine frequency and velocity of stall motion.

Particle Image Velocimetry and High Speed Imaging Particle Image Velocimetry (PIV) studies were conducted to further characterize the flow inside the diffuser. Optical access was obtained through an acrylic window on the top surface of the diffuser, as shown in Fig. 6. A LaVision Imager sCMOS double frame camera with 2560×2160 pixels spatial resolution was used to capture images at 15 Hz. The flow was seeded with micro sized oil droplets and illuminated by a 200 mJ/pulse Nd:YAG laser (Quantel Evergreen) with the illumination plane coincident to the plane of radial flow within the diffuser. Dual frame images were cross-correlated using LaVision DaVis software through a multi-pass algorithm using interrogation windows of size 48×48 pixels for the first two passes and 24×24 pixels for the last four passes and with 75% overlap. Due to the periodic nature of the stalled flow, phase conditioned measurements were required to characterize the flow field. An Endevco pressure sensor was used, where its band-passed signal was fed into a phase-locking delay generator. The LaVision system was triggered using the signal from the delay generator and images were captured at 8 different phases of the stall cycle. Three hundred dual frame images were captured for each phase and processed to obtain velocity fields, where the instantaneous velocity fields were ensemble averaged to obtain each phase-averaged velocity field. Due to the periodicity of the flow, the phases were stitched together, in order to obtain the complete velocity field of the stalled flow. Once stall was eliminated, phase locking was no longer possible since no discrete signal (stall frequency) existed for locking. Conventional non-phase conditioned PIV was also attempted, however the attached/unstalled flow field was found to be unremarkable with the flow exiting radially in a uniform manner.

Flow fields with particle seeding were also imaged using a high-speed Photron FASTCAM SA5 camera at a resolution of 1024 × 1024 pixels and sampling rate of 1280 Hz, with illumination provided using a 10 mJ/pulse Nd:YLF dual head laser (New Wave Research Pegasus-PIV). The high-speed images are used for visualization of stall cells.

3 Results

3.1 Baseline (Uncontrolled) Flow

3.1.1 Radial Pressure Distributions

Steady Pressure Distribution Steady pressure data was collected from the 16 radial taps during compressor operation. Impeller speeds of 10000, 15000, 20000, 24000, and 32000 RPM were tested. For each impeller speed, the mass flow rate was slowly decreased while the pressure was closely monitored. Rotating stall cells were assumed to form at the mass flow rate associated with a sudden onset of strong fluctuations in the unsteady pressure. The mass flow rate was held at the initial position of induced stall, and pressure data were again acquired. Figure 7 shows the normalized pressure distribution in both the unstalled and stalled cases as a function of non-dimensional radial location. For the unstalled case, an initial rapid increase in pressure as the flow exits the impeller ($r/R_{Diffuser} = 0$) was followed by a gradual increase to the diffuser exit pressure (ambient pressure conditions). The deviation from a smooth curve between $0.15 < r/R_{Diffuser} < 0.4$ can be explained by con-

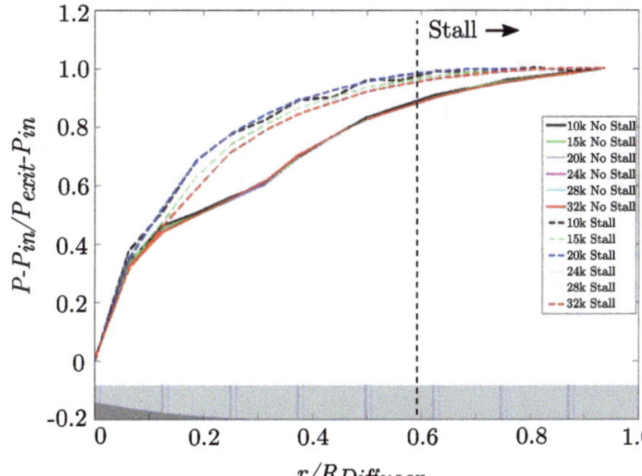

Fig. 7 Pressure distribution in the presence (dashed) and absence (solid lines) of stall

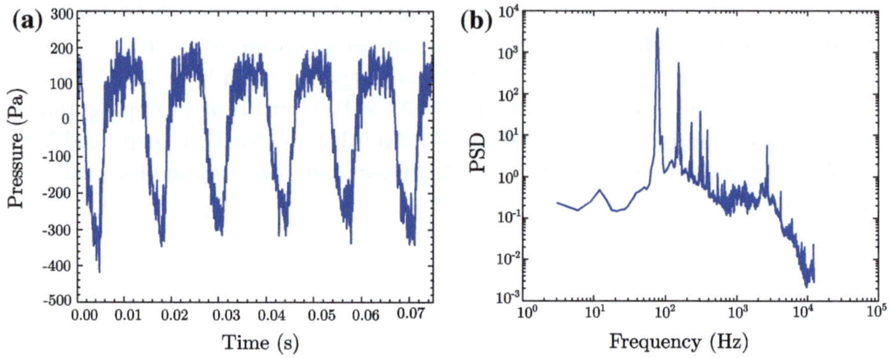

Fig. 8 Unsteady pressure during stalled flow at 20000 RPM. **a** Time series showing periodic oscillations, and **b** PSD with dominant peak at 78 Hz

sidering the diffuser profile contraction, shown in Fig. 4b, which contributes to a change in rate of pressure increase. The stalled pressure distribution shows the same rapid rise as seen in the unstalled case immediately after the impeller exit. Deviation from the unstalled condition occurs at approximately 0.15 $r/R_{Diffuser}$, with rapid pressure rise in the diffuser during stalled flow persisting until approximately 0.6 $r/R_{Diffuser}$. At this point, the non-dimensional pressure approaches ~95% of the exit atmospheric pressure. The region from $0.6 < r/R_{Diffuser} < 1.0$ in the stalled pressure distribution corresponds to the location of the stall cells, where airflow at atmospheric pressure was pulled back into the diffuser.

Unsteady Pressure Distribution Unsteady pressure data along the determined stall radius was recorded during stalled and unstalled conditions. Time series of the pressure from all unsteady pressure sensors during constricted mass flow (stalled) show distinct oscillations associated with rotating stall. This can be seen in Fig. 8a for an impeller speed of 20000 RPM at a mass flow rate of 26.6% unstalled mass flow, along with its associated power spectrum (Fig. 8b). The dominant frequency identified in the spectrum, 78 Hz, is the frequency of the rotating stall cells. Based on the stall frequency, the lag between simultaneous measurements from sequential sensors, the inverse of the sampling rate, and the angular separation of the pressure sensors, the speed and number of stall cells present in the diffuser were calculated. These results show that an increase in impeller speed corresponds to increases in stall cell velocity and percent of baseline mass flow rate at which stall begins. The impeller speeds with the highest mass flow rates at which stall is initiated should have the greatest potential for improvement.

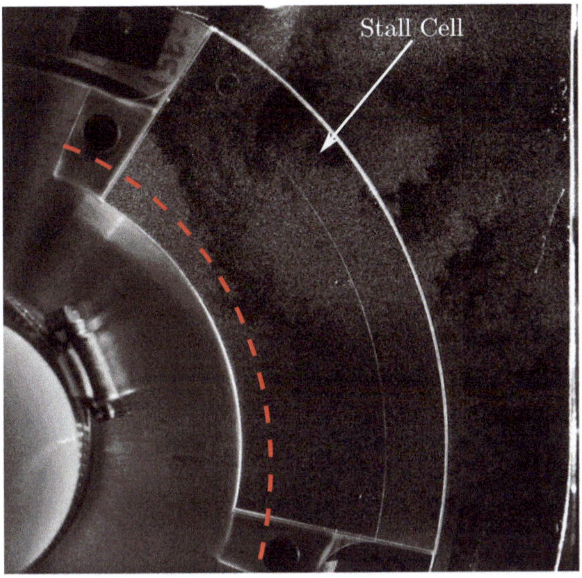

Fig. 9 Instantaneous high-speed image showing stall cell. Dashed line indicates stall boundary

3.1.2 Velocity Field

The velocity field in the radial diffuser was obtained through phase-locked PIV. Stall cells are clearly visible even in instantaneous images (Fig. 9), where the dark regions (marked in Fig. 9) correspond to reverse flow into the diffuser from the unseeded ambient environment. A complete flow field was constructed by shifting the measured quadrant by the proper phase offset. Figure 10 shows a complete stall cycle at 20000 RPM, using the eight phases measured. This was possible due to the rotational nature of the flow. The stall regimes are visible by areas of radially inward flow, as seen in the streamlines shown in Fig. 10. This locally reverse region can also be seen as blue colored contours and are marked in the figure. The velocity field results shown here confirm the radial location of stall cells, beginning approximately 76 mm from the center of the diffuser. This information was used to choose the appropriate radii for microjet injection, as detailed in the following section.

3.2 Effect of Microjet Control

After identifying the onset of stall at different RPMs using mean and unsteady pressure measurements (Figs. 7 and 8) and subsequently characterizing the velocity field, see Fig. 10, we systemically implement steady microjet control. These experiments were conducted over a range of operating and actuation conditions, only a summary is provided here along with some representative results for the sake of brevity. Hence,

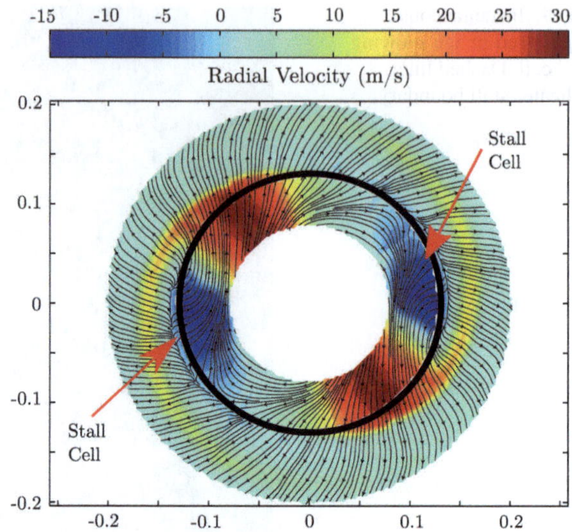

Fig. 10 Contours of radial velocity at impeller speed of 20000 RPM. Spatial dimensions are in units of meter and solid black line represents the edge of the diffuser

although actuator arrays at four radial locations in the diffuser were tested, we only present results for the most effective case, corresponding to a radius of 63.5 mm (see Sect. 2.2). Prior studies on the use of this control strategy identified that the most effective range of blowing ratio (BR) for control of separated flows ranges between 0.5 and 3 [27, 33]. This is the range of blowing ratios examined here, where it is defined as the ratio of the velocity of the jets issuing from the actuator to that of the crossflow. The microjet velocity was estimated by measuring the mass flow rate as well as the stagnation pressure and temperature in the microjet supply chamber. The latter measured properties then provide an estimate of the microjet velocity through the actuator orifices and the total mass flow rate. By comparing the two mass flow rates, a loss coefficient was estimated to account for head losses through the orifices. Following this procedure, the requisite microjet stagnation chamber pressures corresponding to each blowing ratio were determined for all combination of array configurations.

3.2.1 Unsteady Pressures and Flowfield

The innermost circular array of microjets, located 63.5 mm from the diffuser center and ~31.8 mm inward of the stall boundary (based on pressure and velocity field results) proved to be the most effective for elimination of stall cells. The effect of this control is readily evident in the unsteady pressure signals as shown in Fig. 11a.

As seen here, the large-scale fluctuations characteristic of stall cells (in the green trace) were dramatically reduced, essentially to a pressure signature corresponding to unstalled flow (blue trace) through the application of microjets. Figure 11b shows

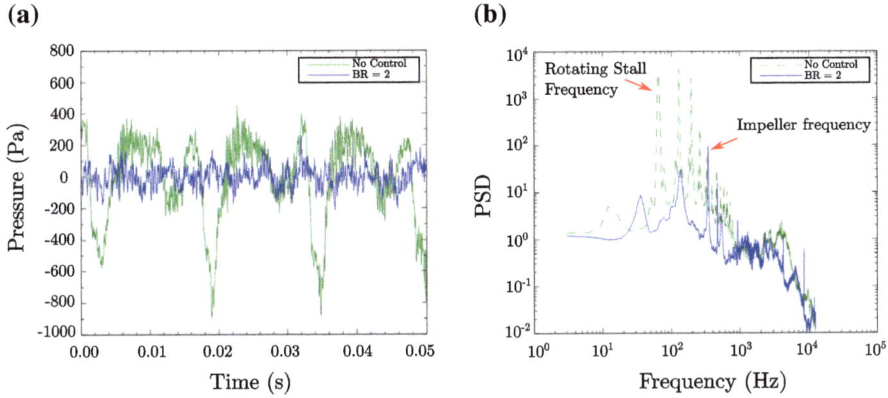

Fig. 11 Unsteady pressures at impeller speed of 28000 RPM with and without control. **a** Time series, and **b** Corresponding spectral content

the same impact of flow control in the spectral domain—the high-amplitude spectral peaks corresponding to stall, in the dashed green line spectrum, have been eliminated in the blue spectrum. The remaining lower amplitude peaks in the controlled case correspond to impeller passage frequency. The results shown here correspond to an inter-microjet spacing of $25d_{jet}$ with the compressor operating at 28000 RPM. This impeller speed is of particular interest, since this speed is most similar to normal compressor operating conditions.

Flow control tests were also conducted at two other RPMs, using two different inter-microjet spacing: $25d_{jet}$, discussed above, and $12.5d_{jet}$. Results of these tests are summarized in Table 1. As seen here, microjets arrays with a $25d_{jet}$ spacing achieved an 11.5% extension of the compressor operating range, while microjets spaced 12.5 diameters apart improved the operating range by 13.9%. Compared to the original stall limit, the $25d_{jet}$ and $12.5d_{jet}$ spacing microjets showed improvements of 40.4% and 49.1%, respectively. Although a higher improvement was achieved using the closer jet spacing this does not necessarily translate to higher efficiency, when accounting for the cost of control. This is discussed in a subsequent section.

Finally, in Fig. 12 we depict the effect of microjet control on the surge limit at the three RPMs corresponding to Table 1. The cases shown here correspond to $25d_{jet}$ spacing only. The surge limit has now been moved to significantly lower mass flow rates with a concomitant increase in the pressure recovery through the diffuser. The pressure ratio between the diffuser inlet and exit is a pivotal measure of compressor performance and microjet control has significantly reduced the adverse impact due to rotating stall. As seen here, the improvement in the operating range and pressure ratio are more significant at higher impeller speeds, translating to greater improvements at normal operating conditions.

For cases where rotating stall is completely eliminated and the radial flow is completely attached due to microjet control, the velocity field contains no unusual features as noted in the discussion of the PIV results in Sect. 2.3. However, the

Table 1 Improvement in stall limit with microjets located 63.5 mm from diffuser center

Microjet spacing	Impeller RPM	Design flow rate (kg/min)	Original stall limit (kg/min)	Percent design flow rate (%)	New stall limit (kg/min)	Percent design flow rate (%)	Extension of operating range (%)	Improvement from original stall limit (%)
$12.5d_{jet}$	20000	7.90	2.10	26.6	1.10	14.0	12.6	47.5
	25000	9.85	2.76	28.0	1.35	13.7	14.3	51.0
	28000	10.84	3.07	28.3	1.56	14.4	13.9	49.1
$25d_{jet}$	20000	7.90	2.10	26.6	1.35	17.1	9.5	35.7
	25000	9.85	2.76	28.0	1.70	17.3	10.7	38.4
	28000	10.84	3.07	28.3	1.83	16.9	11.5	40.4

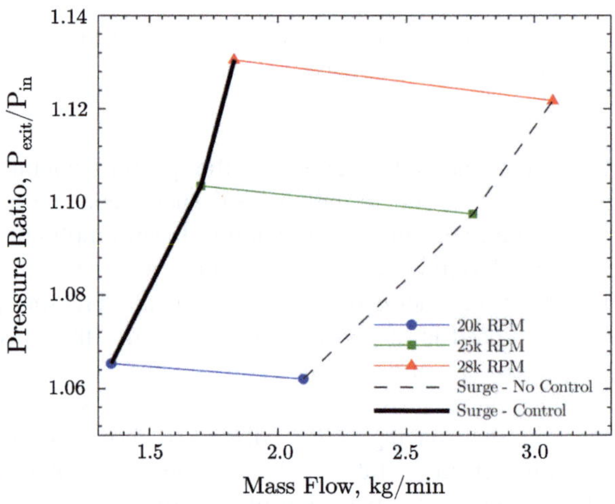

Fig. 12 Effects of control on surge limit and pressure ratio. Shown for impeller speeds of 20000, 25000 and 28000 RPM

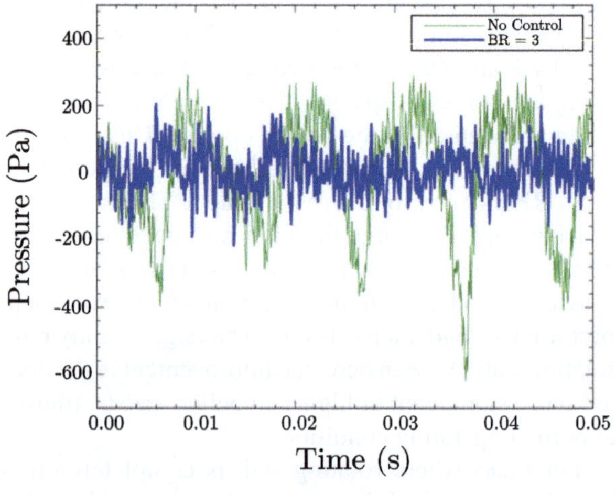

Fig. 13 Time history showing reduction of large-scale pressure fluctuations with control. Impeller speed is 28000 RPM

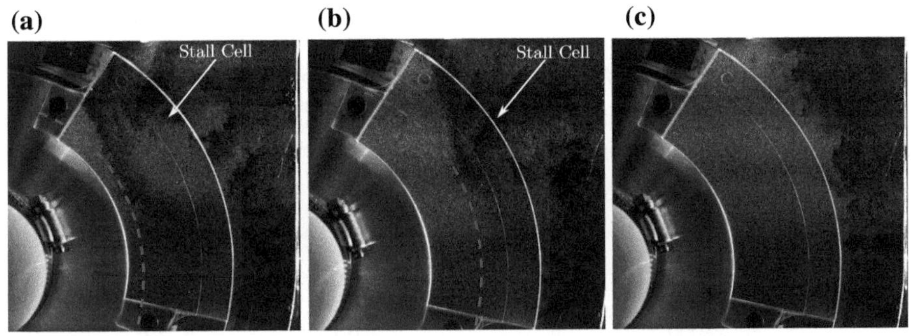

Fig. 14 Elimination of stall cell under microjet control. **a** Stall cell before microjet initiation (same as Fig. 9), **b** stall cell moving outward in the diffuser as microjets take effect, **c** complete flow reattachment. Dashed line indicates stall boundary

progressive reduction in stall as the actuator control authority is gradually increased provides interesting insight. This is seen in the images shown in Fig. 14, which show a succession of high speed, instantaneous images acquired at 1280 Hz. The flow was seeded to enable visualization as microjet actuator flow was progressively increased. Starting from the leftmost image, the presence of stall cells very close to impeller exit is clearly seen—this corresponds to the stalled condition. Moving to the right, one can see the transient effect of actuation where the stall cells—visible as dark, unseeded regions, are progressively 'pushed' radially outward towards the diffuser exit. This culminates in the completely attached flowfield, seen in the last image on the right (Fig. 14c).

3.2.2 Parametric Effects

While the innermost actuator array proved to be most effective, the impact of the other three actuator arrays, located at 76.2, 88.9 and 99.7 mm (from the diffuser center), was also systematically examined. The second, 76.2 mm, row of microjets was significantly less effective in stall control. While the first array led to a significant improvement in the stall limit, stall occurred at the same mass flow rate with and without microjets for the second row. However, there was some beneficial impact at certain conditions in that pressure fluctuations were somewhat attenuated with microjet actuation. This can be seen in the pressure time history shown in Fig. 13. These results are shown for the $25d_{jet}$ spacing at 28000 RPM and 28.3% baseline mass flow rate, the original stall limit. Similar results were found at all blowing ratios and microjet spacings for this radial actuator. Furthermore, control effectiveness is reduced as impeller speed are increased. Finally, the last two microjet arrays located at 88.9 and 99.7 mm were completely ineffective in eliminating or reducing rotating stall cells. This is expected as the most effective control schemes require actuation

Table 2 Effective blowing ratios and microjet flow rates for microjets located 63.5 mm from diffuser center

Microjet spacing	Impeller RPM	Minimum blowing ratio for control	Microjet flow rate (kg/min)	Percent diffuser flow to supply microjets (%)
$12.5d_{jet}$	20000	2.5	0.15	14.0
	25000	2	0.16	12.1
	28000	2	0.22	14.1
$25d_{jet}$	20000	3	0.13	9.6
	25000	2.4	0.14	8.5
	28000	2	0.16	8.6

in the near vicinity of separation whereas these arrays are significantly downstream of where separation is first initiated in the diffuser.

When evaluating the effectiveness of a control strategy one must also consider the cost of control. In the present case, one of the most easily measured and commonly used 'cost' is the mass flow required through the microjet arrays. Similarly, in the context of the present application the relative extension of the stall limit is an appropriate measure of the benefit. The cost-benefit of this control scheme was characterized through parametric studies for the most effective actuator array. Here the blowing ratio was varied at each RPM and for two inter-microjet actuator spacings, the corresponding results are summarized in Table 2.

As the 28000 RPM condition is most relevant for this compressor in the context of its desired operating conditions, this case is highlighted in gray in Table 2. According to Table 1, the closer spacing of microjets improves the stall limit by 49% relative to the $25d_{jet}$ inter-microjet spacing which lead to a 40% improvement. However, the mass flow required for the $12.5d_{jet}$ spacing is considerably higher—14.1% than that for the larger spacing which require a mass flow of 8.6%. Unless the additional increase in the stall limit is critical, the array with the larger spacing is much more effective. The interaction of the CVPs generated by the JICF account for the difference in efficacy due to microjet spacing as discussed in Refs. [27, 33]. These results reaffirm the fact that efficient control requires an understanding of the underlying physical mechanisms and simply 'more control' or ill—conceived control is rarely effective.

4 Conclusion

In this paper we describe our study where we implement microjet-based active flow control to reduce, ideally eliminate, stall in the radial diffuser of a production compressor used for commercial HVAC systems. This research is part of the continuing effort to translate microjet based control out of the laboratory to increasingly realistic and challenging applications. The results of the studies in canonical configurations were used to guide the design of the actuators for the present. In the present application, radial stall occurs in the diffuser at low load conditions where the reduced mass flow rate through the system leads to lower radial momentum which makes it susceptible to separation, i.e. stall, at certain RPMs. This can significantly limit the operational regime of such machines and delaying separation would result in efficiency gains.

Using a systematic approach, we first characterized the baseline, uncontrolled flowfield at unstalled and stalled conditions. Through steady and unsteady pressure as well as velocity field measurements the presence of two rotating stall cells was clearly identified, along with the extent of the separated flowfield and the conditions (RPM, mass flow rate) at which these occur. This was used to design and integrate a number of circular microjet arrays at various radial locations in the diffuser. The flow response to microjet actuation was examined over a range of compressors RPMs and diffuser mass flow rates where different actuator arrays were systematically activated over various blowing ratios. The array closest to the initiation of stall cells was found to be most effective in reattaching the flow, as somewhat anticipated based on our prior studies of separation control. Control of the modified compressor led to a very significant increase in the stall margin, reducing the minimum operational mass flow rate to 14% of the design flow rate. This is nearly half of the original 28% flow rate at which stall occurs before microjet control is implemented. This study clearly demonstrates the potential efficacy of this relatively simple and robust control strategy in a real-world application. The results showed that the parameters found be most effective in the simple configurations proved to be near-optimal for the present surge control application in a much more complex geometry. This provides us confidence that the lessons learned from prior studies can be extended to other applications. The next step in further evaluating this control strategy is to implement it in a closed compressor system using the actual working fluid used therein, R-134a.

Acknowledgements This research was in part supported by the Florida Center for Advanced Aero-Propulsion and Danfoss Turbocor Compressors Inc. We also acknowledge the involvement, support and encouragement of Dr. Joost Brasz of Danfoss Turbocor during this research. He was one of the early driving forces behind this project at Danfoss, unfortunately he passed away soon after the completion of this phase of the research. He is missed by us all.

References

1. Storms, B.L., Ross, J.C.: Experimental study of lift-enhancing tabs on a two-element airfoil. J. Aircr. **32**(5), 1072–1078 (1995)
2. Lin, J.C.: Review of research on low-profile vortex generators to control boundary-layer separation. Prog. Aerosp. Sci. **38**(4–5), 389–420 (2002)
3. Zaman, K.B.M.Q., Bar-Sever, A., Mangalam, S.M.: Effect of acoustic excitation on the flow over a low-Re airfoil. J. Fluid Mech. **182**, 127–148 (1987). https://doi.org/10.1017/S0022112087002271

4. Ahuja, K.K., Whipkey, R.R., Jones, G.S.: Control of turbulent boundary layer flows by sound. In: AIAA Paper 1983, p. 726 (1983)
5. Amitay, M., Pitt, D., Glezer, A.: Separation control in duct flows. J. Aircr. **39**(4), 616–620 (2002). https://doi.org/10.2514/2.2973
6. Glezer, A., Amitay, M.: Synthetic jets. Annu. Rev. Fluid Mech. **34**, 503–29 (2002)
7. Smith, B.L., Swift, G.W.: A comparison between synthetic jets and continuous jets. Exp. Fluids **34**(4), 467–472 (2003)
8. Raju, R., Mittal, R., Cattafesta, L.: Towards physics based strategies for separation control over an airfoil using synthetic jets. In: AIAA Paper 2007, p. 1421 (2007)
9. Raju, R., Aram, E., Mittal, R., Cattafesta, L.: Simple models of zero-net mass-flux jets for flow control simulations. Int. J. Flow Control **1**, 179–97 (2009)
10. Corke, T.C., Enloe, C.L., Wilkinson, S.P.: Dielectric barrier discharge plasma actuators for flow control. Annu. Rev. Fluid Mech. **42**, 505–29 (2010)
11. Samimy, M., Adamovich, L., Webb, B., Kastner, J., Hileman, J., et al.: Development and characterization of plasma actuators for high-speed jet control. Exp. Fluids **37**, 577–88 (2004)
12. Samimy, M., Kim, J.H., Kastner, J., Adamovich, I., Utkin, Y.: Active control of high-speed and high-Reynolds number jets using plasma actuators. J. Fluid Mech. **578**, 305–30 (2007)
13. Popkin, S.H., Cybyk, B.Z., Foster, C.H., Alvi, F.S.: Experimental estimation of SparkJet efficiency. AIAA J., 1831–1845 (2016)
14. Narayanaswamy, V., Raja, L.L., Clemens, N.T.: Characterization of a high-frequency pulsed-plasma jet actuator for supersonic flow control. AIAA J. **48**, 297–305 (2010)
15. Cybyk, B., Grossman, K., Wilkerson, J.: Performance characteristics of the Sparkjet flow control actuator. In: 2nd Conference AIAA Flow Control, AIAA paper 2004, p. 2131, Portland (2004)
16. Cattafesta, L.N., Sheplak, M.: Actuators for active flow control. Annu. Rev. Fluid Mech. **43**, 247–72 (2011)
17. Zhuang, N., Alvi, F.S., Alkislar, B., Shih, C.: Supersonic cavity flows and their control. AIAA J. **44**(9), 2118–2128 (2006)
18. Zhuang, N., Alvi, F.S., Shih, C.: Another look at supersonic cavity flows and their control. In: AIAA 2005, p. 2803 (2005)
19. Bower, W.W., Kibens, V., Cary, A.W., Alvi, F.S., Raman, G., Annaswamy, A., Malmuth, N.: High-frequency excitation active flow control for high-speed weapon release (HIFEX). In: AIAA Paper 2004, p. 2513 (2004)
20. Alvi, F.S., Shih, C. Elavarasan, R., Garg, G., Krothapalli, K.: Control of supersonic impinging jet flows using supersonic microjets. AIAA J. **41**(7) (2003)
21. Alkislar, M.B., Krothapalli, A., Butler, G.W.: The effect of streamwise vortices on the aeroacoustics of a Mach 0.9 jet. J. Fluid Mech. **578**, 139–169 (2007)
22. Upadhyay, P., Gustavsson, J.P., Alvi, F.S.: Development and characterization of high frequency resonance enhanced microjet actuators for control of high speed jets. Exp. Fluids **57**(5), 1–16 (2016)
23. Lou, H., Alvi, F.S., Shih, C.: Active and passive control of supersonic impinging jets. AIAA J. **44**(1), 58–66 (2006)
24. Alvi, F.S., Lou, H., Shih, C., Kumar, R.: Experimental study of physical mechanisms in the control of supersonic impinging jets using microjets. J. Fluid Mech. **613**, 55–83 (2008)
25. Kumar, V., Alvi, F.S.: Towards understanding and optimizing separation control using microjets. AIAA J. **47**(11), 2544–2557 (2009)
26. Ali, M.Y., Alvi, F.S.: Jet arrays in supersonic crossflow—an experimental study. Phys. Fluids **27**(12) (2015). https://doi.org/10.1063/1.4937349
27. Fernandez, E., Alvi, F.S.: Vorticity Dynamics of Microjet Arrays for Active Control. AIAA Science and Technology Forum and Exposition, Maryland (2014)
28. Kreth, P., Alvi, F.S.: Microjet-based active flow control on a fixed wing UAV. J. Flow Control Meas. Vis. **2**(2) (2014)
29. Fernandez, E., Kumar, R., Alvi, F.S.: Separation control on a low-pressure turbine blade using microjets. J. Propul. Power **29**(4), 867–881 (2013)

30. Ljevar, S., De Lange, H.C., Van Steenhoven, A.A.: Two-dimensional rotating stall analysis in a wide vaneless diffuser. Int. J. Rotating Mach. **2006**, 1–11 (2006)
31. Kurokawa, J., Saha, S.L., Matsui, J., Kitahora, T.: Passive control of rotating stall in a parallel-wall vaneless diffuser by radial grooves. J. Fluids Eng. **122**(1), 90–96 (2000)
32. Tsurusaki, H., Kinoshita, T.: Flow control of rotating stall in a radial vaneless diffuser. J. Fluids Eng. **123**(2), 281–286 (2001)
33. Fernandez, E.: On the properties and mechanisms of microjet arrays in crossflow for the control of flow separation. Ph.D. Dissertation, Florida State University (2014)

High Frequency Boundary Layer Actuation by Fluidic Oscillators at High Speed Test Conditions

Valentin Bettrich, Martin Bitter and Reinhard Niehuis

Abstract Detailed investigations of high frequency pulsed blowing and the interaction with the boundary layer at high speed test conditions were performed on a flat plate with pressure gradient. This experimental testbed features the imposed suction side flow of an aerodynamically highly loaded low pressure turbine profile. For actuation, a newly developed coupled fluidic oscillator with an independent mass flow and frequency characteristic was tested successfully. Several oscillator operating points were investigated at one turbine profile equivalent operating point with Reynolds number of 70,000, theoretical outflow Mach number of 0.6, and an inflow free stream turbulence level of 4%. The examined frequency range was between 6.5 and 7.5 kHz and the actuation mass flow rates were varied between 0.68% and 1.32% of the overall passage mass flow. As a result, the flow separation and transition can be controlled and the suction side profile losses even halved. Differences in the interaction with the boundary layer of the different oscillator operating points are also presented and discussed.

Keywords Active flow control · Fluidic oscillator
High frequency pulsed blowing · Boundary layer actuation
Flat plate with pressure gradient · Aerodynamic loading
Boundary layer transition · Flow separation

V. Bettrich (✉) · M. Bitter · R. Niehuis
Institute of Jet Propulsion, Bundeswehr University Munich, Neubiberg,
85577 Munich, Germany
e-mail: valentin.bettrich@unibw.de
URL: https://www.unibw.de/isa

R. King (ed.), *Active Flow and Combustion Control 2018*,
Notes on Numerical Fluid Mechanics and Multidisciplinary Design 141,
https://doi.org/10.1007/978-3-319-98177-2_4

Nomenclature

Symbols

h	[mm] wall-normal height
f	[Hz] frequency
$\dot{m}$	[kg/s] mass flow rate
Ma	[-] Mach number
p	[Pa] pressure (static in case of no subscript)
q	[Pa] dynamic pressure
Re	[-] Reynolds number
s	[mm] surface length
Sr	[-] Strouhal number
t	[mm] spacing between two oscillator outlets
Tu	[%] turbulence level
U	[m/s] velocity
V	[m/s] velocity based on optical measurements
z	[mm] spanwise direction
β	[-] Hartree parameter
$\delta*$	mm displacement thickness
δ_2	mm momentum thickness
δ_{99}	mm boundary layer thickness
γ	[-] ratio of specific heats
ν	[m^2/s] kinematic viscosity
ζ	[-] total pressure loss coefficient

Subscripts

1	Inflow condition
2,th	Theoretical exit condition
∞	Free stream condition
AFC	Active flow control
inst.	Instability
int	Integral value
is	Isentropic
osc	Based on the oscillator
pas	One passage of the cascade / experimental testbed
Pp	Based on Preston probe
PIV	Based on Particle Image Velocimetry
t	Total condition
TEC	Based on the turbine exit casing
tot	Total surface length

T161	Based on the T161 results
Wazzan	Based on Wazzan's charts
Walker	Based on Walker's equation

Abbreviations

AFC	Active Flow Control
CTA	Constant Temperature Anemometry
DEHS	Di-Ethyl-Hexyl-Sebacat
HGK	Hochgeschwindigkeits-Gitterwindkanal (High Speed Cascade Wind Tunnel)
ISA	Institut für Strahlantriebe (Institute of Jet Propulsion)
KH	Kelvin-Helmholtz
LPT	Low Pressure Turbine
PIV	Particle Image Velocimetry
PSD	Power Spectral Density
TS	Tollmien-Schlichting

1 Introduction

Since the general research trend identified active flow control (AFC) to play a key role for efficient aerodynamic applications, many different concepts and methods were established. Depending on the flow problem or system to be controlled, different approaches are possible or reasonable. The important questions for a chosen flow problem arise regarding the most efficient actuator concept, the actuation position, the actuation impact or energy, and actuation frequency.

The latest flow control concepts are all "active". Among them, differences can be found in their actual mechanism principle. Some concepts use sensor feedback for controllers, e.g. closed-loop designs like in the work of King et al. [1]. Other applications are considered to be active if the actuator is switched on and off when needed, as outlined in the review paper by Niehuis and Mack [2]. In some cases, feedback control cannot be applied due to small geometries, high temperatures or frequencies, or if it is just the more expedient approach. The variety of available actuators is also quite high. The most important ones to name are plasma, piezo, and synthetic jet actuators as well as fluidic oscillators. Cattafesta [3] gives a comprehensive overview of the mentioned ones and considers even more actuators. For flow separation problems, many studies on the most efficient position of actuation were already carried out. A suggestion for an optimal actuation according to [4–8] is at the position of highest receptivity.

Considering mass flow investment, Mack et al. [9] used for AFC on the T161, an aerodynamically highly loaded low pressure turbine (LPT) research profile, fluidic oscillators with a mass flow rate related to the passage mass flow of $\dot{m}_{osc.,T161} \approx 1.14\% \cdot \dot{m}_{pas.,T161}$. To improve the efficiency, the latest investigations at the Institute of Jet Propulsion (ISA) prove for an aerodynamically highly loaded LPT exit casing profile that the mass flow investment of the oscillator can be decreased as low as $\dot{m}_{osc.,TEC} = 2.1 \cdot 10^{-4} \cdot \dot{m}_{pas.,TEC}$. The key for this achievement is to consider the AFC in the design process with an optimized position and the right trigger frequency [10–12].

Which concept, actuator, position, impact, or frequency range should be used depends on many parameters, e.g. the velocity distribution, necessary actuation frequency or available reaction time, available sensors, controllers, and actuators as well as the nature of the flow problem—periodic or random. In turbomachinery, one main research area for flow control problems are LPTs. Modern LPTs are aerodynamically highly loaded, usually featuring flow separation on the suction side towards low Reynolds number operating points. Since the high speed investigations at ISA showed many promising results [2, 9, 13, 14] for AFC applied on the T161, this paper focuses on a T161-like suction side flow of a flat plate with pressure gradient. The massive flow separation is controlled with high frequency periodic excitation [15], induced by a newly developed coupled fluidic oscillator [16].

The focus of this paper are investigations of different oscillator operating points and their respective influence on loss behavior, transition and interaction with the boundary layer. According to literature it is most promising to actuate in the range of the natural instabilities or receptivity of the boundary layer. Regarding the discussion whether to trigger the so-called "Tollmien-Schlichting" (TS) or the "Kelvin-Helmholtz" (KH) instabilities, a simple explanation might clarify the topic. The TS waves or instabilities develop naturally in the laminar boundary layer and can therefore triggered [17]. The instabilities developing and growing within the shear layer for separated flows are commonly known as KH instabilities [18]. Consequently, the goal is to trigger with the most effective frequency in the range of the boundary layer receptivity to amplify linear instability effects and therefore use this interaction within the boundary layer for a controlled transition towards a turbulent boundary layer.

2 Experimental Setup

2.1 Actuator—Coupled Fluidic Oscillator

The actuator employed for the investigations of this paper is a coupled fluidic oscillator developed at ISA [16]. In the experimental testbed (Sect. 2.3) an array of actuators enables distributed pulsed blowing at very high frequencies of several kHz. The unique advantage of the coupled oscillator over all common designs is the ability to change frequency and mass flow independently in a certain range, which is fundamental for detailed investigations of the impact and interaction between the boundary layer and the AFC device.

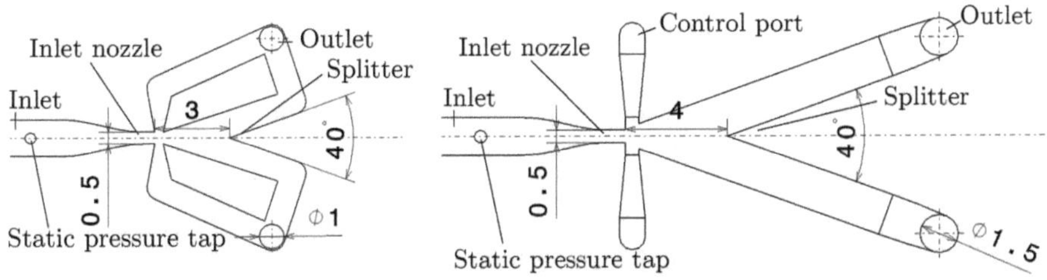

Fig. 1 Master oscillator (left) and slave oscillator (right) (all dimensions in millimeter)

The frequency of the coupled fluidic oscillator can be tuned with a standard feedback oscillator (Fig. 1 left). This design, which has a direct dependency between mass flow rate and frequency, was already used in former studies (see [2, 9, 13, 14]. In contrast to the former applications, the pulsed blowing at the outlet of the feedback oscillator (master) is not directly used for flow control, but for frequency tuning of the oscillator without feedback loops (slave) through the control ports (Fig. 1 right). The excitation of the master forces the additional or secondary mass flow entering the system through the slave's inlet nozzle to flip according to the trigger frequency to each side of the splitter. The combined mass flow leaves the device as high frequency pulses through the outlets of the slave oscillator at the master's frequency. More details on the working principle of the coupled fluidic oscillator and a characterization of the actuator pulses at different operating points can be found in [16].

2.2 *High Speed Cascade Wind Tunnel*

The results presented here are based on experiments performed in the High Speed Cascade Wind Tunnel (HGK) at ISA of the Bundeswehr University Munich. The core of the test facility (Fig. 2) is an open loop wind tunnel which is placed inside a 12 × 4 m pressure tank. A six stage axial compressor is driven

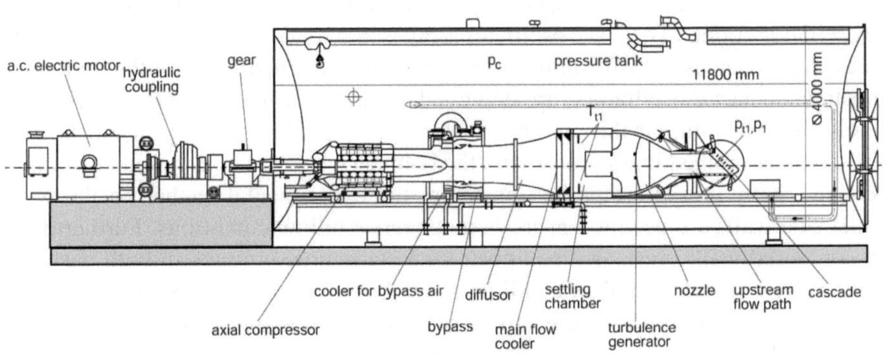

Fig. 2 High speed cascade wind tunnel (HGK) test facility

by an external 1.3 MW a.c. electric motor. It enables compression ratios up to 2.14. The operating point is adjusted with a variation in shaft speed, a variable bypass, and coolers. Engine relevant turbulence levels can be realized with a turbulence generator just upstream the nozzle. In the pressure tank it is possible to vary the absolute pressure level from 35 mbar to approximately 1200 mbar. The independent variation of the Reynolds and Mach number makes the HGK test facility very unique among very few world wide. More details can be found in Sturm and Fottner [19].

For the specific experimental testbed boundary layer suction at the contoured walls (Sect. 2.3) of the configuration was required. This was realized with the secondary air supply system of the test facility. A one-stage radial compressor sucks at the test section inside the pressure tank and re-injects it into the vessel, in order to preserve the over all pressure level.

2.3 Flat Plate with Pressure Gradient

Motivated by several promising investigations regarding passive and AFC concepts on the aerodynamically highly loaded T161 LPT cascade, summarized by Niehuis and Mack [2], the research on this topic with the well-known boundary layer topology was intensified. The T161 low pressure turbine profile features a rather large separation bubble at low and even an open separation at very low Reynolds numbers. Therefore, it is an ideal experimental testbed to investigate flow control concepts in this Reynolds number regime. However, with the use of the coupled fluidic oscillator for the planned fundamental investigations, the T161 blade geometry is physically too small to integrate the new actuator system inside the cascade profile at its recent design stage. That is why a flat plate with pressure gradient, imposed by symmetric contoured walls, was developed in order to generate a closely matching boundary layer topology at the same flow conditions as for the T161 profile. The benefit of the new configuration is also a better accessibility for probes and optical measurement techniques. The experimental testbed integration into the HGK test facility is shown in Fig. 3. The flat plate itself (Fig. 4) consists of three individual plates. The uppermost one (yellow) features a super elliptical leading edge, static pressure taps, and the oscillator outlets. The slave oscillator array is integrated underneath in the middle plate (orange) and is fed with air through the plenum. The actuators for the frequency adjustment (master oscillators) are deployed in the lowermost plate (red), also separately supplied with air through the master's plenum. Through the slave oscillator outlets, the combined mass flow enters the boundary layer as high frequency pulsed blowing. Static pressure taps just upstream the experimental testbed in the wind tunnel side walls are used to ensure homogeneous inflow conditions. Furthermore, the contoured walls were also equipped with static pressure taps in both passages to ensure symmetric main flow conditions. The homogeneous and symmetric inflow conditions are verifiable not affected with activated oscillator on the upper surface of the flat plate only. Thus, equivalent passage inflow conditions can be assumed for all cases. In order to control the main passage flow and to prevent separation on the

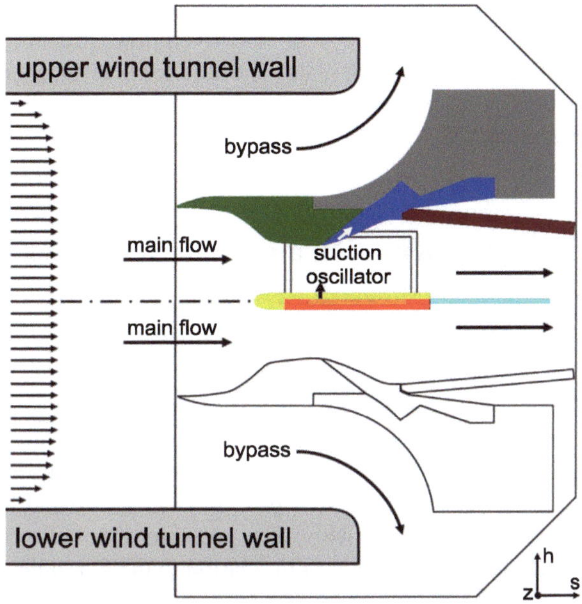

Fig. 3 Experimental testbed for the cascade section of the HGK

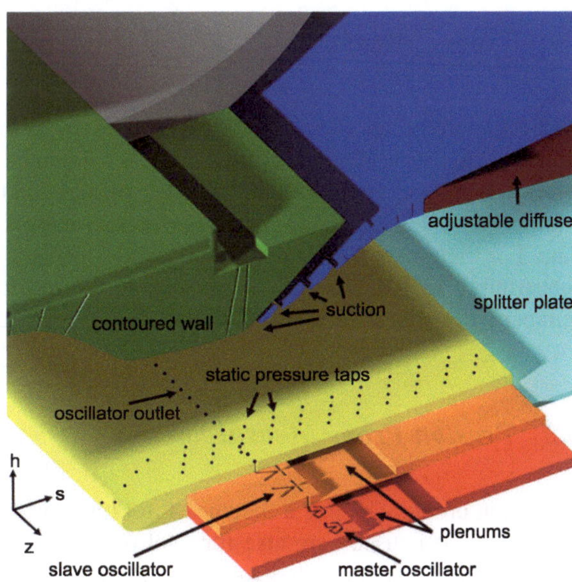

Fig. 4 Detailed view of the experimental testbed with the coupled fluidic oscillators

outer contour, air is sucked by several slots installed on contoured walls, opposing the flat plate. A splitter was assembled downstream to avoid vortex shedding at the trailing edge of the flat plate. The adjustable diffuser is used to fine-tune the flat plate boundary layer characteristic. More details and aspects of the design process can be found in [15].

3 Instrumentation and Data Acquisition

3.1 Surface Pressure Distribution

Compared to the T161 profile, the size of the flat plate is a 1:1 suction side surface length scale. It is equipped with 33 static pressure taps within $s/s_{tot} = 0$ and 1, ensuring twice the resolution of static pressure taps in relation to the T161 cascade, compare [2, 9, 13, 14]. The positions $s/s_{tot} = 0$ and 1 correspond to the leading and trailing edge, respectively. The pressure distribution is presented as the isentropic Mach number distribution in relation to the flat plate's surface length s

$$Ma_{is}(s/s_{tot}) = \sqrt{\frac{2}{\gamma - 1} \cdot \left[\left(\frac{p_{t1}}{p(s/s_{tot})} \right)^{\frac{\gamma-1}{\gamma}} - 1 \right]}. \tag{1}$$

At the inlet of the test section, the total pressure p_{t1} is measured with a pitot tube. The isentropic Mach number along the surface is calculated with local static pressure data $p(s/s_{tot})$, attained from the static pressure taps. As part of the standard measurement equipment, a 98RK rack mounted pressure system is used to acquire the pressure data. For the presented results pressure transducers with 345 mbar full scale range (uncertainty of 0.05%) were used. Hence, the resulting uncertainty of the Mach number is less than $\Delta Ma_{is} \leq 0.02$ for the investigated operating point.

3.2 Preston Probe Measurements

A Preston probe, which is a flattened Pitot tube, was used to traverse closely along the surface. It is used here also for total pressure boundary layer traverses normal to the surface. The local dynamic pressure $q(s/s_{tot}, h)$ is calculated with the local static pressure data $p(s/s_{tot})$ and the total pressure of the probe $p_t(s/s_{tot}, h)$. It is normalized by the local dynamic pressure outside of the boundary layer $q_\infty(s/s_{tot})$ to obtain the dimensionless local dynamic pressure coefficient

$$\frac{q(s/s_{tot}, h)}{q_\infty(s/s_{tot})} = \frac{p_t(s/s_{tot}, h) - p(s/s_{tot})}{p_{t1} - p(s/s_{tot})}. \tag{2}$$

More details on the Preston probe head shape and size, the procedure to determine the boundary layer condition (laminar, turbulent or separated), and the determination of the transition point are given in [15, 22].

Furthermore, the total pressure loss coefficient calculated from the Preston probe data is used to compare different actuator operating points. Therefore, boundary layer traverses were taken at $s/s_{tot} = 1$, representing the trailing edge of the T161 profile. The total pressure loss coefficient is calculated by

$$\zeta = \frac{p_{t1} - p_t(s/s_{tot}, h)}{p_{t1} - p(s/s_{tot})}. \tag{3}$$

In order to cover the entire boundary layer up to the free stream for all operating points, the integral value is calculated for $h = 0$ to 15 mm by

$$\zeta_{int} = \int_{h=0}^{h=15} \zeta(h). \tag{4}$$

A PSI 9116 pressure transducer with a full scale range of 69 mbar and 0.05% full scale range uncertainty was used for all Preston probe measurements. This results in an uncertainty of the total pressure loss coefficient of $\Delta\zeta \leq 0.005$ and a Mach number uncertainty of $\Delta Ma \leq 0.01$ for the traverses at the edge of the boundary layer. It has to be noted that the uncertainty increases towards lower Mach numbers near the surface.

3.3 *Particle Image Velocimetry*

A standard planar two-component particle image velocimetry setup (2D2C-PIV) was realized for the characterization of the boundary layer topology with and without activated flow control mechanism. Two 4 Mpx scientific CCD cameras were used side-by-side to measure the flow field across the full length of the flat plate in parallel. The mid plane of the configuration was illuminated by a Beamtech Vlite 200 double pulse Nd:YAG laser with 200 mJ pulse energy. A carefully polished model surface enabled the investigation of the wall-near velocity field down to sub-millimeter scale as diffuse reflections of the laser light on the model surface were minimized. This approach enables detailed near-wall PIV measurements [23]. The PIV tracer particles were produced from Di-Ethyl-Hexyl-Sebacat (DEHS) oil, generated in a Laskin-nozzle seeding atomizer which typically produces particles with 1 μ in size.

For the data processing, each PIV image was pre-processed with a shift correction to compensate for low frequency vibrations and by subtracting the local image minimum for intensity normalization. A Butterworth temporal filtering combined with a horizontal FFT-filtering allowed a complete suppression of the laser light reflection on the polished metal surface. Unfortunately, some tracer particles always remain stationary on the model surface even after careful cleaning. These particles partly impose spurious vectors to the final result at these positions at the wall. Nevertheless, a common multi-pass PIV with interrogation windows down to 16-by-16 pixel with 50% overlap was possible. As the particles had an imaged diameter of roughly 2 pixels, the final result was not effected by typical PIV bias errors as peak-locking. Finally, 2000 instantaneous vector fields were averaged. This evaluation procedure led to an effective spatial resolution of about 0.3 mm per velocity vector.

3.4 Hot-Wire Boundary Layer Measurements

For time-resolved data, hot-wire measurements were performed with a 1D hot-wire boundary layer probe (Type 55P15) from DANTEC, operated in the constant temperature anemometry (CTA) mode. A velocity calibration was conducted for the expected static pressure and velocity range. At each position the signal was acquired for 10 s at 60 kHz with a 30 kHz hardware low pass filter set to prevent aliasing effects. Based on the velocity data, the power spectral density (PSD) according to Welch's method, as implemented in Matlab with default inputs, is calculated.

The uncertainty of the hot-wire measurements is strongly dependent on the Reynolds number (static pressure within the vessel) and the flow velocity. At the absolute pressure level of the conducted measurements, the uncertainty for the velocity data is around ±2% at the edge of the boundary layer. Due to a low thermal conduction as a consequence of low absolute static pressure of the operating point and the low velocities near the surface, the uncertainty increases within the boundary layer.

4 Design Aspects of the Experimental Testbed

4.1 Validation of the Experimental Testbed and Oscillator Integration

To ensure comparability with former results, the experimental testbed was validated against the T161 cascade measurements. Bettrich et al. [15] proved very good agreement between the flat plate with pressure gradient and the T161 suction side flow. The operating point was chosen at Reynolds number $Re = 70,000$, theoretical outflow Mach number of $Ma_{2,th} = 0.6$, and an inflow free stream turbulence level of $Tu \approx 4\%$. The suction side flow is imposed on a flat plate in 1:1 scale of the surface length. With matching in- and outlet conditions and without actuation, an excellent correspondence with former studies is shown, compare [9]. The Mach number distribution within the area of interest (white shaded) is shown in Fig. 5. The differences near the leading edge are typical for any flat plate setup, as discussed in [15]. Altogether, the design goals of the experimental testbed for the planned investigations is fully achieved.

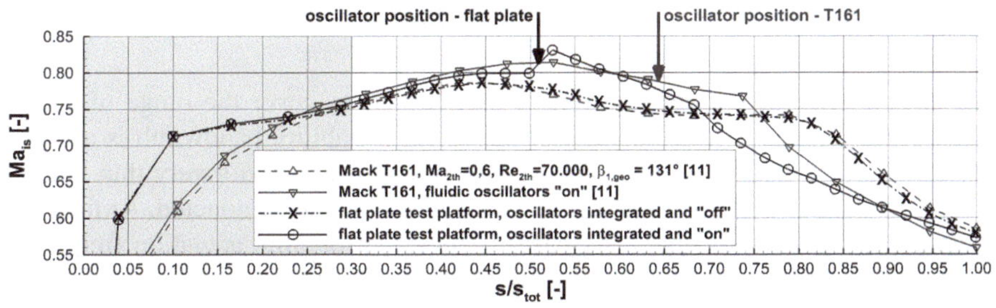

Fig. 5 Mach number distributions from static pressure taps

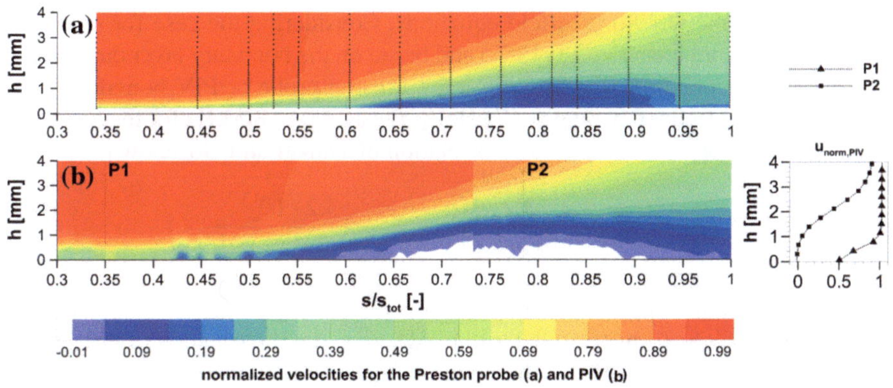

Fig. 6 Non-actuated boundary layer flow with separation bubble of the experimental testbed, based on Preston probe **a** and PIV **b** measurements

In order to resolve the boundary layer flow of the experimental testbed without actuation, Preston probe (Fig. 6a—each dot indicating a measured position) and PIV (Fig. 6b) measurements were performed. Both methods indicate the non-actuated time averaged flow field near the surface. While the overall flow field shows very good agreement, some differences in the separation bubble and the wall near flow are evident. Reason for differences are that only PIV can resolve reverse flow within the separation bubble (grey and white contours) but are quite challenging in high speed applications near the surface. The pneumatic investigations, acquired with the Preston probe, deliver very robust results, but allow only for limited resolution due to the size of the probe head, as discussed in [15].

In contrast to former studies with the T161 cascade, the oscillator outlets were positioned further upstream on the flat plate configuration. This approach was motivated by other results found in literature. It is proven to be beneficial in terms of mass flow investment (reduction of 60%) and control effectiveness compared to the further downstream position, used in the T161 investigations [15].

4.2 Estimation of Receptivity of the Boundary Layer

The design goal for the coupled fluidic oscillator is to cover the range of receptivity. Since a precise determination of the actual instability bandwidth is a rather challenging problem, especially for high speed applications with very thin boundary layers, several empirical approaches were available and conducted. At first, the latest numerical data for the T161 cascade reveal an optimal actuation frequency around 7.6–7.8 kHz. Second, an estimation for the TS instabilities was carried out with the use of the spatial and temporal stability charts according to Wazzan et al. [20]. The charts indicate the temporal amplification rates for TS waves for different flow profiles from stagnation to separation and are attained by a step-by-step integration method of the Orr-Sommerfeld equations. Considering the case for separated flow (Hartree parameter $\beta = -0.1988$) and based on the boundary layer data, which differ slightly depending whether to use the Preston probe or PIV boundary layer data, it delivers a frequency range of $7\,\text{kHz} \leq f_{TS,Wazzan} \leq 8.5\,\text{kHz}$ for the separated flow. The third approach utilizes the correlation of Walker and Gostelow [21]:

$$f_{TS,Walker} = \frac{3.2 \cdot U_{\infty}^{2}}{2\pi\nu(\frac{U_{\infty}\delta^{*}}{\nu})^{1.5}} = \frac{3.2 \cdot U_{\infty}^{0.5}}{2\pi\nu^{-0.5} \cdot (\delta^{*})^{1.5}}. \tag{5}$$

The free stream velocity U_{∞} outside the boundary layer and the kinematic viscosity ν are straight-forward to determine, whereas the displacement thickness δ^{*} is very sensitive to the results of the boundary layer measurements. Calculating δ^{*} with the Preston probe measurements results in $f_{TS,Walker,Pp} \approx 7.5\,\text{kHz}$ and with the PIV data in $f_{TS,Walker,PIV} \approx 6.8\,\text{kHz}$. A slight variation of the boundary layer profile results also in a change in displacement thickness, which has a strong influence on the potential instability frequency (compare Eq. 5). Therefore the results of the different measurement techniques should be considered supplemental rather than only one of the two. Furthermore, the frequency estimation is also quite sensitive to the actual position of the separation point, since its location is used as reference for the variables and their values included in the equation. For details to determine the separation point see [15]. To conclude, the target actuation frequency for the oscillator is determined to be between 6.8 and 8.5 kHz.

4.3 Design of the Fluidic Oscillator

For the design of the coupled fluidic oscillator array, several requirements had to be taken into account. Most important, the frequency range of the oscillator should cover the receptive range of the boundary layer. The entire estimated frequency range at the low absolute pressure conditions with an adequate mass flow variability could not be realized with one actuator. Therefore, the focus was set on the lower half of the estimated range, because it appeared to be the most promising compromise.

The mass flow rate of the coupled actuator was aimed to be comparable to the ones used for the T161 cascade measurements. This allows a comparison of the further upstream outlet position of the oscillator with former results. Furthermore, lower and higher mass flow rates should also be achieved.

The influence of the design parameters on the frequency range, the operating stability, and the mass flow rate was determined in preliminary experimental investigations. The design of the coupled actuator was performed in a two step process. First, the master oscillator was scaled to reach the desired frequency range. Second, the mass flow range was adjusted with the slave oscillator and the same outlet diameter of the actuator was chosen as for the T161 cascade measurements.

5 Results and Discussion

5.1 *Investigated Actuator Operating Points*

Within the coupled oscillator's operating range, several operating points were investigated. The frequency range was varied between 6.5 and 7.5 kHz at a constant mass flow rate of $\dot{m}_{osc.} \approx 1.05\% \cdot \dot{m}_{pas.}$. In addition, the mass flow rate was varied between $\dot{m}_{osc.} \approx 0.78\%$ and $1.32\% \cdot \dot{m}_{pas.}$ for a constant frequency of 7 kHz. An evaluation of the performance of each operating point can be derived from the total pressure loss coefficient (Fig. 7) and transition (Fig. 8, same legend as Fig. 7) behavior. Since the differences in the distribution of the total pressure loss coefficient between all operating points with activated actuation is rather small, the integral values ζ_{int} in wall-normal direction for $h = 0$ to 15 mm are outlined in Fig. 7, too. The distribution

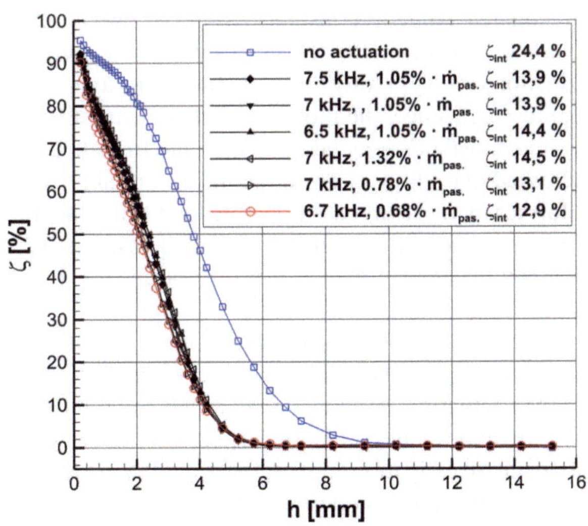

Fig. 7 Distribution and integral value of the total pressure loss coefficient for different operating points at $s/s_{tot} = 1$ from Preston probe

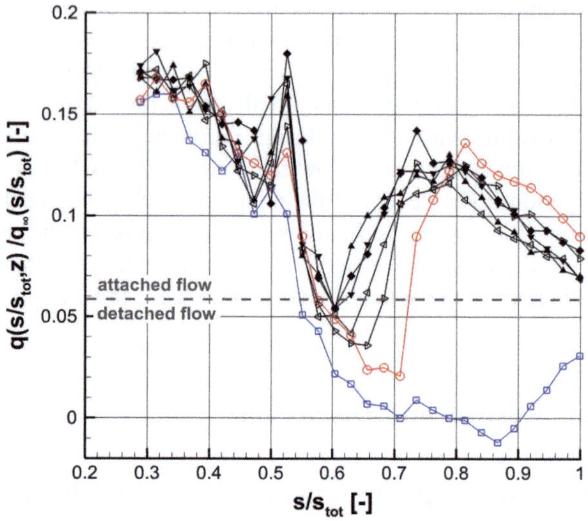

Fig. 8 Actuation effect on the boundary layer state for different actuator operating points (same legend as Fig. 7) from Preston probe

of the dimensionless local dynamic pressure coefficient along the surface (Fig. 8) indicates if separation occurs (values in the range of detached flow). It reveals as well if and where the transition takes place. Decreasing values can either represent a laminar or turbulent boundary layer, whereas a strong increase is an indicator for transition, compare Stotz et al. [22]. According to Fig. 8, for all actuated cases a transition to turbulent flow occurs. For some operating points even a small separation bubble is present. The impact of the oscillator can clearly be seen in all cases with actuation at $s/s_{ges} \approx 0.52$ (Fig. 8). If no actuation is applied, the boundary layer remains detached until the trailing edge.

It can clearly be seen in Figs. 7 and 8 that the frequency variation within the investigated range at constant mass flow rate is of minor importance on both loss behavior and transition. The key factor, however, is the mass flow investment. A reduction results in lower integral losses, as long as the open flow separation can be suppressed. Hence, the momentum itself is not the driving factor for loss reduction. For the constant frequency of 7 kHz the transition point first moves upstream with increased mass flow rate and the integral total pressure losses increase accordingly. A further increase in mass flow rate then shifts the transition point downstream with a further increased loss characteristic. Thus, another mechanism aside the injected momentum has an influence on transition and loss behavior, which will be highlighted in Sect. 5.2.

Based on the observations drawn from the loss and transition behavior, the actuator operating point at the lowest overall mass flow rate ($\dot{m}_{osc.} \approx 0.68\% \cdot \dot{m}_{pas.}$) and 6.7 kHz actuation frequency was found to be most efficient in terms of mass flow investment and integral losses. Therefore, it was of special interest for a more fundamental investigation. It features a small and closed separation bubble, the furthest downstream transition point, and therefore the shortest turbulent boundary layer in

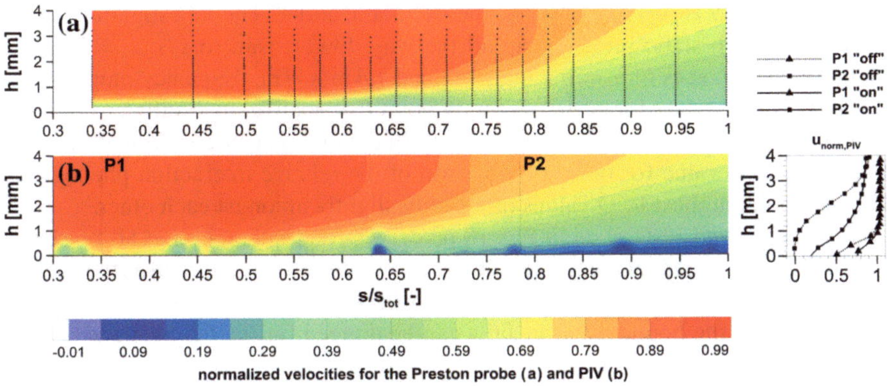

Fig. 9 Actuated boundary layer flow from Preston probe and PIV measurements for the 6.7 kHz and $\dot{m}_{osc.} \approx 0.68\% \cdot \dot{m}_{pas.}$ operating point

streamwise extent. In contrast to the case without any AFC (Fig. 6), the controlled boundary layer topology is very different (Fig. 9). The large separation bubble is successfully suppressed by the actuation. Both Figs. 6 and 9 were normalized with the same scales. Therefore, the peak Mach number increases with actuation applied, due to the reduced blocking effect. This can also clearly be seen in the surface Mach number distribution for both the flat plate with pressure gradient and the T161 cascade measurements, depicted in Fig. 5. The Preston probe and the PIV measurements were performed in streamwise direction in the centerline between two oscillator outlets. The steady and unsteady interaction phenomena of the 6.7kHz actuation with the boundary layer flow is already discussed in a previous paper [15]. A comparison of the interaction of different oscillator operating points with the boundary layer will be outlined below.

5.2 *High Frequency Pulsed Blowing Boundary Layer Interaction*

For the different oscillator operating points, 1D hot-wire measurements were carried out to investigate the respective interaction. Interaction is defined to an influence of the actuation pulse on the boundary layer flow. It takes place when the coherent structures, induced by the high frequency actuation, are not immediately damped away but can develop downstream and in wall-normal direction. If interaction takes place, the design goal of the AFC concept is reached and an increased power level in the spectrum must become visible. The results are presented in Fig. 10. It shows normalized PSD intensity plots in streamwise direction, parallel to the surface, for different wall-normal distances. The path for the hot-wire probe in streamwise direction was chosen to be in line with an oscillator outlet. The most upstream measurement position $s/s_{ges} \approx 0.5$ is just upstream of the oscillator outlet. The investigations cover positions at each static pressure tap all the way downstream to $s/s_{ges} = 1$ and in wall normal distances of $h = 0.2, 0.6, 1.0$, and 1.4 mm. The frequency range shown is 500 Hz each, shifted for every operating point to visualize the oscillator influence and boundary layer interaction at a high level of detail. Each case is normalized with its highest PSD value for better comparison of the relative interaction phenomena. The individual highest PSD values differ only slightly amongst each other.

The normalized PSD plots in Fig. 10 reveal a strong interaction at all presented wall-normal distances for the 6.7 kHz and $\dot{m}_{osc} \approx 0.68\% \cdot \dot{m}_{pas}$ as well as the 7.0 kHz and $\dot{m}_{osc} \approx 0.78\% \cdot \dot{m}_{pas}$ operating points. A decreased interaction of the actuation frequency with the boundary layer flow can be noticed for the 6.5 kHz and $\dot{m}_{osc} \approx 1.05\% \cdot \dot{m}_{pas}$ as well as for 7.5 kHz and $\dot{m}_{osc} \approx 1.05\% \cdot \dot{m}_{pas}$. A weak or damped interaction downstream the oscillator outlet and in wall-normal direction is present for the 6.5 kHz and $\dot{m}_{osc} \approx 1.05\% \cdot \dot{m}_{pas}$ as well as the 7.0 kHz and $\dot{m}_{osc} \approx 1.32\% \cdot \dot{m}_{pas}$. The reason for the different degrees of interaction can either be a matter of the frequency, the initial momentum, or a feature of the oscillator itself. Since the pulsed flow of the oscillator can be considered constant flow superimposed by a (temporal) sinusoidal flow portion, higher mass flow rates at constant frequency can either result in a higher constant flow with relatively lower sinusoidal flow amplitudes or vise versa. However, these observations correspond well with the integral of the total pressure loss coefficients, presented in Fig. 7. Besides the fact of the different transition points, the lowest integral total pressure losses correspond to the strongest interaction with the boundary layer (6.7 kHz and $\dot{m}_{osc} \approx 0.68\% \cdot \dot{m}_{pas}$ case).

With the interaction being identified as the second key factor of the AFC besides the induced momentum, the discrepancy of the transition point of the three different mass flow rates at 7 kHz of Sect. 5.1 can be explained. For the $\dot{m}_{osc} \approx 0.78\% \cdot \dot{m}_{pas}$ case the mass flow rate is low but the interaction strong. Thus a further downstream but controlled transition can be achieved. Increasing the mass flow rate to $\dot{m}_{osc} \approx 1.05\% \cdot \dot{m}_{pas}$, some interaction within the boundary layer can be observed. The higher momentum in combination with the boundary layer interaction consequently moves the transition point upstream. Increasing the momentum even further to $\dot{m}_{osc} \approx 1.32\% \cdot \dot{m}_{pas}$, barely any interaction is noticed. Considering the delayed transition point compared to the $\dot{m}_{osc} \approx 1.05\% \cdot \dot{m}_{pas}$ case leads to the conclusion that the interaction has also a strong influence on transition control. A variation of the frequency (6.5, 7, and 7.5 kHz) at the same mass flow rate of $\dot{m}_{osc} \approx 1.05\% \cdot \dot{m}_{pas}$ shows that the interaction with the boundary layer for constant mass flow rates increases with increasing frequency. The differences in the integral total pressure losses confirm, that the profile losses depend not only on momentum but also on the degree of interaction of the boundary layer with the actuation. Among theses three oscillator operating points, the 6.5 kHz case with the weakest interaction has higher integral losses than the other two cases.

To investigate the interaction in more detail, boundary layer traverse measurements were carried out for the most efficient oscillator operating point (6.7 kHz and $\dot{m}_{osc} \approx 0.68\% \cdot \dot{m}_{pas}$) in terms of mass flow investment, loss characteristic, and interaction with the boundary layer. The results are presented in Fig. 11. It shows PSD

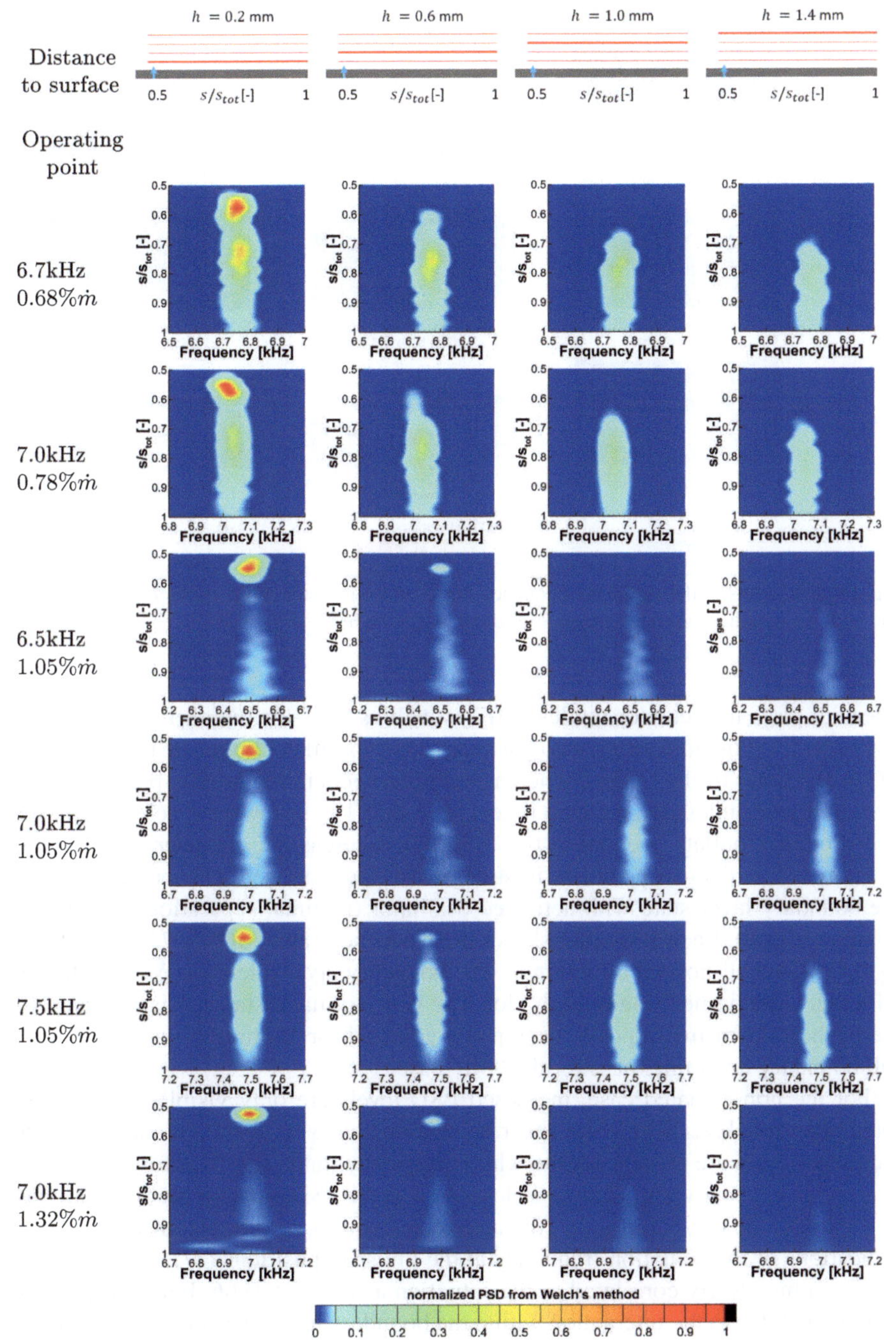

Fig. 10 Normalized PSD plots of the oscillator frequency from 1D hot-wire probe measurements: streamwise development at different distances to the surface

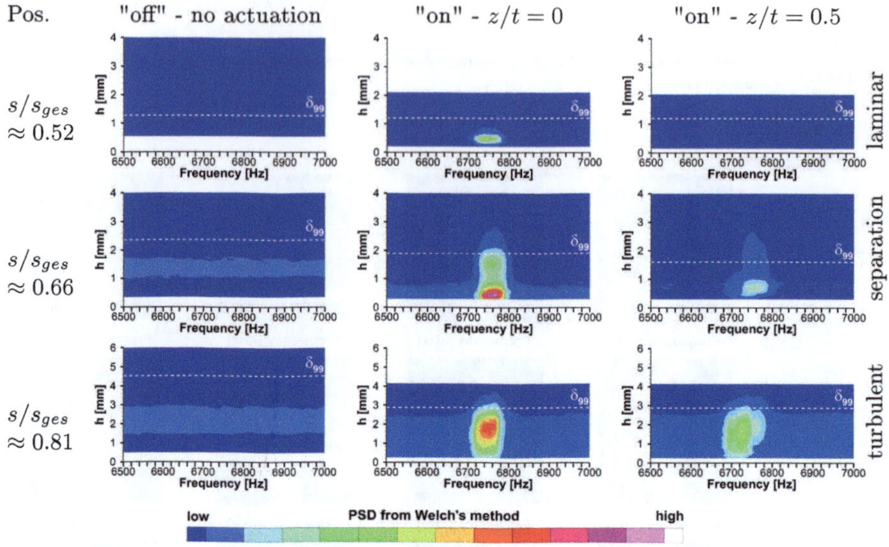

Fig. 11 PSD plots in the receptive frequency range of the investigated boundary layer flow with and without actuation: invested case 6.7 kHz and $\dot{m}_{osc} \approx 0.68\% \cdot \dot{m}_{pas}$

intensity plots for actuator modes "off" as well as "on" for two different spanwise positions z/t. Positions straight downstream of one oscillator outlet are indicated with $z/t = 0$ (center of the outlet), whereas measurements in the middle plane between two oscillator outlets are referred to as $z/t = 0.5$. The spanwise spacing t between two oscillator outlets is equal to $t = 5.33$ mm, whereas the diameter of one outlet is 1 mm. Three distinct positions in streamwise direction were chosen according to the boundary layer state with activated oscillator (laminar, separation, turbulent—compare Fig. 8). Each PSD plot shows the frequency range of 6.5–7 kHz analogue to Fig. 10 in wall-normal direction h. The boundary layer thickness δ_{99} is indicated with the white dashed line in each plot. The wall-normal distance h is re-scaled for the furthest downstream positions, equally adjusted for all three cases, to cover the full boundary layer thickness for the "off" case.

For the non-actuated case, increased PSD levels are only visible for the separated boundary layer within the shear layer (positions $s/s_{ges} \approx 0.66$ and 0.81). The increased levels are present for the whole frequency range and therefore not related to any specific bandwidth. Taking a closer look at the PSD plots for the "on" case just downstream the oscillator outlet, the actuation frequency is visible only at $z/t = 0$. Indications of the actuation frequency at the midspan positions ($z/t = 0.5$) can for the first time clearly correlated to the actuation at $s/s_{ges} \approx 0.66$. Furthermore, it is evident that due to the interaction within the boundary layer, the actuation frequency spreads throughout the whole boundary layer height. At the same time it is also restricted to it. At $s/s_{ges} \approx 0.81$ the character and extension in wall-normal direction are similar, the intensity, however, is slightly lower at the $z/t = 0.5$ position.

The interaction phenomena and the mixing process are very complex phenomena and are discussed in more detail in another paper by Bettrich et al. [15].

The development of the boundary layer thickness is strongly dependent on the state of the boundary layer. For the non-actuated case the boundary layer thickness increases rapidly when flow separation occurs, resulting in high total pressure losses. Since the momentum thickness increases as well, a thicker boundary layer can be correlated to an increase in loss behavior, too. This is the case for the boundary layer without actuation. However, with AFC activated, the boundary layer thickness remains thin with a moderate increase up to $s/s_{ges} \approx 0.66$. Further downstream, when transition over the small separation bubble occurs, the boundary layer thickness slightly increases. To preserve a thin boundary layer, AFC in the right frequency bandwidth and small induced momentum but strong interaction of the high frequency pulsed blowing with the boundary layer turns out to be favorable in terms of loss reduction. A target oriented AFC design cannot only control flow separation but also allows to control the boundary layer development, including the location of transition. Allowing a small separation bubble with defined and delayed turbulent reattachment turns out to be beneficial to reduce profile losses. The later transition occurs, the lower are the overall all losses, as long as the flow reattaches. Based on these results it is expected that there is further potential for loss reduction by reducing the mass flow. Decreased mass flow could delay the transition even further and will therefore potentially reduce turbulent boundary layer losses if interaction with the boundary layer still occurs. A mass flow reduction can be achieved either with scaled oscillators in size with the same frequency and/or an increased spacing of the oscillator outlet holes.

6 Conclusions and Outlook

The investigations on high frequency boundary layer actuation presented in this paper are carrying forward several promising investigations on flow control with the T161 cascade at the Institute of Jet Propulsion (ISA). With the use of fluidic oscillators, the aerodynamically highly loaded T161 low pressure turbine research profile showed a very significant reduction in the overall profile losses by 40%. An even greater mass flow reduction (as low as $\dot{m}_{AFC} = 2.1 \cdot 10^{-4} \cdot \dot{m}_{pas}$) was achieved for a turbine exit casing, which was specifically designed for utilization of active flow control (AFC). Motivated by these promising results, much potential for further improvements was expected. However, for investigations with the focus on the fundamental understanding of the interaction phenomena between high frequency pulsed blowing and the boundary layer, a new approach became necessary. For this purpose, a new experimental testbed was developed. In order to continue the previous work on the T161, its suction side flow was successfully reproduced on a flat plate with the same pressure distribution, induced by opposed contoured walls. All investigations were carried out at the High Speed Cascade Wind Tunnel at ISA. The results presented here are for the T161 equivalent operating point with Reynolds number of $Re = 70,000$ (based on chord lenght), theoretical outflow Mach number of $Ma_{2,th} = 0.6$ and inflow free stream turbulence level of $Tu \approx 4\%$.

The actuator used here for the flow control application is a specifically adapted and scaled design of the coupled fluidic oscillator, developed at ISA. It enables independent mass flow and frequency variations in a certain range. Based on preliminary studies, the receptive range of the boundary layer was estimated in order to ensure an effective actuator design. The frequency of the oscillator was varied between 6.5 and 7.5 kHz and the mass flow rate between $\dot{m}_{osc} \approx 0.68$ and $1.32\% \cdot \dot{m}_{pas}$. The different oscillator operating points are evaluated and compared according to their respective integral total pressure loss coefficients, transition behavior along the surface, and interaction with the boundary layer flow. Among these operating points the one with the lowest mass flow rates (6.7 kHz and $\dot{m}_{osc} \approx 0.68 \cdot \dot{m}_{pas}$) turned out to be most efficient in terms of loss reduction and air consumption. Consequently it was chosen for more detailed investigations applying Preston probe, particle image velocimetry, and 1D hot-wire measurements.

The main findings can be summarized as follows:

1. A further upstream position of the oscillator outlets compared to the previous work on the T161 cascade is very beneficial in terms of effectiveness of the flow control concept. The invested mass flow could be reduced by 60%.
2. Reducing the invested mass flow turns out to be beneficial, as long as the flow separation is under control. The reason for the lower integral losses can be explained by a delayed but controlled transition, which leads to a shorter streamwise extent of the turbulent boundary layer.
3. A change in actuation frequency within the estimated receptive range shows only minor impact on the boundary layer development and the associated loss generation. The control effect within the investigated range is found to be primarily dependent on the mass flow rate. However, actuation frequencies with weak boundary layer interaction show higher integral total pressure losses compared to cases with higher interaction and same mass flow rates.
4. Actuation within the estimated receptive range of the boundary layer indicates an increased interaction of the AFC with the boundary layer for decreased mass flow rates. High AFC mass flow rates turned out to be disadvantageous for the high frequency actuation to interact with the boundary layer. Higher mass flow rates are less effective in terms of transition control and loss reduction.
5. The key for further substantial increase in AFC effectiveness is to decrease the boundary layer thickness. Based on the results of this work this can be achieved most efficiently with an actuation in the receptive range to take advantage of the interaction between actuation and boundary layer flow. At the same time the momentum should be reduced to the extent that a controlled transition occurs as far downstream as possible. A Promising strategy would be down scaling the oscillators or increase their spacing.

With the current investigations on AFC with fluidic oscillators, substantial progress was achieved. The valuable results in the presented degree of details became possible with the newly developed experimental testbed, which will be used in upcoming investigations on high frequency boundary layer actuation even more extensively. The oscillator design in the estimated receptive frequency range shows for some cases a strong interaction with the boundary layer flow with promising loss characteristics. When all the governing effects are investigated even further, the authors are convinced that there is great potential to further reduce the mass flow investment, while keeping the control aspect effective, resulting in a very efficient fluidic oscillator design for AFC.

Acknowledgements The authors gratefully acknowledge the financial support of the German Research Foundation (Deutsche Forschungsgemeinschaft, DFG), which funded this research project (NI 586/9-1) on fundamental investigations of fluidic oscillators. The authors are also very grateful to the reviewers. Their valuable and critical comments contributed significantly to the quality of the final paper.

References

1. King, R., Heinz, N., Bauer, M., Grund, T., Nitsche, W.: Flight and wind-tunnel tests of closed-loop active flow control. AIAA. J. Aircr. **50**(5), 1605–1614 (2013). 10 pages. https://doi.org/10.2514/1.C032129
2. Niehuis, R., Mack, M.: Active boundary layer control with fluidic oscillators on highly-loaded turbine airfoils. In: King, R. (ed.) Active Flow and Combustion Control. Springer Series Notes on Numerical Fluid Mechanics and Multidisciplinary Design, vol. 127 (2014). (Invited Paper). https://doi.org/10.1007/978-3-319-11967-0_1
3. Cattafesta, L.N., Sheplak, M.: Actuators for active flow control. Ann. Rev. Fluid Mech. **43**(1), 247–272 (2011). https://doi.org/10.1146/annurev-fluid-122109-160634
4. Dovgal, A.V., Kozlov, V.V., Michalke, A.: Laminar boundary layer separation: instability and associated phenomena. Progr. Aerosp. Sci. **30**(1), 61–94 (1994). https://doi.org/10.1016/0376-0421(94)90003-5
5. Bons, J.P., Hansen, L.C., Clark, J.P., Koch, P.J., Sondergaard, R.: Designing low-pressure turbine blades with integrated flow control. In: ASME Turbo Expo GT2005-68962, Reno, pp. 1079–1091 (2005). 13 pages. https://doi.org/10.1115/GT2005-68962
6. Bons, J.P., Reimann, D., Bloxham, M.: Separated flow transition on an LP turbine blade with pulsed flow control. ASME J. Turbomach. **130**(2), 021014 (2008), 8 pages. https://doi.org/10.1115/1.2751149
7. Bons, J.P., Pluim, J., Gompertz, K., Bloxham, M. und Clark, J.P.: The application of flow control to an AFT-loaded low pressure turbine cascade with unsteady wakes. ASME J. Turbomach. **134**(3), 031009 (2011). 11 pages. https://doi.org/10.1115/1.4000488
8. Rist, U.: Instability and transition mechanisms in laminar separation bubbles. In: VKI/RTO-LS, Low Reynolds Number Aerodynamics on Aircraft Including Applications in Emerging UAV Technology, Rhode-Saint-Genese (2003). http://citeseerx.ist.psu.edu/viewdoc/download?doi=10.1.1.638.6411&rep=rep1&type=pdf
9. Mack, M., Brachmanski, R., Niehuis, R.: The effect of pulsed blowing on the boundary layer of a highly loaded low pressure turbine blade. ASME Turbo Expo GT2013-94566, San Antonio, p. V06AT36A015 (2013). 10 pages. https://doi.org/10.1115/GT2013-94566
10. Kurz, J., Hoeger, M. und Niehuis, R.: Design of a highly loaded turbine exit case airfoil with active flow control. In: ASME IMECE2016-66008, ASME IMECE, Phoenix, p. V001T03A049 (2016). 11 pages. https://doi.org/10.1115/IMECE2016-66008

11. Kurz, J., Hoeger, M., Niehuis, R.: Active boundary layer control on a highly loaded turbine exit case profile. In: ETC12, ETC2017-191, Stockholm (2017). http://www.euroturbo.eu/publications/proceedings-papers/etc2017-191/
12. Kurz, J., Hoeger, M., Niehuis, R.: Influence of active flow control on different kinds of separation bubbles. In: Proceedings of the XXIII. ISABE, Manchester, ISABE-2017-22572 (2017)
13. Mack, M., Niehuis, R., Fiala, A.: Parametric study of fluidic oscillators for use in active boundary layer control. In: ASME Turbo Expo GT2011-45073, Vancouver, pp. 469–479 (2011). 11 pages. https://doi.org/10.1115/GT2011-45073
14. Mack, M., Niehuis, R., Fiala, A., Guendogdu, Y.: Boundary layer control on a low pressure turbine blade by means of pulsed blowing. ASME J. Turbomach. **135**(5), 051023 (2013). 8 pages. https://doi.org/10.1115/1.4023104
15. Bettrich, V., Bitter, M., Niehuis, R.: Interaction phenomena of high frequency pulsed blowing in LP turbine-like boundary layers at high speed test conditions. In: ASME Turbo Expo 2018 GT2018-75475, Oslo (2018)
16. Bettrich, V., Niehuis, R.: Experimental investigations of a high frequency master-slave fluidic oscillator to achieve independent frequency and mass flow characteristics. In: ASME IMECE 2016-66782, Phoenix, p. V001T03A061 (2016). https://doi.org/10.1115/IMECE2016-66782
17. Schlichting, H., Gersten, K.: Boundary-Layer Theory. Springer, Heidelberg (2017). https://doi.org/10.1007/978-3-662-52919-5
18. Simoni, D., Ubaldi, M., Zunino, P., Lengani, D., Bertini, F.: An experimental investigation of the separated-flow transition under high-lift turbine blade pressure gradients. Flow Turbul. Combust. **88**(1–2), 45–62 (2012). https://doi.org/10.1007/978-3-319-11967-0_1
19. Sturm, W., Fottner, L.: The high-speed cascade wind tunnel of the German armed forces university Munich. In: 8th Symposium on Measuring Techniques for Transonic and Supersonic Flows in Cascades and Turbomachines, Genova, Italy (1985)
20. Wazzan, A.R., Okamura, T.T., Smith, A.M.O.: Spatial and temporal stability charts for the Falkner-Skan boundary-layer profiles. REPORT NO. DAC-67086, McDonnell Douglas Astronautics Company-Huntington Beach (1968)
21. Walker, G.J., Gostelow, J.P.: Effects of adverse pressure gradients on the nature and length of boundary layer transition. ASME J. Turbomach. **112**(2), 196–205 (1990). 10 pages. https://doi.org/10.1115/1.2927633
22. Stotz, S., Wakelam, C.T., Niehuis, R., Guendogdu, Y.: Investigation of the suction side boundary layer development on low pressure turbine airfoils with and without separation using a Preston probe. In: ASME Turbo Expo GT2014-25908, Duesseldorf, p. V02CT38A025 (2014). 13 pages. https://doi.org/10.1115/GT2014-25908
23. Kaehler, C.J., Scholz, U., Ortmanns, J.: Wall-shear-stress and near-wall turbulence measurements up to single pixel resolution by means of long-distance micro-PIV. Exp. Fluids **41**(2), 327–341 (2006). https://doi.org/10.1007/s00348-006-0167-0

Model Predictive Control of Ginzburg-Landau Equation

Mojtaba Izadi, Charles R. Koch and Stevan S. Dubljevic

Abstract This work explores the realization of model predictive control (MPC) design to an important problem of vortex shedding phenomena in fluid flow. The setting of vortex shedding phenomena is represented by a Ginzburg-Landau (GL) equation model and leads to the mathematical representation given by complex infinite dimensional parabolic PDEs. The underlying GL model is considered within the boundary control setting and the modal representation is considered to obtain discrete infinite dimensional system representation which is used in the model predictive control design. The model predictive control design accounts for optimal stabilization of the unstable GL equation model, and for the naturally present input constraints and/or state constraints. The feasibility of the optimization based model predictive controller is ensured through a large enough prediction horizon. The subsequent feasibility is ensured in a disturbance free model setting. The applicability of an easily realizable discrete controller design is demonstrated using simulation with known parameters from the literature.

M. Izadi
Ferdowsi University of Mashhad, Mashhad, Iran
e-mail: m_izadi@mail.um.ac.ir

C. R. Koch (✉)
Department of Mechanical Engineering, University of Alberta,
Edmonton,
Alberta, AB T6G2G8, Canada
e-mail: bob.koch@ualberta.ca
URL: https://sites.ualberta.ca/ckoch

S. S. Dubljevic
Department of Chemical and Materials Engineering, University of Alberta,
Edmonton, Alberta, AB T6G2V4, Canada
e-mail: stevan.dubljevic@ualberta.ca
URL: https://dpslab.eche.ualberta.ca/

R. King (ed.), *Active Flow and Combustion Control 2018*,
Notes on Numerical Fluid Mechanics and Multidisciplinary Design 141,
https://doi.org/10.1007/978-3-319-98177-2_5

Keywords Vortex shedding · Ginzburg-Landau (GL) model
Model Predictive Control (MPC) · Boundary control
Complex Dissipative Parabolic Partial Differential Equations (PDEs)

1 Introduction

The realization of flow manipulation is an important technological achievement for engineering applications. A variety of applications ranging from drag reduction, lift enhancement, noise suppression and turbulence augmentation are prime examples of efficient flow control realization with direct benefits to operational costs and savings [1]. One, among large number of modelling and control realization extensively explored in practice and theory, is vortex shedding flow phenomena which describes the flow past submerged obstacles for a Reynolds numbers slightly larger than the critical value. Experiments show that feedback, from a suitable sensor can be used to suppress the shedding, at least in a region close to the sensor location at a Reynolds numbers close to the onset of vortex shedding. Examples of these experiments include oscillating a cylinder normal to mean flow [2, 3] or stabilizing the wake by suction and blowing on the surface of the body in wind tunnel [3–5]. Numerical simulation to demonstrate vortex shedding control based on a feedback by fluid injection and fluid suction applied at a cylinder wall is described in [6]. Optimal drag reduction in an open-loop setting based on a discretized Navier-Stokes equations [7, 8], or on the basis of reduced order representation using proper orthogonal decomposition (POD) [9, 10] were also explored.

Optimal control of partial differential equation (PDE) models is a mature area of research [11, 12]. For flow control, optimal control realizations have received much attention [7, 8] where optimal control and adjoint-based suboptimal optimal control is applied to the vortex-shedding suppression via blowing and suction at cylinder wall. The success of these optimal control realization was conditional on the cost function being defined as the difference among the given velocity field and the velocity field of steady laminar flow, and on the optimization time interval duration being larger than the vortex-shedding period. From these studies became clear that the key element of optimal and suboptimal control realizations is the determination of cost function to be minimized. The optimal control of suppressing vortex shedding in the wake of a circular cylinder has been recently revisited in adjoint-based optimal control framework [13]. There, a detailed exploration of the effectiveness of simply increasing the optimization horizon length or the influence of different cost functions with various Reynolds numbers was employed. In addition, the necessity to have a full flow state information available for the optimal control law calculation is a limiting factor in applying flow optimal control in practice. One recent remedy to address the state reconstruction for vortex shedding flow models was provided by the application of backstepping methodology [14–17].

The powerful backstepping transformation based controller and observer design can be easily constructed for the Ginzburg-Landau (GL) equation. In this case the

GL equation is derived for Reynolds numbers close to the critical Reynolds number describing the onset of vortex shedding. Although the backstepping controller and/or observer designs are applicable to large class of PDEs including fluid flow problems, they do not address optimality nor the presence of actuator or possible state imposed constraints in the control problem. In particular, input constraints cannot be accounted for in backstepping designs which are of great interest for fluid flow control.

Exploring implementable, optimal (or suboptimal) design methodologies which can be computationally realizable in the real time and with the degree of robustness in design and implementation are of interest. Recent studies in the realm of fluid flow, see [18, 19] considered the application of the model predictive control (MPC) design in the fluid flow setting. More recently flow separation and vortex shedding suppression by applying numerical methods on a two-dimensional space grid that utilizes a large scale numerical computational scheme to account for the cost function evaluation with the predictive horizon equal to the period of vortex shedding has been considered [20]. However, none of predictive control strategies mentioned account for the input constraints or other constraints [21, 22]. For optimization based design of fluid flows, the success of model predictive control stems from the successful application in the context of distributed parameter systems (DPS) setting—in particular dissipative parabolic PDEs. Discrete optimization based stabilizing controller realizations for linear PDEs in [23, 24] explicitly account for instability in the model and optimality. In addition, they account for input and/or state constraints which are typically found in fluid flow applications.

In this work, MPC design has been explored in the setting of Ginzburg-Landau (GL) equation describing the onset of vortex shedding in fluid flow. A complex parabolic partial differential equation (PDEs) setting that is amenable to modal model predictive control design is used. The design methodologies explore boundary applied actuation in infinite dimensional DPS setting [25], discrete model representation which accounts for quadratic cost function and simultaneous inclusion of input and state/output constraints using convex optimization. The model predictive control accounts for stabilization of the unstable mode by imposing the equality constraint on the evolution of the unstable mode associated with the vortex instability. The input constraints and state constraints are active over the optimization horizon which being feasible calculates the control input. This control input is applied in closed-loop and the process is repeated every discrete time instance with the prediction horizon window moving forward in time. It is important to note that since that disturbance free setting is feasible this guarantees feasibility in the dissipative distributed parameter systems setting. In particular, the finite time horizon over which the optimization is performed is one of design variables which impacts the optimization feasibility of the MPC realization. For a large enough time horizon the optimization problem is feasible (at least for unconstrained linear MPC), however too large a time horizon is impractical in realtime since solving the larger convex optimization problem takes too long.

This paper is organized in sections which are described next. In the mathematical modelling and system representation section, the Ginzburg-Landau equation is presented and transformed from the original geometric setting to a complex-value parabolic PDE with boundary actuation. The required boundary conversion to in-domain exact state transformation is applied to yield the model representation which is amenable to the MPC design. The constraints and model predictive control section provides design of a predictive constrained stabilizing controller which accounts for input and state constraints in an explicit manner. Finally, in the GL numerical simulation with constrained MPC section, representative simulations of GL equation model under MPC control law in the feedback loop provide numerical demonstration of the method.

2 Mathematical Model and System Representation

A model of vortex shedding phenomenon for the flow past a 2-D circular cylinder is given by the Ginzburg-Landau (GL) equation which was derived for Reynolds numbers close to the critical Reynolds number for the onset of vortex shedding, see Fig. 1. This model has been shown to remain accurate for larger Reynolds numbers [17] and is a nonlinear complex partial differential equation (PDE). Since the nonlinearities in GL equation have a damping effect on large states, a linear stabilizing controller is also stabilizing for large initial conditions [27]. The linear GL equation is:

$$\begin{aligned} \frac{\partial A(\tilde{\xi},t)}{\partial t} &= a_1 \frac{\partial^2 A(\tilde{\xi},t)}{\partial \tilde{\xi}^2} + a_2(\tilde{\xi}) \frac{\partial A(\tilde{\xi},t)}{\partial \tilde{\xi}} + a_3(\tilde{\xi}) A(\tilde{\xi},t) \\ A(0,t) &= \tilde{u}(t) \\ A(\xi_d,t) &= 0 \end{aligned} \tag{1}$$

where $\tilde{\xi} \in [0, \xi_d] \subset \mathbb{R}$ is space, $t \in \mathbb{R}^+$ is time and $A(\tilde{\xi}, t)$ is a complex-valued function. Truncating the spatial domain is due to the fact that the upstream flow is approximately uniform and the downstream subsystem can be approximated to any level of accuracy by selecting a sufficiently large ξ_d [28]. Model parameters are real positive constant a_1 and complex-value space dependent functions $a_2(\tilde{\xi})$ and $a_3(\tilde{\xi})$ (see [27] for more modelling details). This complex-value PDE can be unstable in general and $\tilde{u}(t)$ is the stabilizing boundary control to be designed.

Here $A(\tilde{\xi}, t)$ represents a complex-valued function of $(\tilde{\xi}, t)$ which is related to the transverse fluctuating velocity along the flow centerline [16]. The possible unstable

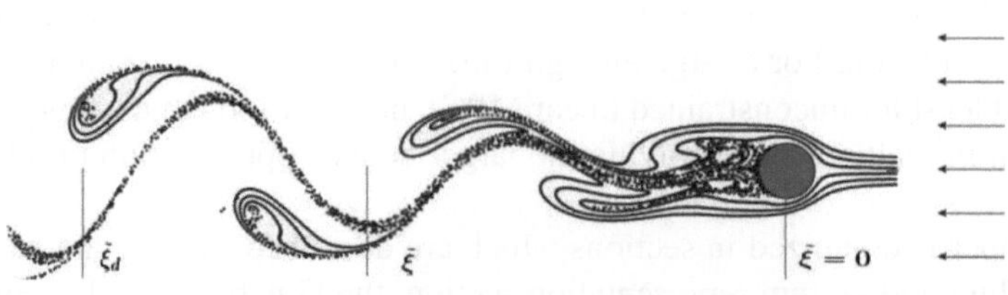

Fig. 1 Schematics of vortex shedding in the 2D flow past cylinder [26]

zero solution of nonlinear GL equation corresponds to the unstable equilibrium state with symmetric vortices above and below the centerline. The actuation $\tilde{u}(t)$ is the transverse velocity applied at downstream end of cylinder $\tilde{\xi} = 0$, which could be physically realized by rotation of the cylinder.

The convective term in linear GL equation can be eliminated by applying the following invertible state transformation $\tilde{x}(\tilde{\xi}, t) = A(\tilde{\xi}, t)g(\tilde{\xi})$, where $g(\tilde{\xi}) = \exp\left(\frac{1}{2a_1}\int_0^{\tilde{\xi}} a_2(\eta)d\eta\right)$. The space is transformed to $[0, 1]$ with the use of $\xi = (\xi_d - \tilde{\xi})/\xi_d$ coordinate transformation, resulting in the following PDE:

$$\begin{aligned}
\frac{\partial \tilde{x}(\xi, t)}{\partial t} &= a\frac{\partial^2 \tilde{x}(\xi, t)}{\partial \xi^2} + b(\xi)\tilde{x}(\xi, t) \\
\tilde{x}(0, t) &= 0 \\
\tilde{x}(1, t) &= \tilde{u}(t) \\
\tilde{x}(\xi, 0) &= \tilde{x}_0
\end{aligned} \tag{2}$$

where $a = a_1/\xi_d^2$ and $b(\tilde{\xi}) = -\frac{1}{2}a_2'(\tilde{\xi}) - \frac{1}{4a_1}a_2^2(\tilde{\xi}) + a_3(\tilde{\xi})$, and a_2'-denotes derivative with respect to space.

We consider the complex Hilbert space $\mathcal{H} = L_2(0, 1)$ with the inner product and norm given by:

$$\langle w_1, w_2\rangle = \int_0^1 w_1(\xi)\overline{w_2(\xi)}d\xi$$

$$\|w_1\| = \langle w_1, w_1\rangle^{\frac{1}{2}}$$

where the over bar represents complex conjugation and w_1, w_2 are any two complex functions in $L_2(0, 1)$. To formulate (2) as an abstract boundary control problem, the state function $x(t)$ on the state-space $\mathcal{H}$ is defined as $x(t) = \tilde{x}(\xi, t)$ for $t > 0$ and $\xi \in (0, 1)$. The system (2) can now be written as:

$$\frac{dx(t)}{dt} = \tilde{\mathcal{A}}x(t) \tag{3}$$

$$\mathcal{B}x(t) = \tilde{u}(t) \tag{4}$$

$$x(0) = x_0 \tag{5}$$

where $\tilde{\mathcal{A}}$ is the spatial differential operator:

$$\tilde{\mathcal{A}} := a\frac{d^2}{d\xi^2} + b(\xi) \tag{6}$$

with domain $\mathcal{D}(\tilde{\mathcal{A}}) = \{\psi \in \mathcal{H} : \psi, \psi' \; abs.\; cont., \psi(0) = 0\}$, while the boundary operator is defined as $\mathcal{B}x(t) := [x(\xi, t)]_{\xi=1} = u(t)$ with domain $\mathcal{D}(\mathcal{B}) \in \mathcal{H}$.

The above equation has non-homogenous boundary conditions and the operator eigenvalue problem cannot be solved in this form. To transform this equation into an equivalent distributed (in-domain) control problem, it is assumed that there exists a new operator as:

$$\mathcal{A}x(t) = \tilde{\mathcal{A}}x(t), \quad for\ all\ \psi \in D(\mathcal{A}) \tag{7}$$

where domain of operator $D(\mathcal{A}) = D(\tilde{\mathcal{A}}) \cap ker(\mathcal{B})$ and that there exists a function $B(\xi)$ such that for all $\tilde{u}(t)$ and $B\tilde{u}(t) \in \mathcal{D}(\tilde{\mathcal{A}})$. This results in

$$\mathcal{B}B\tilde{u}(t) = \tilde{u}(t)$$

It is common [25, 29] to use the state transformation $p(t) = x(t) - B\tilde{u}(t)$ to represent the dynamical system with distributed control. By applying an additional condition $\tilde{\mathcal{A}}B = 0$ to calculate the function $B(\xi)$ a decoupling of boundary applied input and model states is provided. Solution of the following two value boundary value problem given as $\tilde{\mathcal{A}}B = 0$ with the associated boundary conditions $\mathcal{D}(B) \subset \mathcal{D}(\tilde{\mathcal{A}}) \subset \mathcal{D}(\tilde{\mathcal{B}})$, is needed. This is written as:

$$\begin{aligned} &a\frac{d^2B(\xi)}{d\xi^2} + b(\xi)B(\xi) = 0 \\ &B(0) = 0 \\ &B(1) = 1 \end{aligned}$$

Next applying the state transformation $p(t) = x(t) - B\tilde{u}(t)$ to expression (3–5), the system representation is:

$$\begin{aligned} &\frac{dp(t)}{dt} = \mathcal{A}p(t) - B\dot{\tilde{u}}(t) \\ &p(0) = p_0 \end{aligned} \tag{8}$$

where the initial condition is $p_0 = x_0 - B\tilde{u}(0)$ and over-dot represents the time derivative. It can be easily shown that

$$\mathcal{A}^*(\cdot) = a\frac{d^2(\cdot)}{d\xi^2} + \bar{b}(\cdot)$$

with $\mathcal{D}(\mathcal{A}^*) = \mathcal{D}(\mathcal{A})$ being the adjoint operator of $\mathcal{A}$.

The analytical calculation of the spectrum of these operators is not straightforward, because the coefficient $b(\xi)$ is a function of space. However, there exists an analytical solution for a constant b with eigenvalues and eigenfunctions given by:

$$\lambda_n = b - an^2\pi^2$$
$$\phi_n = C_1 \sin(n\pi\xi)$$

with $n = 1, 2, \ldots$. The eigenvalue problem of $\mathcal{A}^*$ has the solution

$$\lambda_n^* = \bar{b} - an^2\pi^2$$
$$\phi_n^* = C_2 \sin(n\pi\xi)$$

In these equations C_1 and C_2 are arbitrary complex constant numbers that must satisfy $C_1\overline{C_2} = 2$. The orthonormality property $\langle \phi_n, \phi_m^* \rangle = \delta_{mn}$ is maintained, e.g., $(C_1, C_2) = (\sqrt{2}, \sqrt{2})$ or $(2 + i, 0.8 + 0.4i)$. Note, that for a complex constant b, eigenvalues of $\mathcal{A}$ are not on the real axis and do not appear in complex conjugate pairs unlike for a real constant b. However, each eigenvalue λ_n is the conjugate complex of the corresponding adjoint eigenvalue λ_n^*. For the case of a real constant b operator $\mathcal{A}$ becomes self-adjoint and $\lambda_n = \lambda_n^*$, as expected.

When the coefficient $b(\xi)$ is a function of space, analytical calculation of the spectrum of $\mathcal{A}$ and $\mathcal{A}^*$ is not possible. Thus in what follows, numerical methods are used to find a solution to these eigenvalue problems.

Consider the ordered (with respect to real parts) eigenvalues λ_n of the operator $\mathcal{A}$. The complex space $\mathcal{H}$ can be decomposed into modal subspaces $\mathcal{H}_s = \text{span}(\phi_1, \phi_2, \ldots, \phi_m)$ and the complement $\mathcal{H}_f = \text{span}(\phi_{m+1}^e, \phi_{m+2}^e, \ldots)$, $\mathcal{H} = \mathcal{H}_s \oplus \mathcal{H}_f$ with $\Re(\lambda_{m+1}) < 0$. Defining the orthogonal projection operators P_s and P_f such that $p_s = P_s p$ and $p_f = P_f p$, the state of the system (8) can be decomposed into

$$p(t) = p_s(t) + p_f(t) = P_s p(t) + P_f p(t)$$

Applying orthogonal projection operators P_s and P_f to (8), results in

$$\begin{aligned}
\frac{dp_s}{dt} &= \mathcal{A}_s p_s - B_s \dot{\tilde{u}} \\
\frac{dp_f}{dt} &= \mathcal{A}_f p_f - B_f \dot{\tilde{u}} \\
p_s(0) &= P_s p_0 \\
p_f(0) &= P_f p_0
\end{aligned}$$

where $\mathcal{A}_s = P_s\mathcal{A}$, $\mathcal{A}_f = P_f\mathcal{A}$, $B_s = P_s B$ and $B_f = P_f B$. Using this decomposition, the dynamics of the system (8) is given by two parts. The first part is the slow and possibly unstable subsystem with $\mathcal{A}_s = \text{diag}(\lambda_1, \lambda_2, \cdots, \lambda_l^m)$, a diagonal matrix of slow eigenvalues. The second part is the exponentially stable fast subsystem with $\mathcal{A}_f$ which is an unbounded exponentially stable (since $\Re(\lambda_{m+1}) < 0$) differential operator.

Introducing a new variable $u(t) = \dot{\tilde{u}}(t)$, the system equations are augmented and rewritten as:

$$\begin{bmatrix} \dot{\tilde{u}}(t) \\ \dot{p}_s(t) \\ \dot{p}_f(t) \end{bmatrix} = \begin{bmatrix} 0 & 0 & 0 \\ 0 & \mathcal{A}_s & 0 \\ 0 & 0 & \mathcal{A}_f \end{bmatrix} \begin{bmatrix} \tilde{u}(t) \\ p_s(t) \\ p_f(t) \end{bmatrix} + \begin{bmatrix} 1 \\ -B_s \\ -B_f \end{bmatrix} u(t) \tag{9}$$

with $\tilde{u}(0) = 0$. Note that when the condition $\tilde{\mathcal{A}}B = 0$ is applied, the state operator in (9) is diagonal with zero off-diagonal elements which implies decoupling of dynamic modes (p_s-slow and p_f-fast).

3 Constraints and Model Predictive Control

The ability to explicitly handle input and state constraints makes MPC widely used in the control community. In this section constraints are applied to the system represented in (9) following [23, 24]. As previously mentioned the physical interpretation of the complex function $A(\tilde{\xi}, t)$ is the real part represents the transverse fluid velocity. The following input and state constraints on the GL equation are considered:

$$U^{\min} \leq \Re\left(\tilde{u}(t)\right) \leq U^{\max} \tag{10}$$

$$A^{\min} \leq \Re\left(\int_0^{\xi_d} A(\tilde{\xi}, t) r(\xi) d\tilde{\xi}\right) \leq A^{\max} \tag{11}$$

The first constraint limits the actuation in terms of the velocity at the downstream end of cylinder (related to the rotation of the cylinder). The second constraint is a limit on the velocity along the centerline. The values $U^{\min}$, $U^{\max}$, $A^{\min}$ and $A^{\max}$ are real numbers representing the lower and upper bounds of the manipulated input and state constraints. The real valued function $r(\xi)$ is a state constraint distribution function and describes how the state constraint is applied in the spatial domain.

To formulate the problem in MPC framework, the dynamical system (9) is discretized. Although, the continuous-time system representation can be discretized exactly for finite dimensional systems, for the infinite dimensional system (9) the system dynamics are approximated by considering the slow and a limited number of fast modes. Applying the Galerkin method and using same notation for the state, system (9) is discretized as:

$$\begin{aligned} \begin{bmatrix} \tilde{u}^{k+1} \\ p_s^{k+1} \\ p_f^{k+1} \end{bmatrix} &= \begin{bmatrix} 1 & 0 & 0 \\ 0 & A_s^d & 0 \\ 0 & 0 & A_f^d \end{bmatrix} \begin{bmatrix} \tilde{u}^k \\ p_s^k \\ p_f^k \end{bmatrix} + \begin{bmatrix} h \\ -B_s^d \\ -B_f^d \end{bmatrix} u^k \\ \begin{bmatrix} \tilde{u}^1 \\ p_s^1 \\ p_f^1 \end{bmatrix} &= \begin{bmatrix} 0 \\ p_s(0) \\ p_f(0) \end{bmatrix} \end{aligned} \tag{12}$$

for a sample interval of h and the sample time is $k = 1, 2, \ldots$. All the matrices appearing in this equation are in the associated complex spaces.

However, for the MPC formulation, specifically the optimization objective function, systems with states belonging to real spaces are needed. Therefore, real and imaginary parts of Eq. (12) are separated to get:

$$\begin{bmatrix} \Re(\pi^{k+1}) \\ \Im(\pi^{k+1}) \end{bmatrix} = \begin{bmatrix} \Re(A^d) & -\Im(A^d) \\ \Im(A^d) & \Re(A^d) \end{bmatrix} \begin{bmatrix} \Re(\pi^k) \\ \Im(\pi^k) \end{bmatrix} + \begin{bmatrix} \Re(B^d) & -\Im(B^d) \\ \Im(B^d) & \Re(B^d) \end{bmatrix} \begin{bmatrix} \Re(u^k) \\ \Im(u^k) \end{bmatrix} \tag{13}$$

$$\begin{bmatrix} \Re(\pi^1) \\ \Im(\pi^1) \end{bmatrix} = \begin{bmatrix} \Re(\pi(0)) \\ \Im(\pi(0)) \end{bmatrix} \tag{14}$$

where

$$\pi^k = \begin{bmatrix} \tilde{u}^k \\ p_s^k \\ p_f^k \end{bmatrix}, \quad \pi(0) = \begin{bmatrix} 0 \\ p_s(0) \\ p_f(0) \end{bmatrix}$$

and A^d and B^d are the state and input matrices in (12), respectively. Input constraints (10) can be readily written in the form of constraints on $\Re(\pi)$. Using (13) and projecting p on the eigenfunctions basis $\{\phi_n\}$, it is straightforward to reduce the state constraints (11) to

$$A^{\min} \leq \Re\left(C\pi^k\right) \leq A^{\max}$$

where the first element of C is $\int_0^1 (B(\xi)r(\xi)/g(\xi))\,d\xi$ and the remaining ones are $\int_0^1 (\phi_n(\xi)r(\xi)/g(\xi))\,d\xi$, $n = 1, 2, \cdots$. Finally, the MPC controller design as minimization of the quadratic cost objective function is formulated.

In general, the discrete form of the MPC controller design allows the quadratic form of the cost function which accounts for the input and state evolution penalties to be defined. The standard discrete MPC controller design takes the form of quadratic optimization functional subjected to a linear model (12), input and state/output constraints:

$$\min_u \sum_{j=0}^{\infty} [\bar{u}^j \; p_s^j \; p_f^j] Q \begin{bmatrix} \bar{u}^j \\ p_s^j \\ p_f^j \end{bmatrix} + u^j R u^j \tag{15}$$

$$s.t. \qquad Eqs.(12), (11), (10) \tag{16}$$

where Q and R are penalties on state and input control evolution. This quadratic programming problem is an infinite dimensional and the problem needs to be transformed into a finite dimensional optimization problem. This is accomplished by consideration of the prediction horizon N. The optimization is realized by considering the finite horizon N with the condition that all unstable modes of the system are stabilized by calculated control input solution of the optimization problem. The cost of the infinite horizon contribution is associated with the evolution of the stable modes and therefore can be expressed as finite cost (or cost to go). The above expression now takes the following form:

$$\min_{u} \sum_{j=0}^{N-1} [\bar{u}^j \; p_s^j \; p_f^j] Q \begin{bmatrix} \bar{u}^j \\ p_s^j \\ p_f^j \end{bmatrix} + u^j R u^j + [\bar{u}^N \; p_s^N \; p_f^N] \bar{Q} \begin{bmatrix} \bar{u}^N \\ p_s^N \\ p_f^N \end{bmatrix} \tag{17}$$

$$s.t. \qquad Eqs.(12), (11), (10) \tag{18}$$

where $\bar{Q}$ is the cost associated with the evolution of the stable dynamics of the linear GL model over the infinite horizon. Now due to the specific form of boundary control realized MPC design, the above cost at time instance k can be expressed in the form of real and imaginary parts of (12). So now the cost associated with the boundary actuation is penalized with the Q_u, the state evolution is penalized with the Q_s while the penalties associated with the terminal state in the model predictive control are given by matrix $\tilde{Q}$. This results in:

$$\min_{\tilde{u}} \sum_{j=0}^{N-1} \left(Q_u \left|\tilde{u}^{k+j}\right|^2 + Q_s \left[\Re(p_s^{k+j})' \; \Im(p_s^{k+j})' \right] \begin{bmatrix} \Re(p_s^{k+j}) \\ \Im(p_s^{k+j}) \end{bmatrix} \right) + \\ \left[\Re(\tilde{u}^{k+N}) \; \Im(\tilde{u}^{k+N}) \; \Re(p_s^{k+N})' \; \Im(p_s^{k+N})' \right] \tilde{Q} \begin{bmatrix} \Re(\tilde{u}^{k+N}) \\ \Im(\tilde{u}^{k+N}) \\ \Re(p_s^{k+N}) \\ \Im(p_s^{k+N}) \end{bmatrix} \tag{19}$$

subject to dynamics of the system (13) and the constraints:

$$U^{\min} \leq \Re\left(\tilde{u}^j\right) \leq U^{\max} \tag{20}$$

$$A^{\min} \leq \left[\Re(C) \; -\Im(C) \right] \begin{bmatrix} \Re(\pi^j) \\ \Im(\pi^j) \end{bmatrix} \leq A^{\max} \tag{21}$$

for $j = 1 \leq j_1, j_1 + 1, \ldots, j_2 \leq N$. The weights associated with control input are $Q_u > 0$, the slow modes are $Q_s > 0$, $\tilde{Q}$ which is a positive definite matrix associated with the terminal penalty, and N is the horizon length. The above optimization problem by discrete MPC design methodology takes the following form of a finite dimensional convex optimization problem:

$$\min_{\tilde{u}} \left[\Re(\tilde{u}^0)\ \Im(\tilde{u}^0)\ \cdots\ \Re(\tilde{u}^{N-1})\ \Im(\tilde{u}^{N-1}) \right] \bar{H} \begin{bmatrix} \Re(\tilde{u}^0) \\ \Im(\tilde{u}^0) \\ \cdots \\ \Re(\tilde{u}^{N-1}) \\ \Im(\tilde{u}^{N}) \end{bmatrix}$$

$$+ \left[\Re(\tilde{u}^0)\ \Im(\tilde{u}^0)\ \cdots\ \Re(\tilde{u}^{N-1})\ \Im(\tilde{u}^{N-1}) \right] \bar{G}\pi(0)$$

$$\bar{A} \begin{bmatrix} \Re(\tilde{u}^0) \\ \Im(\tilde{u}^0) \\ \cdots \\ \Re(\tilde{u}^{N-1}) \\ \Im(\tilde{u}^{N-1}) \end{bmatrix} \leq \bar{B}$$

with $\bar{H}, \bar{G}, \bar{A}$ and $\bar{B}$ being finite dimensional matrices. In this problem, the objective is to minimize the vector $\left[\Re(\tilde{u}^0)\ \Im(\tilde{u}^0)\ \Re(\tilde{u}^2)\ \Im(\tilde{u}^2)\ \cdots\ \Re(\tilde{u}^{N-1})\ \Im(\tilde{u}^{N-1}) \right]'$ of finite length N by solving quadratic optimization problem. If optimization is feasible, then the first control input $\Re(\tilde{u}^0)$ and $\Im(\tilde{u}^0)$ is applied to the system and the horizon is advanced one step forward in time then this is repeated moving forward in time. This procedure is repeated for each time step and it can be shown that initial feasibility implies subsequent feasibility in the case of disturbance free systems. Moreover, it was shown that the control law obtained in this way optimally stabilizes the system providing that the minimization problem is successively feasible [23, 24].

4 GL Numerical Simulation with Constrained MPC

Different values for the parameters of GL equation are reported in the literature due to the different applications of the equation and dimensionality of the problem under consideration. Milovanovic and Aamo considered the same problem presented here, but only reported the real part of $b(\xi)$ [30]. The backstepping method [16] was applied to this problem using explicit forms of the parameters given by Roussopoulos [3]. Since different parameters are used in the literature, here the form given in [3] is used and fitted to $b(\xi)$ and the values given by Milovanovic and Aamo [30], which are given in Table 1 are used.

Table 1 GL equation parameters

Parameter	Value
a_1	0.01667
$a_2(\xi)$	$(0.1697 + 0.04939i)\xi^2 - (0.1748 + 0.06535i)\xi - 0.09061 + 0.001485i$
$a_3(\xi)$	$(0.1563 - 0.001352i)\xi^4 + (-1590 + 0.6278i)\xi^3 +$ $(0.3958 - 1.8577i)\xi^2 + (-1.6852 + 1.6759i)\xi + 1.2645 - 0.2489i$

The eigenvalue problem of operator $\mathcal{A}$ is given by

$$a\frac{d^2\phi(\xi)}{d\xi^2} + b(\xi)\phi(\xi) = \lambda\phi(\xi)$$
$$\phi(0) = \phi(1) = 0 \tag{22}$$

Since $b(\xi)$ is space dependent, no analytic expressions for eigenvalues and eigenvectors is possible in general. However, standard numerical methods can be used to solve this problem. A finite element method results in eigenvalues and eigenvectors shown in Figs. 2 and 3. Note that the first eigenvalue $\lambda_1 = 0.0963 + 0.0993i$ has a relatively small positive real part which makes it unstable. This instability is very sensitive to the parameters. Also note that when system (8) is extended to (9) a zero eigenvalue is added to the set.

The function $B(\xi)$ satisfying all the requirements is given by the solution to the following ordinary differential equation:

$$a\frac{d^2B(\xi)}{d\xi^2} + b(\xi)B(\xi) = 0 \tag{23}$$
$$B(0) = 0 \tag{24}$$
$$B(1) = 1 \tag{25}$$

The solution to this equation is found numerically and is shown in Fig. 4.

For MPC control of the GL equation, the sampling interval is chosen to be $h = 0.1$ which ensures capturing the fastest dynamics in the discretization and $Q_u = Q_s = 1$. The optimization horizon is chosen to be $N = 15$ and limits on actuation are $-U^{\min} = U^{\max} = 0.2$. It is assumed that $r(\xi)$ is a smooth function being nonzero in a finite spatial interval of the form $[\xi_r - \mu, \xi_r + \mu]$, where μ is a small positive real number, and zero elsewhere. Hence, the constraint (11) is applied at a single point $\xi_r = 0.6667$ with limits $-A^{\min} = A^{\max} = 0.5$. In the following, by a constrained result, we mean that both input and state constraints are applied to the control problem in the form of (10) and (11), respectively. The solution to the optimization problem is applied

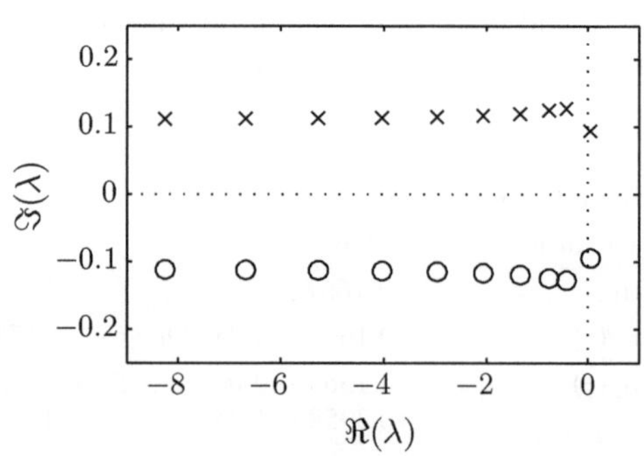

Fig. 2 First ten eigenvalues of $\mathcal{A}$ and $\mathcal{A}^*$ (circles and crosses represent complex eigenvalue (+) with its conjugates (∘)

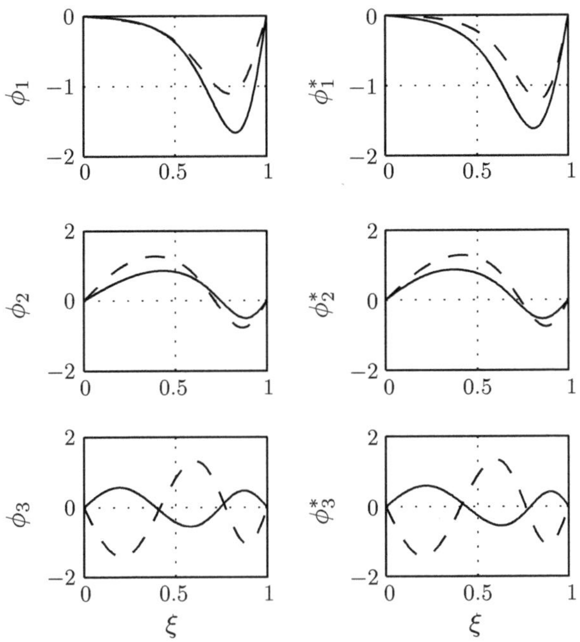

Fig. 3 Real (solid) and imaginary (dashed) parts of the first three eigenfunctions of $\mathcal{A}$ and $\mathcal{A}^*$

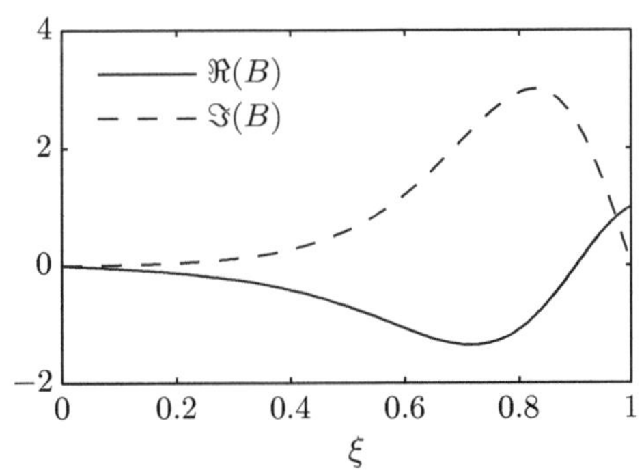

Fig. 4 The function $B(\xi)$ obtained as solution to expression (23)

to a finite difference discretization of the original PDE and the system is solved numerically for each of the constrained case and the unconstrained case.

The real and imaginary parts of the system input for the unconstrained and constrained problem are shown in Fig. 5. Both control laws are stabilizing the unstable system and the constraints on the real part are satisfied. Also, Fig. 6 shows the system response in terms of $\Re(A)$ at the constrained point which shows the satisfaction of state constraint at this point. State evolution of the original PDE is shown in Fig. 7 where MPC is applied at $\xi = 1$. In the vicinity to the constrained point state evolution is shown in Fig. 8 which demonstrates satisfaction of the state constraints under implementation of MPC control law.

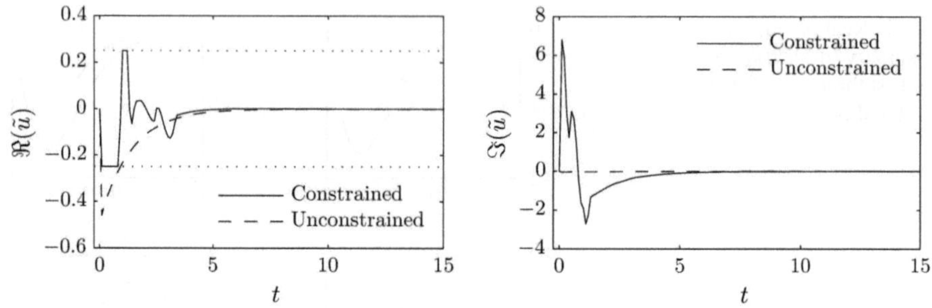

Fig. 5 Real and imaginary parts of the system inputs

Fig. 6 System response at the constrained point ξ_r

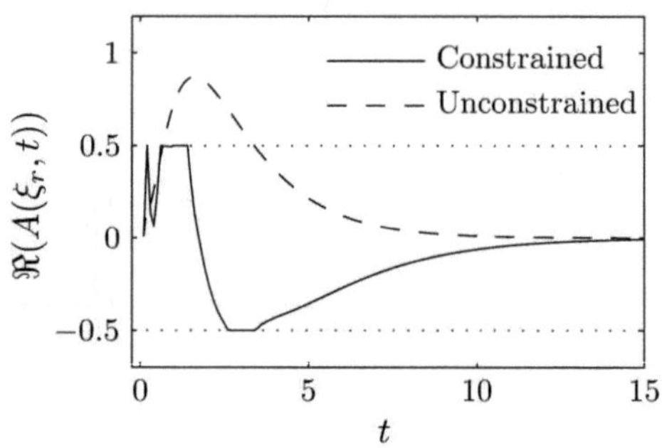

Fig. 7 Evolution of the real and imaginary parts of the state of GL equation under MPC control law

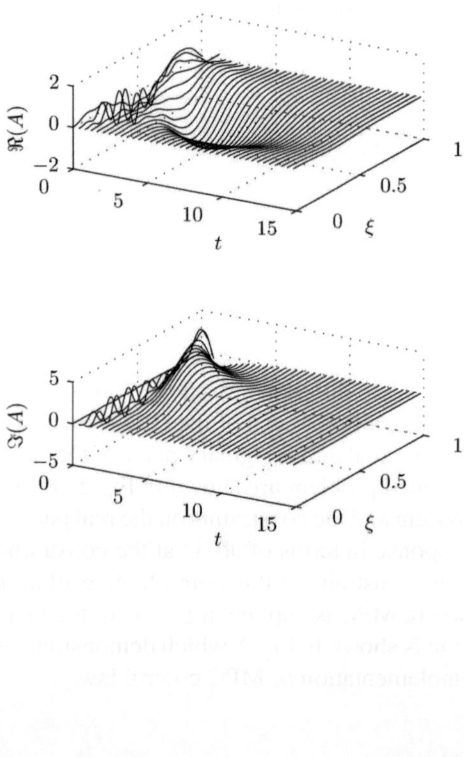

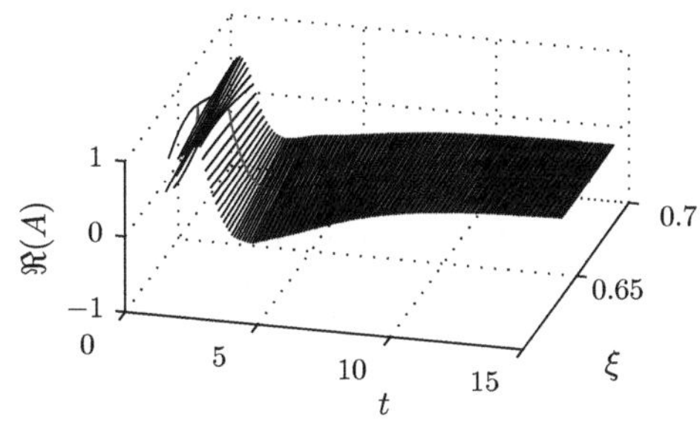

Fig. 8 Evolution of the state of GL equation close to the contained point under MPC control law

5 Conclusions

A linear Ginzburg-Landau equation is used as a model of vortex shedding instabilities of the wake of a bluff body. An MPC formulation is presented for the control of the Ginzburg-Landau equation. The boundary control problem is represented in a complex abstract space as a standard state space formulation for which the available MPC synthesis can be used. The proposed boundary controller achieves stabilization of unstable GL equation and enforces both input and state of PDE constraints which is demonstrated by numerical simulation. Finally, in our future work the experimental application of model based MPC design will be used to demonstrate the application of well known and recognized MPC methodology to fluid flow control problems.

Acknowledgements The authors gratefully acknowledge financial support from the Natural Sciences Research Council of Canada Grant 2016-04646.

References

1. Gad-el Hak, M.: Flow control: the future. J. Aircr. **38**(3), 402–418 (2001) 2018/01/24
2. Berger, E.: Suppression of vortex shedding and turbulence behind oscillating cylinders. Phys. Fluids **10**(9), S191–S193 (1967)
3. Roussopoulos, K.: Feedback control of vortex shedding at low Reynolds numbers. J. Fluid Mech. **248**, 267–296 (1993)
4. Huang, X.Y.: Feedback control of vortex shedding from a circular cylinder. Exp. Fluids **20**(3), 218–224 (1996)
5. Pastoor, M., Henning, L., Noack, B.R., King, R., Tadmor, G.: Feedback shear layer control for bluff body drag reduction. J. Fluid Mech. **608**, 161–196 (2008)
6. Gunzburger, M.D., Lee, H.C.: Feedback control of Karman vortex shedding. J. Appl. Mech. **63**(3), 828–835 (1996)
7. He, J.-W., Glowinski, R., Metcalfe, R., Nordlander, A., Periaux, J.: Active control and drag optimization for flow past a circular cylinder: I. oscillatory cylinder rotation. J. Comput. Phys. **163**(1), 83–117 (2000)
8. Li, Z., Navon, I.M., Hussaini, M.Y., Le Dimet, F.-X.: Optimal control of cylinder wakes via suction and blowing. Comput. Fluids **32**(2), 149–171 (2003)
9. Bergmann, M., Cordier, L., Brancher, J.-P.: Optimal rotary control of the cylinder wake using proper orthogonal decomposition reduced-order model. Phys. Fluids **17**(9), 097101 (2005)

10. Bergmann, M., Cordier, L.: Optimal control of the cylinder wake in the laminar regime by trust-region methods and POD reduced-order models. J. Comput. Phys. **227**(16), 7813–7840 (2008)
11. Curtain, R.F., Zwart, H.: An Introduction to Infinite-Dimensional Linear Systems Theory. Springer, New York (1995)
12. Bensoussan, A., Prato, G., Delfour, M.C., Mitter, S.K.: Representation and Control of Infinite Dimensional Systems. Birkhauser, Boston (2007)
13. Thibault, L.B.F., Colonius, T.: Optimal control of circular cylinder wakes using long control horizons. Phys. Fluids **27**(8), 087105 (2015)
14. Aamo, O.M., Krstic M.: Flow Control by Feedback: Stabilization and Mixing. Springer (2003)
15. Aamo, O.M., Krstic, M.: Global stabilization of a nonlinear Ginzburg-Landau model of vortex shedding. Eur. J. Control **10**(2), 105–116 (2004)
16. Aamo, O.M., Smyshlyaev, A., Krstic, M.: Boundary control of the linearized Ginzburg-Landau model of vortex shedding. SIAM J. Control Optim. **43**(6), 1953–1971 (2005)
17. Eric, L., Bewley, T.R.: Performance of a linear robust control strategy on a nonlinear model of spatially developing flows. J. Fluid Mech. **512**, 343–374 (2004)
18. King, R., Aleksic, K., Gelbert, G., Losse, N., Muminovic, R., Brunn, A., Nitsche, W., Bothien, M., Moeck, J., Paschereit, C., Noack, B., Rist, U., Zengl, M.: Model Predictive Flow Control-Invited Paper. American Institute of Aeronautics and Astronautics (2008)
19. Muminovic, R., Pfeiffer, J., Werner, N., King, R.: Model predictive control for a 2D bluff body under disturbed flow conditions. In: King, R. (ed.) Active Flow Control II, pp. 257–272. Springer, Berlin (2010)
20. Sasaki Y., Tsubakino, D.: Model predictive control of a separated flow around a circular cylinder at a low Reynolds number. In: 2017 56th Annual Conference of the Society of Instrument and Control Engineers of Japan (SICE), pp. 226–229 (2017)
21. Azmi, B., Kunisch, K.: On the stabilizability of the Burgers equation by receding horizon control. SIAM J. Control Optim. **54**(3), 1378–1405 (2016)
22. Grune, L.: Analysis and design of unconstrained nonlinear MPC schemes for finite and infinite dimensional systems. SIAM J. Control Optim. **48**(2), 1206–1228 (2009)
23. Dubljevic, S., Christofides, P.D.: Predictive control of parabolic PDEs with boundary control actuation. Chem. Eng. Sci. **61**(18), 6239–6248 (2006)
24. Dubljevic, S., El-Farra, N.-H., Mhaskar, P., Christofides, P.D.: Predictive control of parabolic PDEs with state and control constraints. Int. J. Robust Nonlinear Control **16**(16), 749–772 (2006)
25. Curtain, R.: On stabilizability of linear spectral systems via state boundary feedback. SIAM J. Control Optim. **23**(1), 144–152 (1985)
26. Aamo, O.M., Krstic, M.: Global stabilization of a nonlinear Ginzburg-Landau model of vortex shedding. Eur. J. Control **10**(2), 105–116 (2004)
27. Aamo, O.M., Smyshlyaev, A., Krstic, M., Foss, B.A.: Output feedback boundary control of a Ginzburg-Landau model of vortex shedding. IEEE Trans. Autom. Control **52**(4), 742–748 (2007)
28. Krstic, M., Smyshlyaev, A.: Boundary Control of PDEs: A Course on Backstepping Designs. Society for Industrial and Applied Mathematics, Philadelphia, USA (2008)
29. Fattorini, H.O.: Boundary control systems. SIAM J. Control **6**(3), 349–385 (1968)
30. Milovanovic, M., Aamo, O.M.: Attenuation of vortex shedding by model-based output feedback control. IEEE Trans. Control Syst. Technol. **21**(3), 617–625 (2013)

A Qualitative Comparison of Unsteady Operated Compressor Stator Cascades with Active Flow Control

Marcel Staats, Jan Mihalyovics and Dieter Peitsch

Abstract Currently, the influence and scaling of active flow control by means of pulsed jet actuators applied to a two-dimensional compressor cascade flow are well understood. However, the presence of a transverse pressure gradient in a 3D annular cascade configuration causes additional effects which need a more profound consideration. The objective of this study is to compare results from the linear cascade setup to the annular one and transfer the AFC technology respectively.

Keywords Compressor cascade · Active flow control · Experiment
Flow mapping · Flow visualization · Pulsed jets

1 Introduction

Improving the overall efficiency of a gas turbine has always been an objective for researchers. One promising possibility is the implementation of a constant volume combustion (*CVC*). Stathopoulus et al. [1] report a potential efficiency increase of up to 20% when passing from the Joule to the Humphrey cycle. Beneficial effects with respect to the efficiency are also reported by Schmidt and Staudacher [2] when introducing the *CVC* cycle. A pressure gain combustion may be envisioned utilizing a pulse detonation combustor (*PDC*). One way of implementing a *PDC* into a gas turbine is the use of multiple combustion tubes arranged in an annular pattern that close subsequently whenever combustion takes place [3]. Such tubes are opened and

M. Staats (✉) · J. Mihalyovics (✉) · D. Peitsch
Technische Universität Berlin, Institute of Aeronautics and Astronautics,
Chair for Aerodynamics, D-10587 Berlin, Germany
e-mail: Marcel.Staats@tu-berlin.de

J. Mihalyovics
e-mail: Jan.Mihalyovics@tu-berlin.de

R. King (ed.), *Active Flow and Combustion Control 2018*,
Notes on Numerical Fluid Mechanics and Multidisciplinary Design 141,
https://doi.org/10.1007/978-3-319-98177-2_6

refilled with fresh gaseous air-fuel mixture and then closed for ignition. Multiple tubes would be operated in a pulse detonation engine (*PDE*), introducing strong unsteady effects to all turbomachinery components [4, 5].

In a turbocharged *PDE*, the compressor will specially suffer from disturbances originating from the *PDC* combustion tubes. In a worst case configuration, the combustion tubes are installed very closely downstream of the last compressor stator without providing a plenum. This configuration would surely be beneficial for the overall length of the gas turbine or aero engine but also maximizes the intermittent pressure fluctuations to the compressor caused by the *PDC*. Investigations on the unsteady flow-field that would be expected under the regime of pressure gaining combustion show severe flow separation phenomena occurring on the compressor stator [6].

These unsteady three-dimensional flows enhance secondary flow structures such as the corner vortex and possible flow separation. Indeed, it has been proven that the highest pressure losses occur specifically within these regions of separated flow [7, 8]. Within this framework, Gbadebo et al. [9] showed that the size and the characteristics of the corner vortices are remarkably increased by periodic blade incidence changes, which can also be associated with a *PDC*.

Therefore, some type of flow control is required to ensure the aerodynamic operability, especially for such unsteady types of gas turbines. Active flow control (*AFC*) opportunities for compressor stators were investigated in [10–14]. Research indicated that pulsed jet actuators (*PJA*) are more effective for *AFC* application compared to steady blowing actuators, as shown by Seifert [15] and Hecklau et al. [16]. The feasibility of suppression of periodic flow separation phenomena in a unsteady operated compressor stator cascade using *PJA* was shown by Staats et al. [17]. Further research regarding unsteady compressor flows was done by Steinberg et al. [18, 19]. Here emphasis was given to apply iterative learning control to an unsteady compressor stator flow-field.

The objective of this work is the transfer of the technology of a two dimensional cascade setup (linear cascade) to a three dimensional setup (annular cascade). The research work emphasizes the qualitative comparability of the two configurations, as the geometric characteristics have mostly been transferred to the annular setup. Furthermore, results with active flow control (using *PJA*) in the annular test setup are compared to the linear compressor test setup. The qualitative comparison of the results showed good consistency among the two test setups.

2 Experimental Setup

In this contribution, results conducted at two separate compressor stator test rigs are presented. Both experiments, the linear- and annular cascade, were equipped with identical highly loaded controlled diffusion airfoils (*CDA*) and operated at a chord-length based REYNOLDS-number of $Re = 6 \times 10^5$. The blades had a design flow

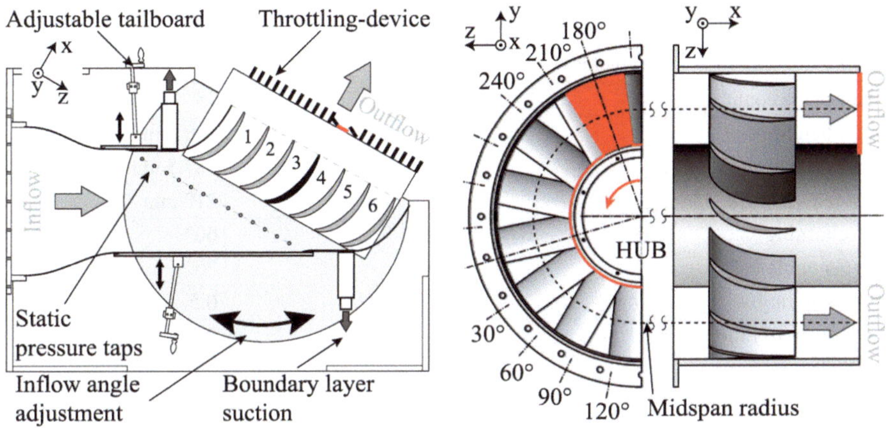

(a) 2D cascade test section (b) 3D cascade test section

Fig. 1 Experimental setup for both cascade test setups (linear- and annular cascade)

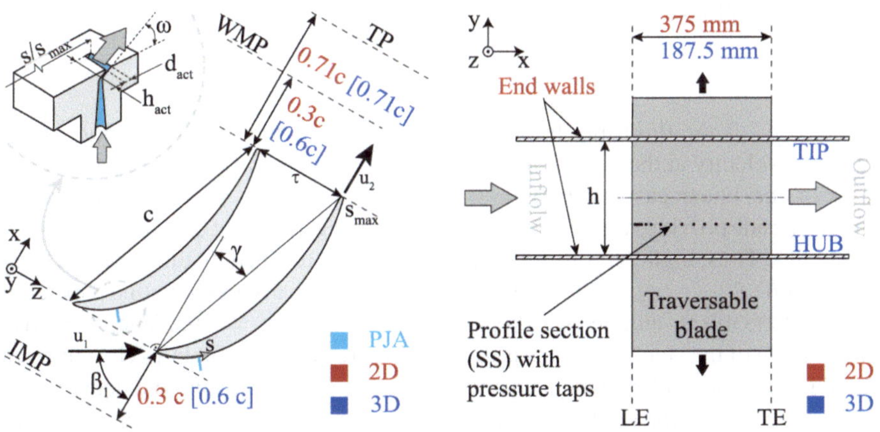

(a) 2D and 3D cascade blade geometry (b) Sketch of a moveable blade used in the 2D and 3D cascade

Fig. 2 Blade geometry for both cascade test setups (linear- and annular cascade)

turning angle of $\Delta\beta = 60°$ and were arranged in an airfoil stagger angle of $\gamma = 20°$. The two test sections used for the experiments are depicted in Fig. 1a and b.

Furthermore, throttling-devices were installed in both configurations that simulate the periodic unsteady effect, similar to the one expected when operating a pulse detonation combustor (*PDC*) downstream the last compressor stator stage. In both cases, these devices were mounted in the throttling plane (*TP*), located $0.71 \cdot c$ downstream of the compressor blades trailing edges (*TE*). Due to a geometric scaling in the cascade designs the dimensionless frequency (STROUHAL-number *St*) is introduced

Table 1 Geometric data of the linear (2D) and the annular (3D) compressor cascade

Name	Parameter	$2D_c$	$3D_c$
Stator blade count	n	7	15
Stator height	h	300 mm	150 mm
Stator chord length	c	375 mm	187.5 mm
Total length of SS	s_{max}	420 mm	210 mm
Stator turning	$\Delta\beta$	60°	60°
Stagger angle	γ	20°	20°
Hub to tip ratio	y_H/y_T	–	0.5
Pitch to chord ratio midspan	τ/c	0.4	0.5
De Haller	u_2/u_1	0.5	0.5

that corresponds to the frequency (f) the throttling devices are operated with. It is calculated as follows:

$$St_{throttling} = \frac{f_{throttling} \cdot c}{u_1}, \tag{1}$$

where $f_{throttling}$ is the throttling frequency, c is the blade chord length and u_1 is the freestream velocity at the inlet of the cascade. The operational limit of the throttling device of the linear cascade is $St = 0.0525$ ($f_{throttling,linear} = 3.5$ Hz), whereas the annular cascade provides the periodic disturbance using a rotating disk, equipped with two paddles. Thus, higher throttling frequencies can be reached. In this case, the limit is $St = 0.06$ ($f_{throttling,linear} = 16.0$ Hz). In the experiments discussed in this paper, the throttling STROUHAL-number was kept constant at $St_{throttling} = 0.03$. The geometric configuration of the stator passages is given in Figure 2a and in Table 1. The numbers printed in red color indicate the measures for the linear cascade and the data printed in blue color denote the geometry used in the annular cascade experiments. Further details on the two individual test setups are presented in the following subsections.

2.1 Linear Cascade Setup and Instrumentation

The linear cascade was attached to a low speed open wind tunnel facility operated at the Chair of Aerodynamics of Technische Universität Berlin (*TUB*) (Fig. 1a). The test section was equipped with seven compressor stator blades forming six two dimensional passages. The test section was mounted to a rotatable disk that allows for inflow angle variations ranging from $\beta_1 = 55°$ to $\beta_1 = 65°$. In the presented experiments within this contribution the design flow turning was used ($\Delta\beta = 60°$). Additional geometric reference data for the linear cascade configuration are shown in Table 1. Furthermore, the test setup featured adjustable tail boards and a boundary layer suction, allowing for highly symmetric flow conditions. The inflow static pressure was measured at the inflow measurement position (*IMP*) at $0.3 \cdot c$ upstream of the leading edges (*LE*). At the operational speed ($u_1 = 25$ m/s) the turbulence level at the inflow plane was below $Tu \leq 1\%$.

The periodic disturbances were introduced by a throttling-device located at $0.71 \cdot c$ downstream of the *TE*s, consisting of 21 throttling blades that were closed one after the other, blocking approx. 90% of one passage at a time. This leads to a periodic disturbance to every stator passage. The passages were choked in the sequence $4 - 5 - 6 - 1 - 2 - 3 - 4 - etc.$ The red shaded throttling-blade (see Fig. 1a) indicates the reference blade. Whenever this blade is closed, a new cycle starts. For such an event, a phase-angle of $\varphi = 180°$ is defined. Further details regarding the throttling device of the linear cascade are given in [6, 20].

2.1.1 Instrumentation

The suction sided surface pressure measurements on the linear cascade test rig were performed on a traversable blade that was equipped with 44 flush mounted miniature differential pressure sensors [First Sensor: HCL12X5], mounted along the suction- (*SS*) and pressure side (*PS*) of the measurement blade (see Fig. 2b). Those pressure data were evaluated by means of the static pressure coefficient ($c_p = \frac{p_{blade}(\varphi) - p_1}{q_1}$). The measurement blade was traversed from $y/h = 3.33\%$ to $y/h = 96.67\%$ in increments of $\Delta y = 5 \cdot 10^{-3}$ m.

Five-hole-probe measurements were performed in the wake measurement plane (*WMP*) at $0.3 \cdot c$ downstream the *TE*s. Five First Sensor: HCLA12X5 (−12.5 to 12.5 mbar) differential pressure sensors were used for the investigations. The traversed grid consisted of 16 × 15 equidistant grid points in the y-z-plane. The acquired data were evaluated in terms of static pressure rise coefficient (C_p, Eq. 2) and total pressure loss coefficient (ζ, Eq. 3).

$$C_p = \frac{p_2(y, z, \varphi) - p_1}{q_1} \ , \tag{2}$$

$$\zeta = \frac{p_{t,1} - p_{t,2}(y, z, \varphi)}{q_1} \ . \tag{3}$$

2.2 *Annular Cascade Setup and Instrumentation*

A low speed, open circuit wind tunnel operated at the flight propulsion laboratory of *TUB* was used for the annular cascade experiments. Figure 1b shows a schematic depiction of the test section, introduced by Brück et al. [21].

The diameter of the casing of the test section measures 0.6 m. In order to create an inflow angle to the annular cascade of $\beta_1 = 60°$ at midspan 19 variable inlet guide vanes (*VIGV*s) produce the swirl needed for the stator inlet conditions.

Moreover, these *VIGV*s provide the option of changing the incidence to the stator by $\pm 5°$. A sketch of the profile geometry used in the 3D annular cascade is shown in Fig. 2a and is supplemented by Table 1. The axial distance between the *VIGV*s and the stator inlet plane measures three chord lengths to ensure for sufficient mixing of the *VIGV* blade wakes and thus produce a homogeneous inlet turbulence of less than $Tu \leq 5.0\%$. In the annular cascade design, the blades from the linear cascade setup were downscaled and adopted to match all stator passage features of the two test setups, such as aspect ratio of the blade, pitch to chord ratio and flow turning, at mid-span.

The application of periodically throttling of the stator passages was realized by a rotating disk mounted at a distance of $0.71 \cdot c$ downstream of the *TE*s of the measurement section. This rotatable disk was equipped with two paddles blocking each stator passage in a given sequence. Hence one cycle of the rotating disk blocks every passage twice. When the measurement passage is fully blocked the phase-angle $\varphi = 180°$ is reached by definition. When both paddles are at the furthest distance from the measurement passage (passage flow is least disturbed) the phase-angle is defined to be $\varphi = 360°$. Thus one full revolution of the throttling device amounts to 720°. The resulting positions of the throttling-device in the annular test setup are illustrated in Fig. 1b), with respect to the phase-angles.

2.2.1 Instrumentation

The annular cascade was also equipped with a traversable stator blade moving in y-direction, allowing for areal surface pressure measurements on the *SS* of the blade, as depicted in Fig. 2b. Here, a total of 24 static pressure ports were equidistantly distributed on the *SS* of the blade profile. The static pressures from these locations were measured using differential pressure sensors [First Sensor: HDOM050] with a calibrated pressure range of −50 to 50 mbar. The static pressure distribution was measured at 99 span-wise locations.

Wake plane measurements were taken at $0.6 \cdot c$ downstream of the stators *TE* (see Fig. 2a). In order to get a detailed wake pressure distribution and velocity profile, a miniature five-hole probe was traversed in a circumferential based polar grid, measuring mean and phase resolved values at each location. The grid covered one passage and had $N = 20$ equidistant radial lines. On each radial line, grid points were distributed equidistantly along the circumference. The radial line at hub side held $M_{N1} = 10$ grid points, while the radial line at tip side held $M_{N20} = 20$ grid points.

2.3 *Actuator Design for Active Flow Control*

In the active flow control experiments a *PJA* system was used for the investigations. The actuator system consisted of a rectangular outlet orifice, measuring $h_{act}/c =$

0.0533 (slot height) and $d_{act}/c = 0.001066$ (slot width) in relation to the chord length. The slot width in the annular cascade experiments was chosen slightly differently due to manufacturing constraints and measured $d_{act}/c = 0.002133$. The outlet orifices had a blowing angle of $\omega = 15°$ relative to the passage end-wall and were perpendicular with respect to the blade's surface. The linear cascade was equipped with twelve actuators located on each passage end-wall (two actuators per passage) at $s/s_{max} = 0.145$ (relative suction surface coordinate). In the annular cascade setup, highly asymmetric flow separation phenomena govern the passage flow field. Accounting for this circumstance only the hub-sided end-wall was equipped with actuators for active flow control at a relative suction surface coordinate of $s/s_{max} = 0.129$. In the annular cascade, the mass flow used for the *AFC* was controlled using mass flow meter [Festo-SFAB-1000] and a proportional directional valve [Festo-MPYE-5-3/8-010-B]. In the linear cascade setup, a mass-flow controller [Bronkhorst-F-203AV-1M0-ABD-55-V] was used to ensure for a constant actuation mass-flow rate. Furthermore, the pulsed blowing was realized by solenoid valves of the type Festo: MHE2-MS1H-3/2G-QS-4-K. The switching frequency, the actuators were operated with, is accounted for by the dimensionless frequency:

$$St_{afc} = \frac{f_{afc} \cdot c}{u_1} \quad . \tag{4}$$

3 Results

3.1 Comparative Investigations on the Steady Flow Fields

In this chapter, results of the undisturbed base flow are presented (no throttling device active). Since the airfoils are highly loaded, the passage flow field was dominated by strong secondary flow phenomena. Figure 3 shows two oil-flow visualizations of the investigated stator blades suction surfaces. In Fig. 3a the flow structures, measured in the linear cascade become evident [13]. A strikingly high symmetry in blade-height direction was detected.

A laminar separation bubble forms at approx. $s/s_{max} = 0.2$ in the 2D case. After the turbulent reattachment the flow separates again, due to the strong adverse pressure gradient and the enhancement of the secondary flow structures (corner separation), from $s/s_{max} = 0.6$ on. Further details on the steady flow field in the linear cascade are provided in Zander et al. [22] and Hecklau et al. [12].

The oil-flow visualization derived from the blade operated in the annular cascade is shown in Fig. 3b. The symmetric separation of the linear cascade cannot be found on the annular cascade blade, however, a massive corner separation occurs on the hub and expands up to 75% of the blade height h at the *TE*. This hub-sided corner separation dominates the entire flow field. The tip region only shows a small area of this phenomenon with a spanwise extension of about 10% h at the *TE*. Between

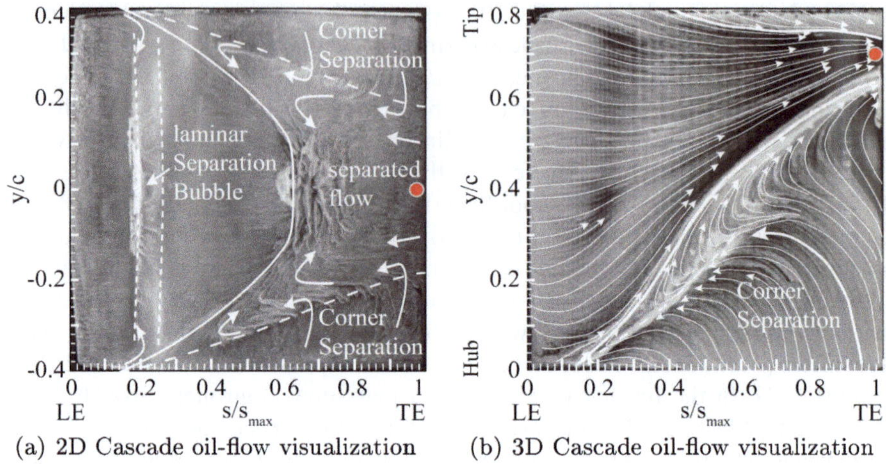

Fig. 3 Comparison of oil-flow visualizations for both cascades

these corner separations a zone of attached flow is visible that cannot be seen in the linear cascade. The corner separation on the annular cascade blade starts at about $s/s_{max} = 0.15$ at the sidewall, expands along the blade span and blocks a huge part of the blade passage in the hub region. The flow is redirected because of this blockade and also accelerated due to the smaller relative flow passage. The stator passages expand in height-wise direction, which leads to a radial pressure gradient causing this increased growth of the hub-sided corner separation. Further investigations on that type of stator flow were performed by Beselt et al. [23].

3.2 Comparative Investigations on the Unsteady Flow Fields

In the following, the unsteady base flow without active flow control with respect to the two compressor stator test rigs (linear- and annular cascade) is discussed. In these cases, the throttling frequency was chosen such that the dimensionless frequency (STROUHAL-number St) was constant in all cases and adjusted to $St_{throttling} = 0.03$ (linear cascade: $f_{throttling} = 2$ Hz; annular cascade: $f_{throttling} = 8$ Hz). The compressor stator performance was evaluated in terms of the Eqs. 2 and 3. The resulting values are depicted in Fig. 4. Here, the compressor stators were operated in the unsteady regime but time averaged data are shown in the figure. It was found that the linear cascade operated at lower total pressure losses with higher static pressure recoveries, compared to the annular cascade.

Figure 5 shows key results obtained from individual measurement campaigns. The figure subdivides into three columns. The plots arranged in the left column include information on the static pressure coefficient fluctuations, measured on the suction

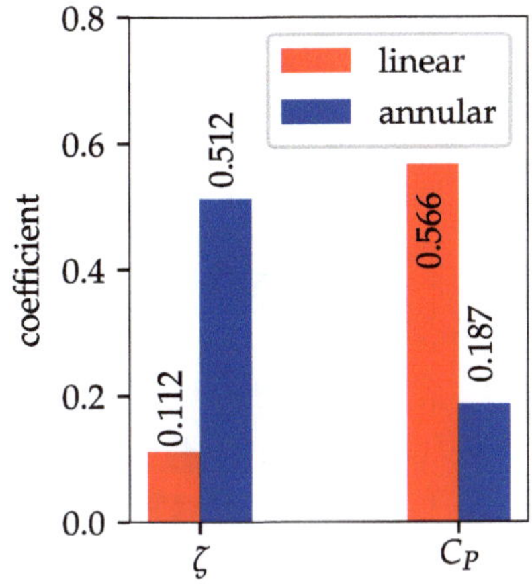

Fig. 4 Comparison of the compressor stator performances at STROUHAL number $St_{throttling} = 0.03$ evaluated for both cascades

side of the measurement blade (between passage three and four; see Fig. 1a) on the linear cascade setup. The mid-column shows the corresponding data obtained from the annular cascades measurement blade. The data shown in the right column of Fig. 5 were taken from wake measurements and indicate the fluctuations in static pressure rise downstream of the stator passage. All data were phase-averaged and arranged row-wise in the figure. The sampling frequency was chosen to exceed the NYQUIST SHANNON sampling theorem with respect to the actuation frequency, which was higher than the throttling frequency, in order to gain sufficient fidelity. The corresponding phase-angles to each plot are found on the line plots ordinate. Here, the static pressure fluctuations on the suction surface of the blades were calculated by means of the REYNOLDS decomposition:

$$c_p{}' = c_p(\varphi) - \overline{c_p} \ . \tag{5}$$

The phase-averaged static pressure rise through one passage, depicted by the line-plot, was divided by its mean value. The phase-angle, where the measurement passage (passage four in the linear cascade) of the cascades were fully blocked is marked in the line plot of Fig. 5 (phase-angle $\varphi = 180°$). In both cascades, the surface pressure distribution oscillated around a mean value. In the linear cascade the oscillation amplitude was $c_p{}' = \pm\, 0.1$. In the annular test setup, lower amplitudes occurred and static pressure fluctuations of up to $c_p{}' = \pm\, 0.05$ were measured. That value corresponds to approximately half the magnitude measured in the linear cascade flow. The oscillation in static pressure rise coefficient $C_P/\overline{C_P}$ through one passage reveal that the amplitudes measured in the annular cascade exceeded the values that originated from the linear cascade flow. The reason for that is twofold. The wake-measurement plane in the annular cascade was located at $0.6 \cdot c$ downstream the trailing edges, whereas the wake measurement plane in the linear cascade was located closer to the trailing edges, at $0.3 \cdot c$. It should be noted that in the annular cascade the wake measurement plane is affected earlier by the upstream moving

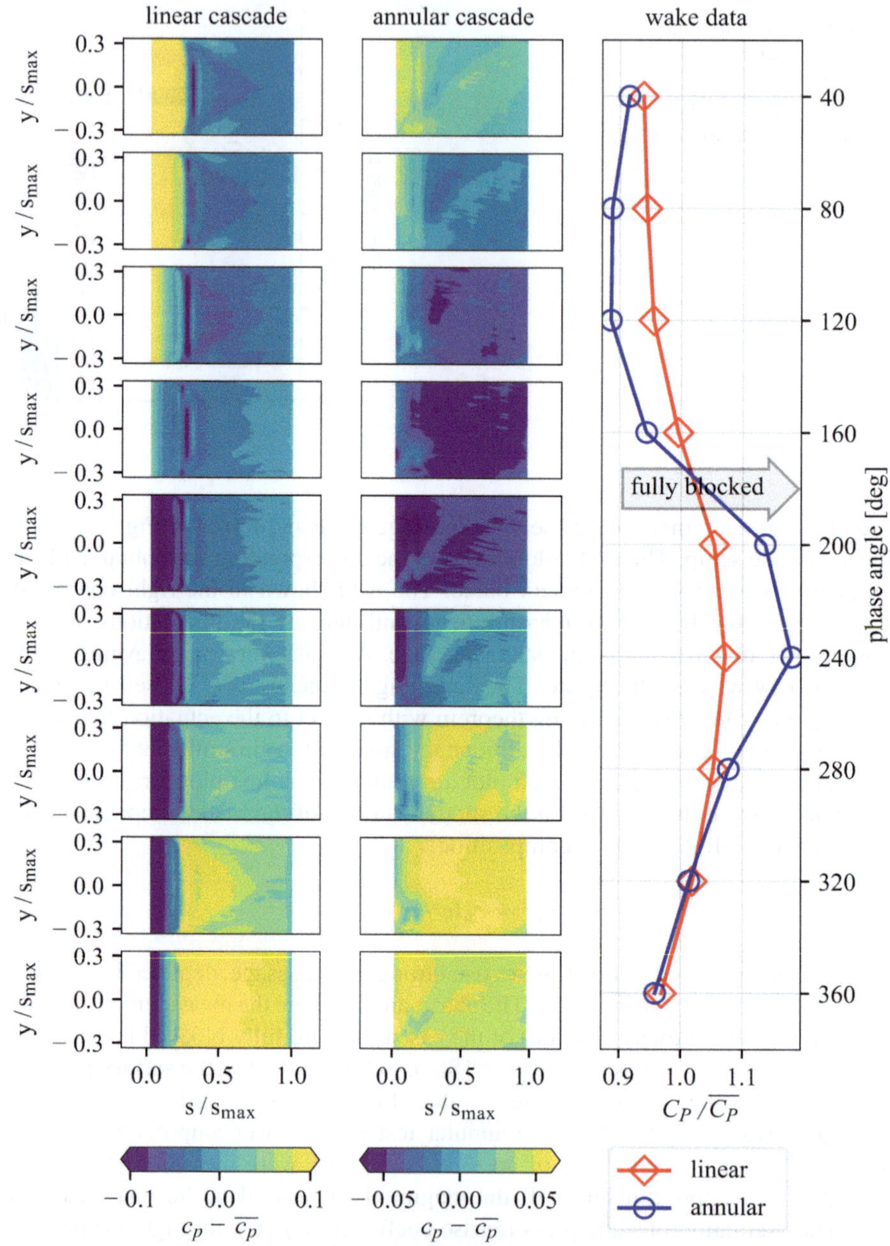

Fig. 5 Comparison of suction side static pressure distribution and wake pressure rise coefficient for both cascades at different throttling phase angles for unsteady base flow at $St = 0.03$

pressure wave caused by the throttling device, than in the linear cascade. This causes a decrease in measured pressure amplitude due to dampening effects. Secondly, the annular- and linear cascade flows indicated major differences in the passage flow field, thus leading to individual reactions of the passage flow, due to the unsteady operation of the two test rigs. However, the qualitative behavior of both cascades under unsteady loading conditions is comparable even though the two throttling devices are operated by different working principles (linear cascade: 21 periodically closing throttling flaps; annular cascade: rotating disk with two oppositely arranged paddles that choke the passages). Despite those differences, striking similarities in the data were found. The phase-angles with highest static pressure recovery measured in the *WMP* were found at approximately $\varphi = 240°$. The highest static pressures on the stator blade suction sides occurred at $\varphi = 320°$ (annular cascade) and $\varphi = 360°$ (linear cascade). With respect to the highest values of the static pressure field on the suction surfaces of the blades and the maximum in static pressure rise through the measurement passage (measured in the *WMP*), a phase-angle shift in the same order of magnitude could be observed in both cascade setups. This shift equaled to approximately $\Delta\varphi = 80°$ to $100°$ and is attributed to a certain inertia of fluid. The resulting phase-angle, with the minimum static pressure fluctuations on the blade, is in good correspondence among the two compressor test setups ($\varphi = 160°$ to $200°$). Due to the periodic throttling of the passages the operating point of one passage is constantly shifted (e.g. inflow angle variations that occur) [20].

3.3 *Comparison of Active Flow Control Results*

The general effect of using a side-wall actuator is that the passage is de-blocked by the actuation of the passage vortex that is reduced in size but not in strength [13, 17, 24].

As mentioned earlier, the linear cascade was equipped with two side-wall actuators per passage, whereas the annular cascade only uses one side-wall actuator to account for the non-symmetric flow structures. The increase of the trailing edge pressure was used to evaluate the impact of the actuation to the compressor stator flow field. The respective positions where $c_{p,TE,afc}$ has been measured is marked with a red circle in Figure 3. These positions have been chosen because the maximum static pressure at the *TE* is observed there, as shown in [25, 26]. For the actuated case the maximum increase in pressure occurs at the same location. Figure 6 shows the increased time-averaged static pressure recovery ($\Delta c_{p,TE,afc}$) of the stator blade on the ordinate. This parameter was calculated by the following equation:

$$\Delta c_{p,TE,afc} = c_{p,TE,afc} - c_{p,TE,ref} \quad . \tag{6}$$

The reference trailing edge pressure coefficient is subtracted from the trailing edge pressure coefficient measured in a given actuation scenario. High trailing edge pressure recoveries indicate higher pressure recoveries throughout the whole passage,

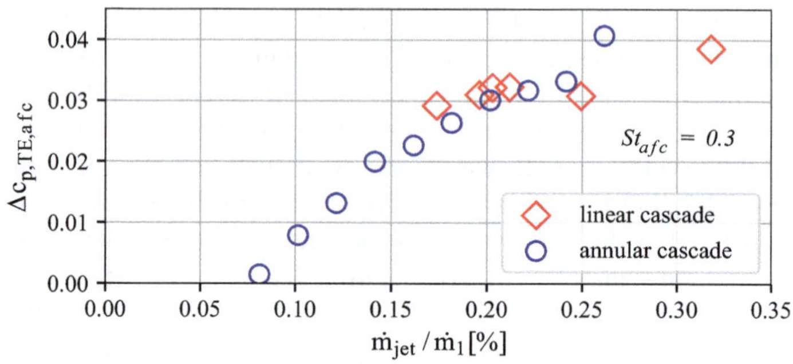

Fig. 6 Impact of *AFC* to the trailing edge pressure coefficient in the linear and annular test setup

due to reduced passage blocking. The abscissa in Figure 6 shows the mass flow ratios of the active flow control. In both test setups, the dimensionless frequency of the actuation was kept constant at a STROUHAL number of $St_{afc} = 0.3$. Here, the actuation frequency in the linear cascade case was $f_{afc} = 30$ Hz and the annular cascade was actuated with a frequency of $f_{afc} = 81$ Hz.

In the discussed cases, the time-averaged trailing edge pressure recoveries, measured in the linear and the annular cascade, were in the same order of magnitude when a mass-flow ratio ranging from $\dot{m}_{jet}/\dot{m}_1 \approx 0.15\%$ to $\dot{m}_{jet}/\dot{m}_1 \approx 0.32\%$ was applied. By applying $\dot{m}_{jet}/\dot{m}_1 \approx 0.2\%$ of the passage mass-flow rate through the actuators a trailing edge pressure increase of $\Delta c_p = 0.03$ was achievable in both configurations. Higher mass-flow rates lead to slightly increased trailing edge pressure recoveries. In the annular cascade case, the maximum investigated actuation mass-flow ratio was $\dot{m}_{jet}/\dot{m}_1 \approx 0.27\%$, where the trailing edge pressure was increased by $\Delta c_p = 0.04$. In the linear cascade case. the maximum investigated actuation mass-flow ratio of $\dot{m}_{jet}/\dot{m}_1 \approx 0.32\%$ led to a comparable gain in trailing edge pressure recovery.

4 Conclusion

In this contribution, results obtained from two unsteady operated compressor stator cascades were shown. The unsteady outflow conditions were imposed by throttling-devices that simulated the condition expected in a pulse detonation engine. In sub-project B01 of the CRC1029, at TUB, two compressor test setups are operated (linear cascade and annular cascade), where such flows are investigated. The present paper contributes to a better understanding of the flow separation phenomena expected in a PDE and compares results from AFC experiments from such flow phenomena, measured in two different test setups. It was shown that the basic flow structures in the passage (e.g. corner separation) formed differently in both configurations. The impact of a periodic disturbance on the other hand was in good agreement for both investigated cases. The static pressure oscillations on the stator blading of the annular cascade reached half the magnitude, when compared to the linear cascade. Preliminary results with active flow control indicated a comparable effect of the pulsed jet actuation with respect to the increase of static pressure recovery for both investigated configurations. Increasing the time-averaged static pressure recovery by $\Delta c_{p,TE,afc} = 0.03$, with the use of $\dot{m}_{jet}/\dot{m}_1 \approx 0.32\%$ of the passage mass-flow rate, was feasible and identical for the investigated stator flows.

Acknowledgements The authors gratefully acknowledge support by the Deutsche Forschungsgemeinschaft (DFG) as part of collaborative research center SFB 1029 "Substantial efficiency increase in gas turbines through direct use of coupled unsteady combustion and flow dynamics" on project B01.

References

1. Stathopoulos P., Vinkeloe, J., Paschereit, C.O.: Thermodynamic evaluation of constant volume combustion for gas turbine power cycles. In: Proceedings of IGTC Tokyo: 11th International Gas Turbine Congress, November, 15th–20th, Tokyo, Japan, WePM1G.2 (2015)
2. Schmidt F., Staudacher S.: Generalized thermodynamic assessment of concepts for increasing the efficiency of civil aircraft propulsion systems. In: Proceedings of the ASME Turbo Expo 2015: Turbine Technical Conference and Exposition (GT2015), June, 15th–19th, Montreal, Canada, GT2015-42447 (2015)
3. Rouser K.P., King P.I., Schauer F.R., Sondergaard R., Hoke J.L. Goss, L.P.: Time-resolved flow properties in a turbine driven by pulsed detonations. J. Propuls. Power **30**(6), 1528–1536 (2014)
4. Lu J., Zheng L., Wang Z., Peng C., Chen X.: Operating characteristics and propagation of back-pressure waves in a multi-tube two-phase valveless air-breathing pulse detonation combustor. Exp. Therm. Fluid Sci. **61**, 12–23 (2015)
5. Fernelius M.H., Gorrell S.E.: Predicting efficiency of a turbine driven by pulsing flow. In: Proceedings of the ASME Turbo Expo 2017: Turbine Technical Conference and Exposition (GT2017), June, 26th–30th, Charlotte, USA, GT2017-63490 (2017)
6. Staats M., Nitsche W.: Active control of the corner separation on a highly loaded compressor cascade with periodic non-steady boundary conditions by means of fluidic actuators. In: Proceedings of the ASME Turbo Expo 2015: Turbine Technical Conference and Exposition (GT2015), June, 15th–19th, Montreal, Canada, GT2015-42161 (2015)
7. Cumpsty N.A.: Compressor Aerodynamics, reprint ed. w/new preface, introduction and updated bibliography ed. Krieger Publishing Company, Malabar, USA (2004)
8. Wei M., Xavier O., Lipeng L., Francis L.: Intermittent corner separation in a linear compressor cascade. Exp. Fluids **54**(6), 25 (2013)
9. Gbadebo, S.A., Cumpsty, N.A., Hynes, T.P.: Three-dimensional separations in axial compressors. J. Turbomach. **127**(2), 331–339 (2005)
10. Peacock R.E.: Boundary-layer suction to eliminate corner separation in cascades of aerofoils. Technical report, Ministry of Defense, Aeronautical Research Council, Her Majesty's Stationery Office. Reports and Memoranda No. 3663 (1965)
11. Gbadebo, S.A., Cumpsty, N.A., Hynes, T.P.: Control of three-dimensional separations in axial compressors by tailored boundary layer suction. J. Turbomach. **130**(1), 011004 (2008)
12. Hecklau, M., Wiederhold, O., Zander, V., King, R., Nitsche, W., Huppertz, A., Swoboda, M.: Active separation control with pulsed jets in a critically loaded compressor cascade. AIAA J. **49**(8), 1729–1739 (2011)

13. Zander, V., Hecklau, M., Nitsche, W., Huppertz, A., Swoboda, M.: Active flow control by means of synthetic jets on a highly loaded compressor cascade. Proc. Inst. Mech. Eng. Part A J. Power Energy **225**(7), 897–908 (2011)
14. Wang, X., Zhao, X., Li, Y., Wu, Y., Zhao, Q.: Effects of plasma aerodynamic actuation on corner separation in a highly loaded compressor cascade. Plasma Sci. Technol. **16**(3), 244–250 (2014)
15. Seifert A.: Evaluation criteria and performance comparison of actuators for bluff-body flow control. In: Proceedings of the 32nd AIAA Aviation Forum: Applied Aerodynamics Conference, June, 16th–20th, Atlanta, USA, AIAA 2014-2400 (2014)
16. Hecklau M., Zander V., Peltzer I., Nitsche W., Huppertz A., Swoboda M.: Experimental afc approaches on a highly loaded compressor cascade. In: King, R. (ed.) Active Flow Control II. Notes on Numerical and Fluid Dynamics, vol. 108, pp. 171–186. Springer (2010)
17. Staats M., Nitsche W.: Experimental investigations on the efficiency of active flow control in a compressor cascade with periodic non-steady outflow conditions. In: Proceedings of the ASME Turbo Expo 2017: Turbine Technical Conference and Exposition (GT2017), June, 26th–30th, Charlotte, USA, GT2017-63246 (2017)
18. Steinberg S.J., Staats M., Nitsche W., King R.: Comparison of conventional and repetitive MPC with application to a periodically disturbed compressor stator vane flow. In: Proceedings of the IFAC World Congress: The 20th World Congress of the International Federation of Automatic Control, July, 9th–14th, Toulouse, France, vol. 50(1), pp. 11107–11112 (2015)
19. Steinberg S.J., King R., Staats M., Nitsche W.: Constrained repetitive model predictive control applied to an unsteady compressor stator vane flow. In: Proceedings of the ASME Turbo Expo 2016: Turbine Technical Conference and Exposition (GT2016), June, 13th–17th, Seoul, South Korea, GT2016-56002, p. V02AT37A001 (2016)
20. Staats M., Nitsche, W.: Active flow control on a non-steady operated compressor stator cascade by means of fluidic devices. In: Notes on Numerical Fluid Mechanics and Multidisciplinary Design, vol. 136, pp. 337–347 (2016)
21. Brück C., Tiedemann C., Peitsch, D.: Experimental investigations on highly loaded compressor airfoils with active flow control under non-steady flow conditions in a 3D-annular low-speed cascade. In: Proceedings of the ASME Turbo Expo: Turbine Technical Conference and Exposition-2016, The American Society of Mechanical Engineers, p. V02AT37A027 (2016)
22. Zander V., Hecklau M., Nitsche W., Huppertz A., Swoboda M.: Experimentelle methoden zur charakterisierung der aktiven strömungskontrolle in einer hoch belasteten verdichterkaskade. Deutscher Luft- und Raumfahrt Kongress (DLRK2008-081322) (2008)
23. Beselt C., Eck M., Peitsch, D.: Three-dimensional flow field in highly loaded compressor cascade. J. Turbomach. **136**(10), 101007 (2014)
24. Hecklau M.: Experimente zur aktiven Strömungsbeeinflussung in einer Verdichterkaskade mit pulsierenden Wandstrahlen: Zugl.: Berlin, Techn. Univ., Diss., 2012. Aerospace Engineering. mbv Mensch-und-Buch-Verl., Berlin (2012)
25. Staats M., Nitsche W., Peltzer, I.: Active flow control on a highly loaded compressor cascade with non-steady boundary conditions. In: King, R. (ed.) Active Flow and Combustion Control 2014. Notes on Numerical Fluid Mechanics and Multidisciplinary Design, vol. 127, pp. 23–37. Springer (2015)
26. Brück C., Mihalyovics J., Peitsch, D.: Experimental investigations on highly loaded compressor airfoils with different active flow control parameters under unsteady flow conditions. In: Proceedings of GPPS Montreal: Global Power and Propulsion Conference North America-2018, May 7th–9th, Montreal, Canada, GPPS-2018-0054 (2018)

Transitioning Plasma Actuators to Flight Applications

David Greenblatt, David Keisar and David Hasin

Abstract Pulse-modulated dielectric barrier discharge plasma actuators are applied to the problem of flow separation on a Hermes 450 unmanned air vehicle V-tail panel. Risk-reduction airfoil experiments were conducted followed by full-scale wind tunnel tests. Silicone-rubber based actuators were calibrated and subsequently retrofitted to both the airfoil and the panel. A lightweight (1 kg), flightworthy high-voltage generator was used to drive the actuators. Airfoil and full-scale panel wind tunnel experiments showed a mild sensitivity to actuation reduced frequencies and duty cycles. On the panel, actuation produced a significant effect on post-stall control authority: for $17° < \alpha < 22°$ a 100% increase in the post-stall lift coefficient was achieved; leading edge separation was prevented up to angles of attack of 30°; and hysteresis was virtually eliminated. Future research will focus on integrating the actuators into the panel geometry, implementing thicker dielectric materials and flight-testing.

Keywords Plasma · Actuators · Dielectric barrier discharge · Flight applications
Unmanned air vehicle

D. Greenblatt (✉)
Faculty of Mechanical Engineering, Technion, Israel Institute of Technology, Haifa, Israel
e-mail: davidg@technion.ac.il
URL: https://www.flowcontrollab.com

D. Keisar
Grand Technion Energy Program, Technion, Israel Institute of Technology, Haifa, Israel

D. Hasin
Elbit Systems Ltd., Herzliya, Israel

R. King (ed.), *Active Flow and Combustion Control 2018*,
Notes on Numerical Fluid Mechanics and Multidisciplinary Design 141,
https://doi.org/10.1007/978-3-319-98177-2_7

1 Introduction

V-tail configurations are common on unmanned air vehicles (UAVs), but the tail panels suffer from flow separation, resulting in loss of control during crosswind take-off and landing [1, 2]. A potential solution to the problem is the application of plasma actuators at the leading-edges of the panels. Several studies have indicated that significant improvements to airfoil post-stall lift coefficients can be achieved, in some cases doubling the post-stall value [3–11]. Furthermore, leading-edge perturbations on vertical axis wind turbine blades dramatically increase turbine performance [12–15]. The actuators introduce perturbations corresponding to the separated shear layer instabilities. These perturbations grow and roll up into spanwise vortices that transport high-momentum flow to the panel surface [16]. This overcomes or ameliorates stall, exemplified by increases in maximum lift, significant increases in post-stall lift, elimination of hysteresis and drag reduction. In particular, single dielectric barrier discharge (SDBD, or simply DBD) plasma actuators are well-suited to typical takeoff and landing speeds [3].

Recently, DBD plasma actuators were demonstrated in-flight for the purpose of transition control [17]. A flightworthy system must fulfill a number of demanding requirements. Firstly, all components of the system must add insignificant mass to the payload and must require negligible power, as a fraction of propulsor power, for operation. The system must be operable on both sides of each control surface and normal operation should not compromise conventional flight control operation. If possible, initially, the system should not require complex feedback control and should be operable under open-loop or feedforward control. The system must be robust: namely, it must be operable for long periods without failure; if failure occurs, it must not compromise control of the vehicle relative to its original baseline configuration; and finally, the system must be easily manufactured, maintained, and repaired or replaced if necessary.

The global objective of this research is to implement DBD plasma actuators on the tail of a Hermes 450 unmanned air vehicle and conduct flight tests. This phase of the research has two main objectives: the first is to conduct wind tunnel experiments on a two-dimensional profile (airfoil) at takeoff speeds with different DBD plasma actuator configurations (risk-reduction experiments); the second is to conduct full-scale wind tunnel tests on a tail panel. The risk-reduction experiments are performed as a precursor to full-scale tests. A major challenge of this phase is to develop a flightworthy actuation system capable of producing sufficiently high-amplitude perturbations at typical takeoff and landing conditions. Here we consider a target takeoff speed of 43 kts or 22 m/s.

2 Airfoil Risk Reduction Experiments

2.1 *Airfoil Design*

An airfoil identical to the nominal panel section geometry, with a 350 mm chord length (c) and a 610 mm span (b), was designed and 3D printed (Fig. 1). The airfoil components comprise: (1) the main element; (2) the lower cover; (3) a removable leading-edge module; (4) a recessed removable leading-edge module. The main element is designed to carry the aerodynamic loads and removal of the lower cover facilitates access to the internal volume of the model. The recessed leading-edge module was designed for the purpose of integrating the DBD actuators into the airfoil geometry with minimum distortion of the original profile. The airfoil has 76 pressure ports (41 on the main body and 35 on the non-recessed leading-edge module) and these are close-coupled with two 32-port ESP pressure scanners (piezo-resistive transducers) mounted inside the model. The airfoil was installed and tested in the Technion's Unsteady Low-Speed Wind Tunnel (UWT) 610 mm × 1004 mm test section [18]. It embodies a pair of circular Plexiglas® windows which are held in place by aluminum rings (Fig. 2). The airfoil model was firmly connected to both windows and pitched about the quarter-chord position by rotating both rings synchronously via a servomotor and belt drives.

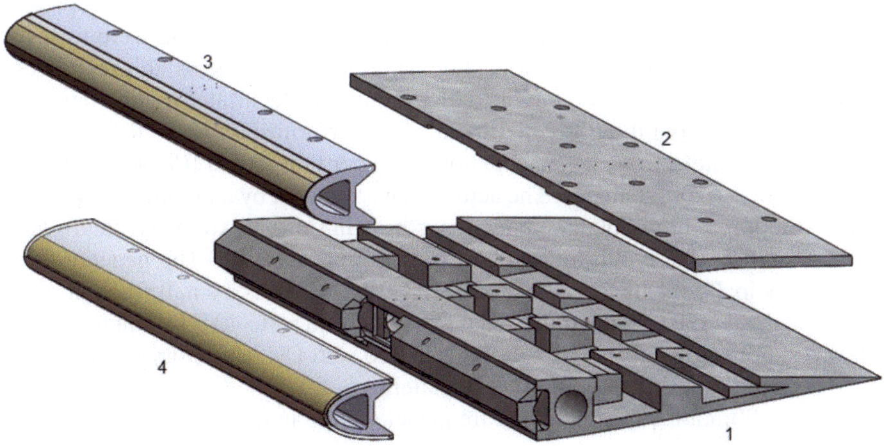

Fig. 1 Expanded schematic of the tail-panel airfoil for two-dimensional wind tunnel testing, showing: the main element (1); the lower cover (2); the removable leading edge modules [without recess (3), with recess (4)]

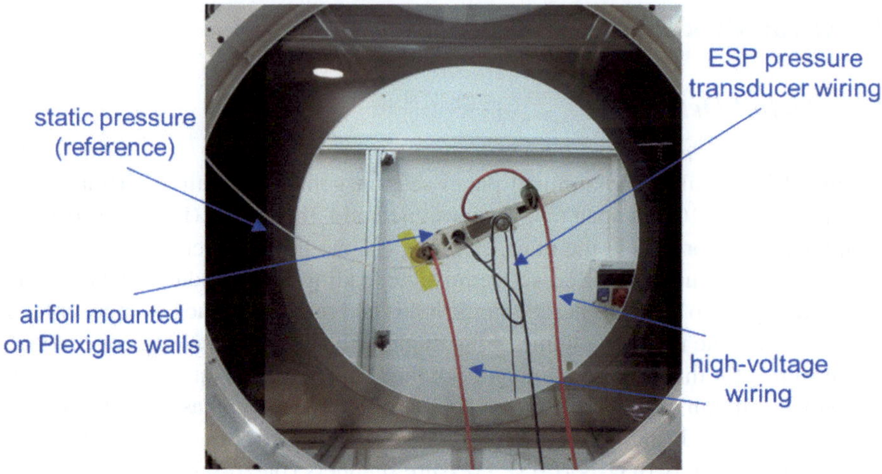

Fig. 2 Photograph of the airfoil with plasma actuator mounted in the tunnel. Pitch-down direction is defined as positive

2.2 DBD Plasma Actuators

In our previous wind turbine related research [12–15], DBD plasma actuators with upper (exposed) and lower (encapsulated) electrodes (both 70 μm thick) separated by three layers of 50 μm thick Kapton® tape were employed. These were wrapped around the leading-edge of the airfoils. For the present experiments, thicker silicone rubber dielectric material was employed (0.3–3 mm) that facilitated higher ionization voltages. Bench-top calibration experiments were performed for both the Kapton and silicone rubber dielectrics, where actuator thrust per unit length $|\mathbf{F}_b|$ was estimated using a Vibra AJ-200E balance. The actuators were driven by a modified GBS Elektronik Minipuls 2 high-voltage generator, consisting of an externally controllable transistor half-bridge and a high voltage transformer cascade. The generator was chosen principally for its low mass, namely 1.0 kg, which is a small fraction of the vehicle payload (150 kg). It requires an input signal and up to 40 V DC input voltage, that was supplied by either a CPx400D-Dual 420 watt DC laboratory power supply or a stack of lithium-ion polymer (LiPo) batteries.

For all calibrations, the ionization frequencies were in the range 8 kHz $\leq f_{\text{ion}} \leq$ 20 kHz; in the separation control study described below this signal was pulse-modulated at frequencies f_p. The power input was calculated from the measured DC voltage and the current supplied to the system: $\Pi_{\text{in}} = V_{\text{in}} \cdot I_{\text{in}}$. A summary of results is presented in Fig. 3, where the input power is referenced to the actuator length b_a.

Based on previous data [12–15], effective separation control was achieved at turbine blade relative wind speeds of 12 m/s. With a target free-stream velocity of 22 m/s, using dimensional analysis, it can easily be seen that the target actuator thrust

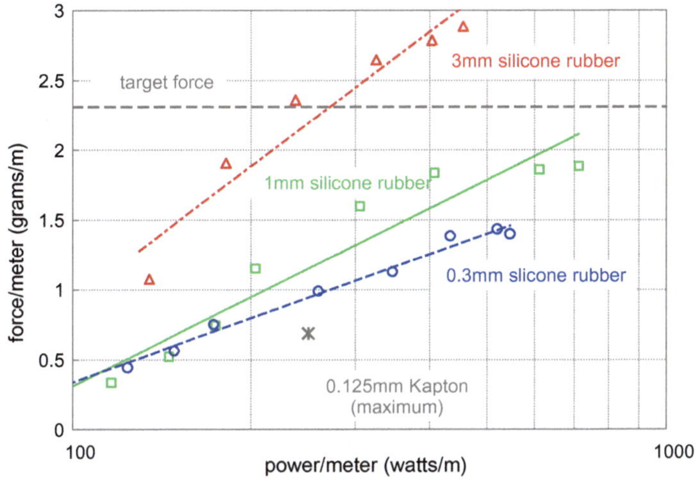

Fig. 3 Measured DBD plasma actuator thrust developed, at d.c. = 100%, as a function of measured input power to the high-voltage generator

must be $(22/12)^2 \cdot |\mathbf{F}_b|_{\text{Kapton}}$. Using silicone rubber as a dielectric material, the target force required for effective separation control at $U_\infty = 22$ m/s, corresponding to mid-span $Re = 7 \times 10^5$, can easily be obtained with a 3 mm thickness. However, in order to minimize changes to the nominal panel geometry, all experiments were performed with thickness 1 mm.

2.3 *Airfoil Results*

Preliminary baseline experiments at free-stream velocities $U_\infty = 19$ m/s and 29 m/s (corresponding to $Re = 4.3 \times 10^5$ and $Re = 6.5 \times 10^5$) were conducted without the actuator present, revealing excellent correspondence with the well-known prediction methods. Static stall occurred at 16° with a $C_{l,\max}$ of 1.3. After validating the fidelity of the baseline experimental setup, different experiments were conducted for the investigation of separation control at different Reynolds numbers, angles of attack, actuator configurations and power input. These were designated as risk-reduction experiments, conducted prior to the full-scale experiments described in Sect. 3. All experiments were performed with the actuator wrapped around the leading-edge of the airfoil, with the encapsulated and exposed electrodes in-line at the $x/c = 0$ location. Both 0.5 and 1.0 mm thick silicone rubber actuator dielectrics were evaluated.

Two key parameters employed for characterizing separation control studies [16] are the momentum coefficient, defined here as:

$$C_\mu \equiv b_a |\mathbf{F}_b| / (q_\infty S) \tag{1}$$

and

$$F^+ \equiv f_p c / U_\infty \tag{2}$$

where b_a, q_∞ and S are the actuator length, free-stream dynamic pressure and planform area respectively. For actuator calibrations $b_a \approx 20\,\text{cm}$, while for the airfoil and panel experiments b_a was equal to the span length. Typical values for effective leading-edge separation control are $C_\mu = O(0.1)\%$ and $F^+ = O(1)$. When the actuators are pulse-modulated, we can also define the net momentum flux that is directly proportional to the duty cycle, namely:

$$\langle C_\mu \rangle \equiv \text{d.c.} \times C_\mu \tag{3}$$

When the plasma ionization frequency is pulse-modulated, d.c. represents the fraction of the modulation period that the plasma is activated. From an applications point of view this is important because d.c. can be reduced to approximately 1%, without loss of airfoil or wing performance, but with a significant reduction in input power.

Since the actuator blocked most of the airfoil leading-edge surface, and the pressure ports with it, it was not possible to compare C_l changes with and without the plasma actuation (see Fig. 4). Therefore, to assess the *relative* changes in performance, three metrics were evaluated, namely: (i) $\Delta C_{p,\text{min}}$—the change in the minimum pressure coefficient; (ii) $\Delta C_{p,\text{TE}}$—the change in the pressure coefficient at the trailing edge of the airfoil; and (iii) $\Delta C_{l,\text{press}}$—the change in the lift coefficient contribution on the high pressure surface of the airfoil. The changes in the high-pressure surface of the airfoil are sensitive to overall circulation (or lift) and elimination of the ports near the leading-edge has only a small effect on the changes.

A summary of the three metrics is shown in Fig. 5 for the post-stall angle $\alpha = 24°$ employing a 0.5 mm thick dielectric. The changes in minimum pressure and lower surface pressure show similar dependence on reduced frequency, while the trailing-edge recovery shows a greater frequency sensitivity. However, the peak is not sharp and it can be concluded that a range of frequencies around $0.75 \leq F^+ \leq 1.5$ will produce positive and comparable increases to post-stall C_l. This is consistent with a number of other studies [16] and is a welcome result, in particular because for a given pulsation frequency, the full-scale panel F^+ varies as a function of the local chord-length (see Sect. 3). To illustrate this, Fig. 5 also shows the reduced frequency range, between root and tip, that would be encountered on the full-scale panel assuming $F^+ = 1$ at the mid-span. On the basis of this observation we project that the panel span-dependent modulation frequency, described in Sect. 3, will produce a positive beneficial result.

Figure 6 shows the variation of all metrics as a function of duty cycle (d.c.) and indicates similar effects for values between 1 and 10%. Duty cycle is a parameter of fundamental importance because the fraction of plasma activation determines the power input to the system [4]. Thus pulse-modulation at low duty cycles has a dual benefit because it can be configured to excite the most effective instability frequency at very low input power. These data are consistent with lift coefficient data acquired at

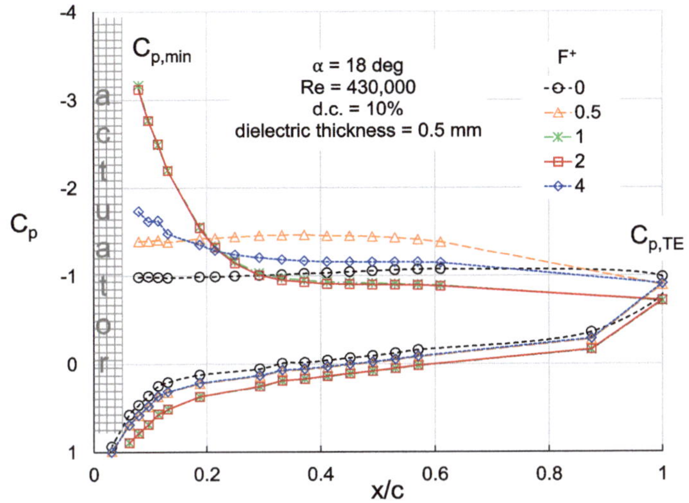

Fig. 4 Pressure coefficient distribution on the airfoil for different reduced frequencies. Actuation conditions: 0.5 mm thick silicone rubber, d.c. = 10%, $f_{\text{ion}} = 9,300\,\text{Hz}$, $\Pi_{\text{in}} = 14\,\text{W/m}$

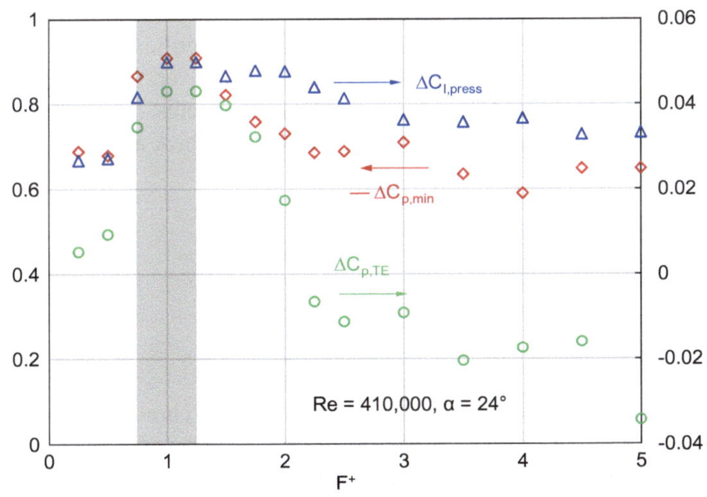

Fig. 5 $\Delta C_{p,\text{min}}$, $\Delta C_{p,\text{TE}}$ and $\Delta C_{l,\text{press}}$, for $0 \le F^+ \le 5$. Actuation conditions: 0.5 mm thick silicone rubber, d.c. = 10%, $f_{\text{ion}} = 9,300\,\text{Hz}$, $\Pi_{\text{in}} = 14\,\text{W/m}$. The shaded area indicates the F^+ range corresponding to the panel

lower Reynolds numbers, where a reduction of the duty cycle from 50 to 1% showed a lift insensitivity similar to [9, 11]. No attempt was made to reduce the d.c. further, although it should be noted that the lower limit should not be reduced to less than one full cycle, namely, d.c. $\ge f_p/f_{\text{ion}}$.

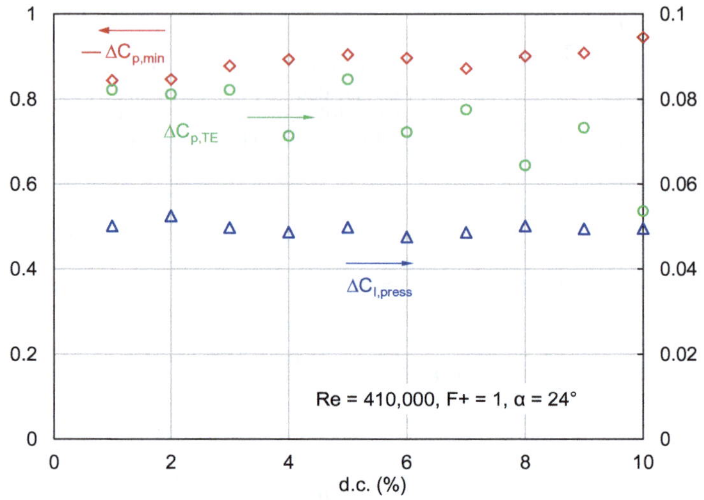

Fig. 6 Variation of the metrics $\Delta C_{p,\min}$, $\Delta C_{p,\mathrm{TE}}$ and $\Delta C_{l,\mathrm{press}}$ as a function of duty cycle, with $F^{+}=1$

Finally, it was noted that the 1.0 mm thick silicone rubber-based actuator produced slightly superior results to those presented above. Moreover, no "burn-through" of the actuator was encountered during any of the experiments. Thus all experiments performed on the full-scale panel employed the 1.0 mm thick actuator.

3 Preliminary Tail-Panel Experiments

3.1 Experimental Setup

The Hermes 450 tail panel has a span of 1.6 m, root and tip chord lengths of 0.6 m and 0.35 m respectively, and a surface area of 0.747 m^2. Experiments were performed in Israel Aircraft Industry's (IAI's) closed-return low speed atmospheric wind tunnel, with test section dimensions 2.6 m × 3.6 m. The panel was mounted on a ∅1.2 m circular end-plate, and fastened to a six-component external aerodynamic balance by means of a clamp and flange (see Fig. 7). The balance operates on the multi-beam principle, employing stepper-motors to drive the riders along the beams to the null setting under each loading condition. The actuator was wrapped around the leading-edge of the panel and attached using double-sided tape in an identical manner to the airfoil application. For purposes of flow visualization, 28 mm fluorescent tufts were fixed to the panel, with 40 mm spacing between them. In order to achieve a strong contrast, the tail panel was painted matt-black and viewed under ultraviolet illumination. Smoke-base flow visualization was also performed.

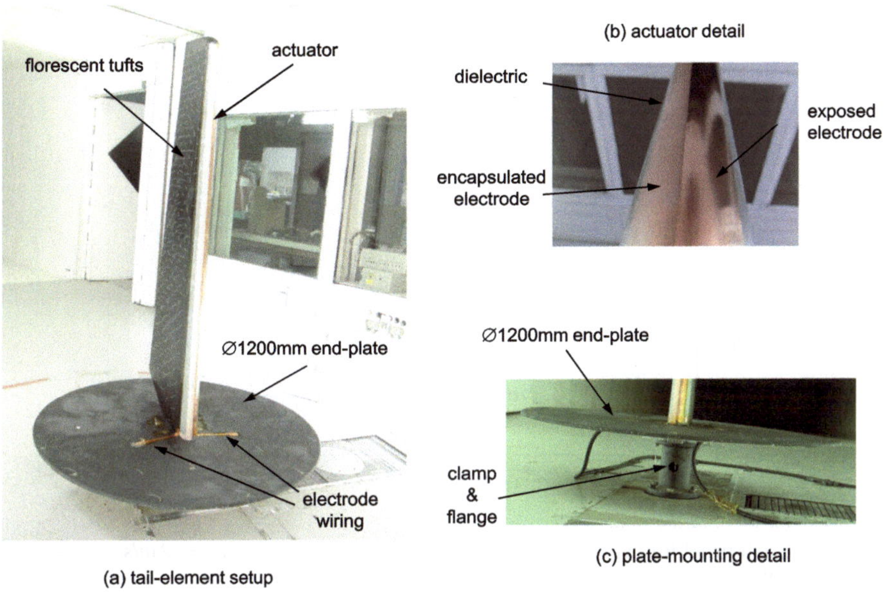

Fig. 7 Photographs of the full-scale panel experimental setup showing the assembly, actuator detail and mounting

3.2 *Preliminary Results and Discussion*

When pulsed perturbations are introduced, the reduced frequency is not uniquely defined because the chord-length is a function of the spanwise location. Here, we simply use the mean panel chord-length (475 mm) in the definition of F^+. Similar to the airfoil experiments, the panel was set at three post-stall angles of attack and for each angle, the pulsation frequency was swept corresponding to $0.25 \leq F^+ \leq 2.5$ at d.c. = 10% and $\Pi_{\text{gross}} = 8.7\,\text{W}$. As before, experiments were performed by measuring the baseline value, followed by initiation of the pulsations, and a subsequent baseline measurement. These data are summarized in Fig. 8. The greatest increases in lift are observed close to the static stall angle at $\alpha = 20°$, where ΔC_L exceeds 0.6 (or 100%) and these data are consistent with prior airfoil investigations. There does not appear to be a significant dependence on reduced frequency and this is broadly consistent with trailing-edge pressure changes and lower surface lift contributions observed on the airfoil. This indicates that these metrics are probably the most reliable for assessing changes in airfoil performance when leading-edge pressure ports are not accounted for. There also may be an averaging effect as the reduced frequency varies across the span. Notwithstanding, this near independence on F^+ bodes well for applications in which it is difficult to accurately determine the crosswind speed. Indeed, even an error on the order of 100% will still produce a substantial, although not necessarily optimum, result.

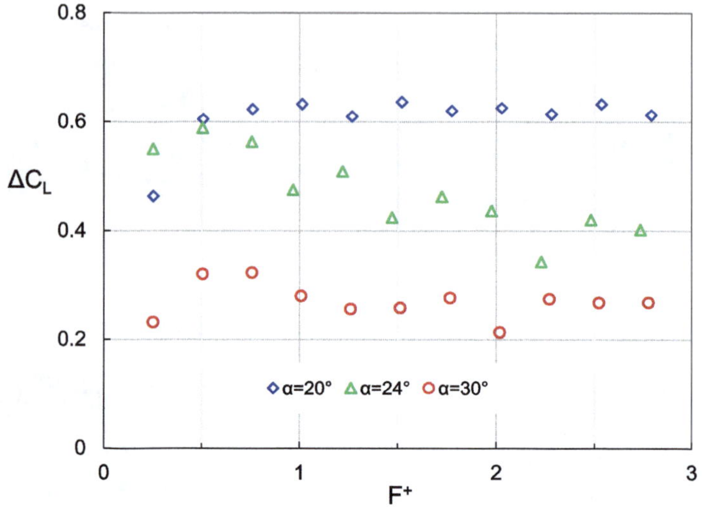

Fig. 8 Post stall panel lift dependence on reduced frequency scan at $U_\infty = 22\,\mathrm{m/s}$: d.c. = 10%, $\Pi_{\mathrm{gross}} = 8.7\,\mathrm{W}$

Baseline and controlled tuft flow visualization at $\alpha = 20°$, under conditions corresponding to Fig. 8 ($F^+ = 0.75$ d.c. = 10%), are shown in Fig. 9. Baseline orientation of the tufts, also visible in video recordings, show apparently random motion. When actuation is applied, the flow appears to attach fully both near the root and tip. However, slightly inboard from the tip and close to the trailing-edge, there exists a flow component towards the root that increases further inboard. At approximately the mid-span position the leading-edge flow has a tip-wise component and the result is a vortical flow with its axis approximately normal to the panel surface. Close inboard, the flow has a component towards the root near the trailing-edge. However, further outboard a similar but opposite-signed vortical structure is evident on the surface and the net result appears to be a stall-cell. However, video recordings show that this structure is not stationary and tends to meander inboard in a wave-like manner along the span.

An example of the lift coefficient versus angle of attack is shown in Fig. 10 for baseline and actuation cases at $U_\infty = 22\,\mathrm{m/s}$. In addition to significant post-stall lift increases, actuation is also clearly capable of almost eliminating hysteresis associated with the panel. However, actuation is not capable of materially increasing $\Delta C_{L,\max}$, due to the fact that the actuator thrust (or body force) is too low. To increase $\Delta C_{L,\max}$ by approximately 0.1, significantly greater plasma thrust, typically an order of magnitude increase, will be required. On the basis of other investgations, this certainly appears to be attainable [19].

When a V-tail configuration is subjected to a crosswind, the panels experience different conditions depending upon whether they are on the windward or leeward side of the vehicle. On the windward and leeward sides, the angle-of-attack will

Fig. 9 Panel flow visualization at $U_\infty = 22\,\text{m/s}$ and $\alpha = 20^{\text{o}}$: left baseline; right $F^+ = 0.75$, d.c. = 10%

increase and decrease respectively. Furthermore, the crosswind also produces an effective sweep-back or sweep-forward depending on whether the panel is leeward or windward respectively. It is important to note that sweep has a non-negligible effect on the mechanism and effectiveness of leading-edge separation control [20] and will be considered in the next phase of this research effort.

To illustrate the effect of plasma-based flow control on takeoff performance, estimates were made by accounting for the effect of sweep [20]. Well-known vehicle performance stability and control software [21] was employed, subject to the assumptions that rotation occurs at 1.15 V_{stall} and downwash in ground effect is accounted for. Based on the experimental data, it was seen that plasma actuation increased the allowable crossflow wind speed from 7.7 m/s (15 kts) to 12.9 m/s (25 kts). This is a meaningful result because, in many locations, wind speeds in excess of 12 m/s are highly improbable.

Furthermore, to better understand the practical weight and power requirements for flight applications, consider, for example, a stack of three typical 12 Volt LiPo batteries (dimensions: 25 × 34 × 104 mm; mass: 183 grams; and capacity 2.2 Ah). These specifications should be compared to the vehicle gross weight (450 kg), payload (150 kg) and endurance (20–30 h). The three batteries add less than 600 g, negligible volume and can operate continuously on two panels for approximately nine hours.

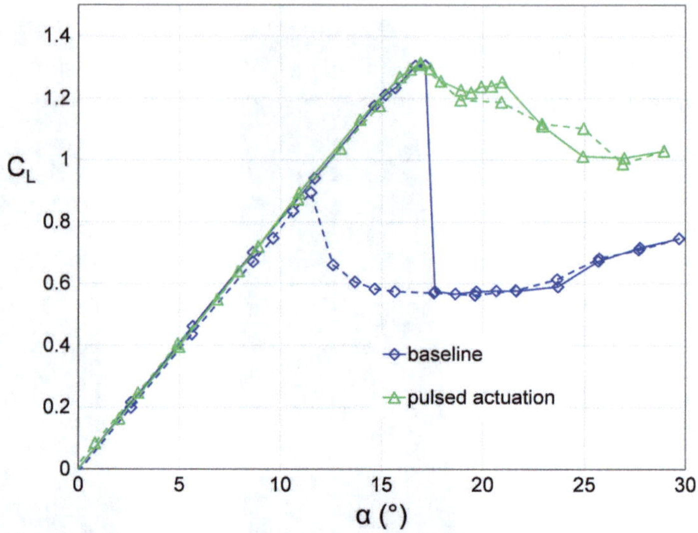

Fig. 10 Full-scale panel lift coefficient as a function of angle of attack at $U_\infty = 22\,\text{m/s}$. Actuation parameters: $F^+ = 0.75$, $f_p = 36\,\text{Hz}$, d.c. = 10%, $f_{\text{ion}} = 5500\,\text{Hz}$, $\Pi_{\text{gross}} = 8.7\,\text{W/m}$, $V_{\text{pp}} = 16.1\,\text{kV}$. Solid line—increasing α; dashed line—decreasing α

Clearly, these numbers can be improved upon, but they illustrate that the weight, volume and power requirements of the plasma actuation system are well within achievable bounds.

4 Concluding Remarks

The major conclusion of this study is that pulsed DBD plasma actuators are a viable and practical solution to the problem of separation control on V-tail panels, resulting from crosswinds during takeoff and landing. In terms of performance, post-stall lift coefficient increases of 0.6 (or 100%) were observed and bi-stable behavior (hysteresis) was eliminated even under deep stall ($\alpha = 30^{\text{o}}$) conditions. Low power requirements (<10 W) can easily be fulfilled off-line, using batteries (<1 kg); the high-voltage generators are a small fraction of the payload (typically 1.0 kg); the actuator themselves are lightweight (around 100–300 g). The relative insensitivity to reduced frequency in the range examined here, between 0.5 and 2.5, also renders the system very robust. Throughout all experiments, no burn-through and no failures whatsoever were encountered and no oxidation or degradation of the silicone rubber dielectric was observed after completion of the experiments.

Future investigations should integrate the actuator into the geometry of the tail element. Thus the only disturbance on the element will be a <25 μm external electrode. Removal of this electrode with an integrated dielectric results in the original clean element configuration. The external electrode is easy to replace or remove. Total failure of the DBD plasma system causes the vehicle to perform in its baseline configuration. Prior to flight testing, greater body forces must be generated using thicker dielectrics, up to O(10 mm), with a target $C_\mu = O(0.1)\%$. This will not materially increase input power, but will require significantly higher V_{rms}. Finally, conditions of forward- and backward-sweep must be fully evaluated prior to flight-testing and actuation must be achievable on both sides of the panel.

Acknowledgements This research was supported, in part, by the Nancy and Stephen Grand Technion Energy Program (GTEP). The authors also wish to thank the staff of Israel Aircraft Industry's (IAI's) Low Speed Wind Tunnel for assistance in performing the full-scale panel experiments.

References

1. Purser, P.E., Campbell, J.P.: Experimental Verification of a Simplified Vee-Tail Theory and Analysis of Available Data on Complete Models with Vee-Tails, NACA Rept. 823 (1945)
2. Abzug, M.J.: V-Tail stalling at combined angles of attack and sideslip. AIAA J. Aircr. **36**(4), 729–731(1999). https://doi.org/10.2514/2.2500
3. Corke, T.C., Post, M.L., Orlov, D.M.: SDBD plasma enhanced aerodynamics: concepts, optimization and applications. Prog. Aerosp. Sci. **43**, 193–217 (2007)
4. He, C., Corke, T.C., Patel, M.P.: Plasma flaps and slats: an application of weakly ionized plasma actuators. AIAA J. Aircr. **46**(3), 864–873 (2009)
5. Sosa, R., Artana, G., Moreau, E., Touchard, G.: Stall control at high angle of attack with plasma sheet actuators. Exp. Fluids **42**, 143–167 (2007). https://doi.org/10.1007/s00348-006-0227-5
6. Kelley, C.L., Bowles, P., Cooney, J., He, C., Corke, T.C.: High mach number leading-edge flow separation control using AC DBD plasma actuators. In: 50th AIAA Aerospace Sciences Meeting Including the New Horizons Forum and Aerospace Exposition, 9–12 January 2012, Nashville, Tennessee. AIAA Paper No. 2012-0906 (2012)
7. Roupassov, D.V., Nikipelov, A.A., Nudnova, M.M., Starikovskii, AYu.: Flow separation control by plasma actuator with nanosecond pulsed-periodic discharge. AIAA J. **47**(1), 169–185 (2009)
8. Little, J., Takashima, K., Nishihara, M., Adamovich, I., Samimy, M.: High lift airfoil leading edge separation control with nanosecond pulse driven DBD plasma actuators. In: 5th AIAA Flow Control Conference, 28 June–1 July 2010, Chicago, Illinois. AIAA Paper No. 2010-4256 (2010)
9. Greenblatt, D., Göksel, B., Rechenberg, I., Schule, C., Romann, D., Paschereit, C.O.: Dielectric barrier discharge flow control at very low flight Reynolds numbers. AIAA J. **46**(6), 1528–1541 (2008)
10. Greenblatt, D., Schneider, T., Schule, C.Y.: Mechanism of flow separation control using Plasma Actuation. Phys. Fluids **24**, 077102 (2012). https://doi.org/10.1063/1.4733399
11. Göksel, B., Greenblatt, D., Rechenberg, I., Nayeri, C.N., Paschereit, C.O.: Steady and unsteady plasma wall jets for separation and circulation control. In: 3rd AIAA Flow Control Conference, 5–8 June 2006, San Francisco. California, USA. AIAA Paper 2006-3686 (2006)
12. Greenblatt, D., Schulman, M., Ben-Harav, A.: Vertical axis wind turbine performance enhancement using plasma actuators. Renew. Energy **37**, 345–354 (2012). https://doi.org/10.1016/j.renene.2011.06.040
13. Ben-Harav, A., Greenblatt, D.: Feed-forward dynamic stall control on a vertical axis wind turbine. Wind Energy **19**(1), 3–16 (2016). https://doi.org/10.1002/we.1814
14. Greenblatt, D., Ben-Harav, A., Mueller-Vahl, H.: Dynamic stall control on a vertical-axis wind turbine using plasma actuators. AIAA J. **52**(2), 456–462 (2014). https://doi.org/10.2514/1.J052776

15. Greenblatt, D., Lautman, R.: Inboard/outboard plasma actuation on a vertical-axis wind turbine. Renew. Energy **83**, 1147–1156 (2015)
16. Greenblatt, D., Wygnanski, I.: The control of separation by periodic excitation. Prog. Aerosp. Sci. **36**(7), 487–545 (2000)
17. Duchmann, A., Simon, B., Tropea, C., Grundmann, S.: Dielectric barrier discharge plasma actuators for in-flight transition delay. AIAA J. (in press). https://doi.org/10.2514/1.J052485
18. Greenblatt, D.: Unsteady low-speed wind tunnels. AIAA J. **54**(6), 1817–1830 (2016)
19. Thomas, F.O., Corke, T.C., Iqbal, M., Kozlov, A., Schatzman, D.: Optimization of dielectric barrier discharge plasma actuators for active aerodynamic flow control. AIAA J. **47**(9), 2169–2178 (2009)
20. Greenblatt, D., Washburn, A.E.: Influence of finite span and sweep on active flow control efficacy. AIAA J. **46**(7), 1675–1694 (2008)
21. Williams, J.E., Vukelich, S.R.: The USAF Stability and Control Digital DATCOM, vol. I. Users Manual (1979). ADA086557, www.dtic.mil/dtic/tr/fulltext/u2/a086557.pdf

Part II
Combustion Control

Effect of the Switching Times on the Operating Behavior of a Shockless Explosion Combustor

Fatma C. Yücel, Fabian Völzke and Christian O. Paschereit

Abstract In the past, a wide range of investigations are made in order to increase the efficiency gain in gas turbines by using constant volume combustion. In comparison to detonation-based concepts, such as pulse detonation engine and rotation detonation engine, a new promising way was proposed by Klein and Paschereit and firstly assessed by Bobusch et al. (Combust Sci Technol 186(10–11):1680–1689 (2014), [1]), the so-called shockless explosion combustion (SEC). The principle is based on a quasi-homogeneous auto-ignition process that leads to an approximate constant volume combustion (aCVC). In order to achieve a quasi-homogeneous auto-ignition, it is necessary to achieve constant ignition delay times along the combustor. The combustion process in the SEC is similar to the one in internal combustion engines, namely Homogeneous Charge Compression Ignition (HCCI). This paper focuses on the use of wastegates to actively control filling and flow motion in the combustor dedicated to perform quasi-homogeneous auto-ignition. The results clearly show the ability to actively control the fuel distribution and purging time in the combustor which is an important step in the evolution of the SEC.

Keywords Constant volume combustion · Shockless explosion combustion
Ignition delay time · Purging time

F. C. Yücel (✉) · F. Völzke · C. O. Paschereit
Technische Universität Berlin, Institut für Strömungsmechanik und Technische Akustik, Müller-Breslau-Str. 8, 10623 Berlin, Germany
e-mail: f.yuecel@tu-berlin.de

F. Völzke
e-mail: fabian.voelzke@tu-berlin.de

C. O. Paschereit
e-mail: oliver.paschereit@tu-berlin.de

R. King (ed.), *Active Flow and Combustion Control 2018*,
Notes on Numerical Fluid Mechanics and Multidisciplinary Design 141,
https://doi.org/10.1007/978-3-319-98177-2_8

1 Introduction

Today, the potential of achieving a notable gain in gas turbine efficiency by improving components such as compressor or turbine is only incremental. A leapfrogging step could be achieved by replacing the conventionally utilized Brayton cycle (ideal constant-pressure cycle) by the Humprey cycle (ideal constant-volume cycle) [2]. In the past decades, different concepts on this topic were investigated, e.g., pulsed detonation combustors (PDC) [3] and rotation detonation combustors (RDC) [4]. The PDC and RDC are both based on periodic combustion processes that utilize a detonation wave to achieve an approximate constant volume combustion (aCVC). By igniting a flammable mixture a detonation wave is initiated that propagates with a high velocity into the unburned mixture, the gas has no time to expand and burns quasi-instantaneously. However, detonation waves are inefficient due to sharp pressure peaks that are associated with strong losses. The shockless explosion combustor suggested by Bobusch et al. [1] is a new promising way to implement an aCVC to achieve an increase in the efficiency of a gas turbine cycle. The combustion process itself is similar to the HCCI process used in internal combustion engines [5, 6]. In the HCCI a homogeneous ignitable mixture is compressed until auto-ignition occurs while the SEC achieves a quasi-homogeneous auto-ignition by stratifying the fuel-air mixture along the combustor close to auto-ignition conditions. The fuel stratification is needed in order to compensate the residence time of the fuel such that a constant ignition delay time along the combustor is achieved. One challenge in the HCCI process is the ignition timing. The occurrence of too early or late combustion turned out to be disadvantageously for the HCCI process. For this, closed-loop control is applied to actively control the ignition timing. One similar challenge in the evolution of the SEC process is to control the ignition timing along the combustor to increase homogeneity of the auto-ignition process.

The SEC is based on a periodic combustion cycle that achieves an aCVC by a quasi-homogeneous auto-ignition process as shown in (Fig. 1). At the beginning of the process (Fig. 1, top), the combustor is filled with a well-defined stratified fuel-air mixture. This axial stratification leads to a quasi-homogeneous auto-ignition of the entire gas volume after a certain ignition delay time. Due to the simultaneous combustion, a pressure wave is induced that propagates in downstream direction. At the open end of the combustor, the pressure wave is reflected as a suction wave that travels upstream (Fig. 1, bottom). As the suction wave reaches the tube inlet, the pressure drops below supply pressure and the recharge cycle can start. First, the combustor is purged with pure air creating a buffer to the hot combustion products. After a certain purging time, fuel is injected and the combustor is filled with the axially stratified fuel-air mixture (Fig. 1, left) while the suction wave is reflected and propagates downstream the tube. The suction wave is then reflected at the tube outlet as a pressure wave traveling upstream again. The process is restarted by another simultaneous combustion of the fuel-air mixture.

The ignition delay time of a mixture depends on temperature, pressure, equivalence ratio and the type of fuel and oxidizer. When assuming all other parameters to

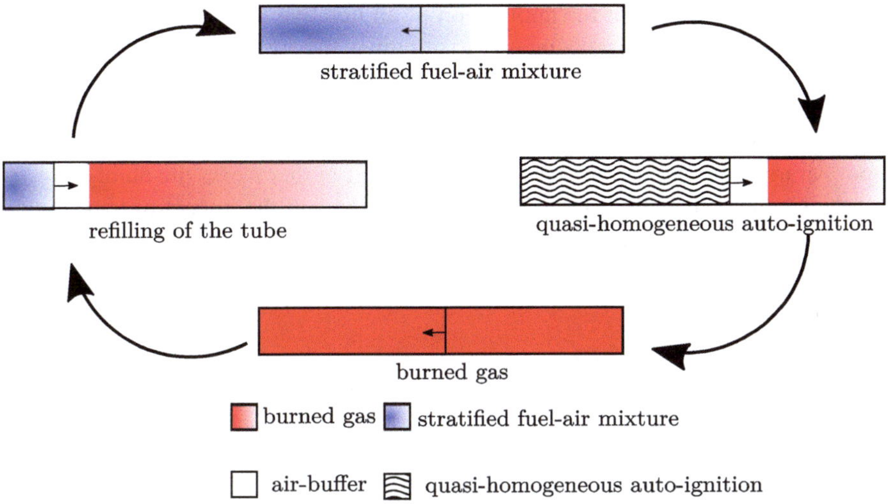

Fig. 1 SEC cycle

be constant, the spatial distribution of the ignition delay time $\tau(x)$ can be controlled by a stratification of the equivalence ratio. The ignition delay time decreases with increasing temperature, pressure and with decreasing deviation from stoichiometric conditions. Neglecting small temperature differences along the tube, a constant fuel injection would lead to an early ignition at the rear end of the combustor which would impede a quasi-homogeneous auto-ignition. In order to achieve a quasi-simultaneous auto-ignition, the ignition delay time must increase from the injection position to the outlet of the combustor. Further investigations have shown that a criteria for a quasi-simultaneous ignition along the combustor is that the difference in ignition delay time $\Delta\tau_{\text{ign}}$ between two neighboring infinitesimal small volumes should not exceed the excitation time τ_{et}, which is a value for the rate of chemical energy release [7, 8]:

$$\Delta\tau_{\text{ign}} \overset{!}{<} \tau_{\text{et}}. \tag{1}$$

One challenging task is the realization of resonant operation. The frequency of the SEC cycle is determined by the acoustic frequency of the combustor—as it relies on the suction wave to initiate the refill process—which is in the order of 100 Hz. For this, depending on tube length and speed of sound, very short ignition delay times are required. For atmospheric pressure conditions this can not be achieved, as the ignition delay times of all known fuels are not small enough. This crucial question gives room for future investigations and is not in the scope of this paper.

Another challenge is maximizing the gain in efficiency by achieving a quasi-homogeneous auto-ignition. Therefore, an increase in homogeneity of the combustion process is aimed since it goes along with an increase in efficiency. This requires efficient control of the axial fuel stratification to realize an operational SEC process.

Investigations showed that the application of an iterative learning controller into a closed-loop control is a promising way to minimize the variance of the ignition time by controlling the spatial fuel distribution [9].

For further improvements of the homogeneity of the auto-ignition, it is important to gain a deeper knowledge of this procedure and to analyze the impact of different input parameters. Two parameters are investigated in this work: the ability to control the spatial distribution of the fuel concentration and the impact of temperature fluctuations due to the purging time of the combustor.

2 Experimental Setup

An atmospheric test rig (Fig. 2) is designed to investigate the behavior of a shockless explosion combustor at non-resonant conditions. The test rig has been used in earlier investigations [9] and was modified in order to increase the experimental reproducibility.

The air is preheated to a maximum temperature of 700 °C using an electrical air heater to realize lowest ignition delay times. Dimethylether (DME) is used as fuel which exhibits an ignition delay time of around 100 ms at the given atmospheric pressure conditions and high temperatures [1, 10]. Additionally, DME has a characteristic small ignition delay time variation with temperature. Fuel lines are preheated to 90 °C preventing liquification of the DME.

The application of the preheater requires a minimal mass flow of 30 kg/h. This results in a combustor flow velocity of about 22 m/s and a residence time in the range of 30 ms. With an ignition delay of 100 ms the ignitable mixture would have already left the combustor before ignition. Enabling also tests under atmospheric pressure conditions a bypass is installed allowing the combustor flow to rest after charging. Currently, valves for turbocharged engine systems that can be controlled actively are used, called wastegates, replacing the previously used fluidic switch and valve

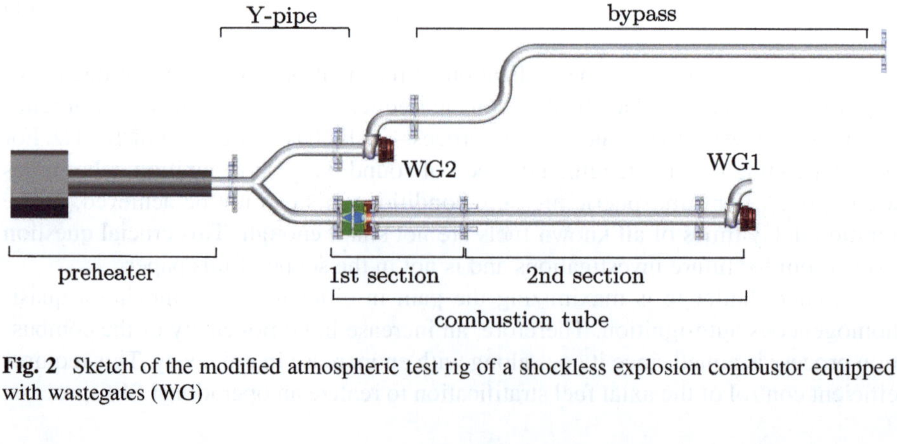

Fig. 2 Sketch of the modified atmospheric test rig of a shockless explosion combustor equipped with wastegates (WG)

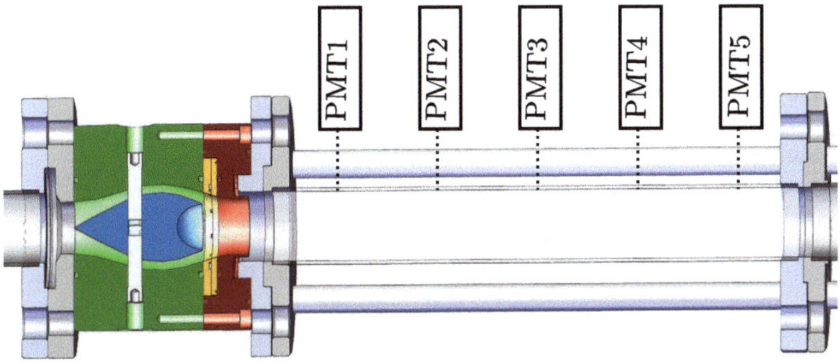

Fig. 3 Detailed view of the combustor with fluidic diode (blue) with oscillators (yellow), quartz tube and PMTs

heads. They have been installed respectively at the combustor (WG1) and at the bypass (WG2). These valves are equipped with several ports where boost pressure can be supplied such that they can be opened and closed actively. The advantage of these wastegates when compared to the fluidic switch in combination with valve heads is that the wastegates prevent backflow more reliably and the opening and closing timings can be adjusted actively. This has not been the case for the valve heads since they closed and opened passively depending on the pressure at the tube outlets. Furthermore, replacing the fluidic switch by a Y-pipe decreased the heat loss upstream of the combustor significantly.

A fluidic diode prevents the backflow of hot gases due to pressure increase after the combustion process and was already successfully used in earlier investigations [11]. The fuel is injected using two identical fuel injector arrays each equipped with four parallel-connected high-speed solenoid valves. The fuel inlet section is equipped with eight radial injection ports and fluidic oscillators to increase the mixing quality in the radial direction.

The combustor has two sections with an inner diameter of $d = 40$ mm. The first section is made from a 0.5 m long quartz tube in order to enable the detection of the ignition event via photomultipliers (PMTs). These PMTs are used to evaluate the ignition delay times at five axial positions in the combustor (Fig. 3). The second part is a steel tube equipped with two water-cooled pressure sensors to detect the combustion-induced pressure wave.

3 Investigation in the Wastegate Behavior

The wastegates used in this work are composed of a main body and an actuator body (Fig. 4). The actuator body contains a piston that is connected to the valve head in the main body. There are two different ways of controlling control the wastegates: passive and active. Equipping the actuator body with springs of different spring constants, the

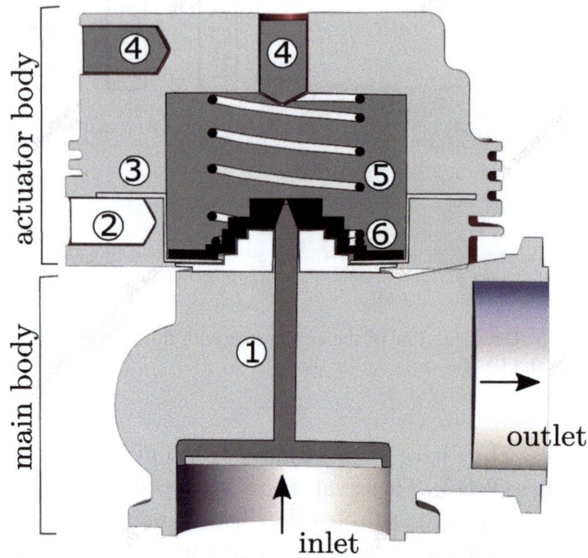

Fig. 4 Sectional view of a wastegate; ① valve head, ② boost pressure port for opening, ③ membrane, ④ boost pressure port for closing, ⑤ spring, ⑥ diaphragm

required input pressure for opening the valve can be controlled passively by setting the reset force of the piston. In this work, the wastegates were controlled actively by regulating the supplied boost pressure. A diaphragm separates the two different areas inside the wastegate for the boost pressure supply. By setting the pressure in the upper area to higher levels, the wastegate closes. When the pressure in the upper region is decreased and the pressure in the lower area is increased, the wastegate opens. 3/2-way valves have been applied to allow for controlling the boost pressure. It is important to note that the wastegate behavior and especially its response times are dependent on the operating speed of the 3/2-way valves. One voltage signal is needed to operate both wastegates simultaneously since the 3/2-way valves are installed on each wastegate at inverted positions.

The switching process and its effect on the combustor flow was investigated using a laser Doppler anemometry system (LDA). This non-intrusive optical measurement technique enables a high spatial and temporal resolution. An aerosol (Bis(2-ethylhexyl) sebacate) was used as seeding which was atomized using the aerosol-generator.

The LDA measurements were done under non-reactive conditions using air only. The air flow has been set to 30 kg/h, which matches the conditions for reacting tests. The laser was positioned 50 mm downstream the injection geometry. Figure 5 displays the voltage signal for controlling the 3/2-way valves and the measured combustor flow velocity as a function of time for the first three periods at a switching frequency of 0.5 Hz. When the air flow is guided to the combustor, the mean velocity measured was $u_{\text{mean}} = 6.3\,\text{m/s}$ with an RMS-value of $u_{\text{rms}} = 0.68\,\text{m/s}$ implying a turbulence intensity of 10 %. When the voltage signal for controlling the 3/2-way valves is set to $U_C = 5\,\text{V}$, WG1 closes and WG2 opens simultaneously. Subsequently,

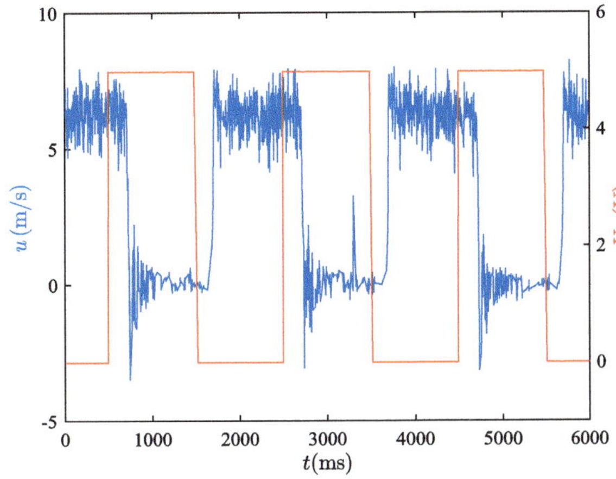

Fig. 5 Measured particle velocity u with Laser Doppler Anemometry and switching signal U_C over time

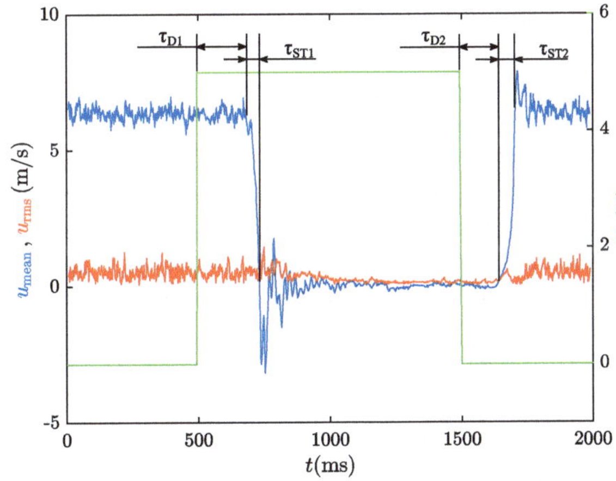

Fig. 6 Mean velocity u_{mean}, RMS-value u_{rms} and switching signal U_C over time

after a certain delay time the velocity decreases fast until the flow comes to rest. The main velocity is overlaid by velocity oscillations due to high turbulence during the opening and closing procedure. When the voltage signal is set back to $U_C = 0\,\text{V}$, the flow velocity increases again until the mean velocity of $\bar{v} = 6.3\,\text{m/s}$ is reached. Both switching events include two time delays respectively. To specify these time delay times, the velocity was averaged over 10 cycles and the delay times were assessed (Fig. 6).

The time delay for closing the WG1 is determined to be $\Delta\tau_{D1} = 215\,\text{ms}$, while the flow needs a time span of $\Delta\tau_{ST1} = 17\,\text{ms}$ to be decelerated to $u = 0\,\text{m/s}$. The values for opening WG1 have been determined to be $\Delta\tau_{D2} = 175\,\text{ms}$ and $\Delta\tau_{ST2} = 31\,\text{ms}$. All values are listed in Table 1.

The RMS-value shown in Fig. 6 displays the stochastic fluctuations of the flow velocity when the air is guided through the combustor. The fluctuations between the cycles that appear when stopping the air flow in the combustor are very low which

Table 1 Wastegate timings

	WG1 closing (ms)	WG1 opening (ms)
Wastegate delay τ_D	215	175
Switching time τ_{ST}	17	31

emphasizes the reproducibility of the switching events. Even though the flow cannot be stopped instantly, the reproducibility of the process enables the consistency of the stratification of the fuel-air mixture for reacting tests.

4 Ignition Timing Measurements

Preliminary investigations showed that the purging time can be actively controlled by triggering the wastegates. For this, the delay times discussed in the previous section (see Table 1) have to be taken into account. In this section, the impact of the actual wastegate timings and the purging time as an input control parameter on the combustion process of the SEC are evaluated under reacting conditions. The success rates for auto-ignition and the spatial distribution of the ignition delay time have been investigated as a function of the switching times. To investigate this behavior, a periodic non-resonant combustion process is applied.

One operational period of the SEC process can be divided into three different parts: charging, homogeneous ignition and purging. During the charging process, the air mass flow is guided through the combustor and a constant fuel mass flow is added for 30 ms. Since this work focuses on the control aspect, the combustible mixture is not stratified as no homogeneous combustion is aimed here. After a well defined time span τ_{WT1}, that is varied between 40 and 120 ms, the air flow is switched into the bypass by closing the wastegate WG1 and opening wastegate WG2. To specify the appropriate timing of the voltage signal, the time delays τ_{D1} and τ_{D2} are taken into account. While the mass flow is directed through the bypass, the ignition event takes place in the combustor. After a second defined time span τ_{WT2} the purging process starts by guiding the air flow through the combustor again. Thus, an increasing residence of the air mass flow in the bypass is equal to a decreasing purging time.

Figure 7 shows an exemplary period of a 1 Hz SEC cycle for $\tau_{WT1} = 40$ ms and $\tau_{WT2} = 300$ ms. During the interval from 0 to 40 ms the air mass flow is guided through the combustor while fuel is injected at the same time from 0 to 30 ms, such that the quartz tube is entirely filled with a flammable mixture. After 40 ms the closing procedure starts such that after the required switching time of 17 ms the wastegate is entirely closed and the air mass flow is guided through the bypass for a duration of 260 ms. The air flow is then switched back into the combustor at 300 ms with an additional time delay of 31 ms until the end of the cycle. In total this induces a purging time of 669 ms.

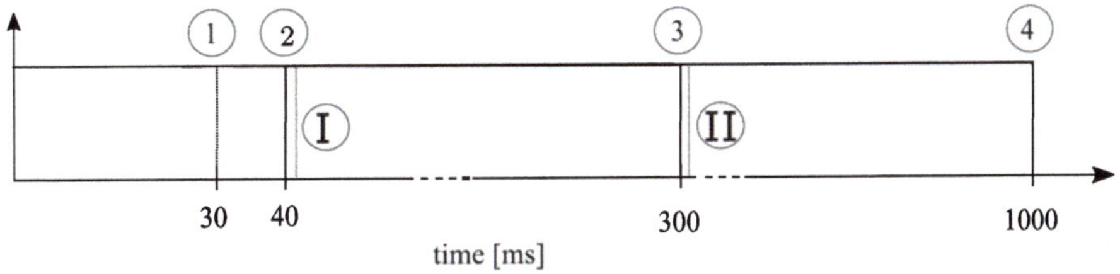

Fig. 7 Simplified representation of the different timings, (1) fuel injection time τ_{inj}, (2) wastegate timing τ_{WT1}, (3) wastegate timing τ_{WT2}, (3) to (4) purging time τ_{pt}, (I) switching time τ_{ST1}, (II) switching time τ_{ST2}

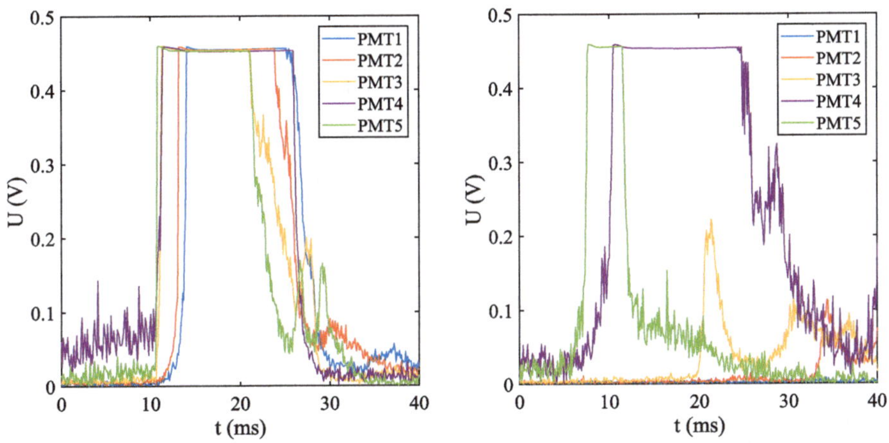

(a) $\tau_{\mathrm{WT1}} = 80\,\mathrm{ms}$, $\tau_{\mathrm{WT2}} = 690\,\mathrm{ms}$ (b) $\tau_{\mathrm{WT1}} = 100\,\mathrm{ms}$, $\tau_{\mathrm{WT2}} = 540\,\mathrm{ms}$

Fig. 8 Output-signal of PMTs

As shown in Fig. 3, PMTs are used to detect the ignition event at five different axial positions. Figure 8 shows two exemplary output signals of the PMTs. The PMTs detect the light emission of the flame due to auto-ignition. The time of ignition is assumed to be the time when the voltage signal of a certain PMT reaches the threshold of 0.45 V. In Fig. 8a the PMT signal displays an output, where a combustion event is detected by all PMTs, while in Fig. 8b combustion event was only detected by PMT4 and PMT5. Although there is a peak in the signal of PMT3, this is not interpreted as an ignition event at this position, since it does not reach the threshold of 0.45 V. It rather can be interpreted as an ignition event close to the position of PMT3 causing some light emission entering the PMT3. As a criteria for auto-ignition only the first PMT signals are considered as actual auto-ignition events since in this paper, it is not aimed for a fuel stratification. Starting from the first ignition it is considered that the remaining PMTs only detect a deflagrative propagating flame front.

Figure 9 illustrates the ignition events at different positions in the combustor for different wastegate timings τ_{WT1} and τ_{WT2}. Every node at the x-axis is representative

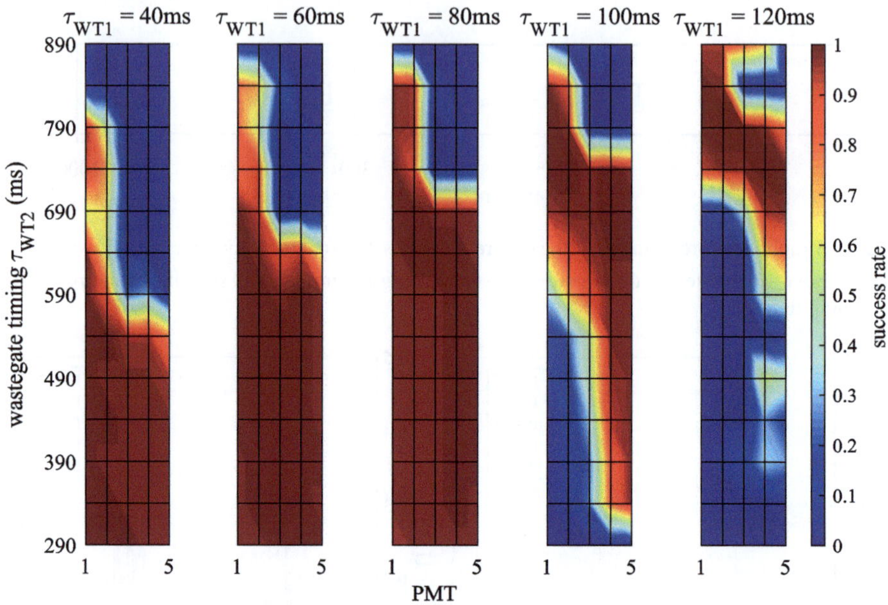

Fig. 9 Investigation on the impact of the purging time on the success rate on every single PMT by varying τ_{WT1} and τ_{WT2}

for one PMT, while the y-axis represents different timings for switching from bypass into the combustor. The colorbar represents the success rate of auto-ignition whereby 40 cycles where evaluated in total for every combination of tested wastegate timings. A success rate of 1 means that the PMT detected an ignition event in every single cycle. The blue areas, representing a value of 0, show that no ignition could be detected, respectively.

It is apparent that for relatively high purging times ($\tau_{WT2} \leq 540$ ms) and 40 ms $\leq \tau_{WT1} \leq 80$ ms, every single PMT has a success rate of 1. It can be assumed that the entire quartz tube is filled with a flammable mixture as visualized in Fig. 10 case B. Thus, this regime contains a periodic auto-ignition event that induces a combustion process at all observed axial positions in the combustor.

Increasing τ_{WT1} results in the injected fuel package traveling further downstream before the flow is stopped due to the closing of WG1 (see Fig. 10). Thus, the ignition event is shifted more and more into the exhaust tube. Therefore, an increase to $\tau_{WT1} = 100$ ms reduces the detected ignition events at the first three PMTs drastically (Fig. 10 case D). For $\tau_{WT1} = 120$ ms, no PMT detects a stable combustion for $\tau_{WT2} < 690$ ms as the success rate is zero at every position since the entire burnable mixture has been moved outside of the quartz tube, where the combustion event cannot be detected by the PMTs (Fig. 10 case E). This observation demonstrates the ability to control the spatial fuel distribution by using actively controlled wastegates.

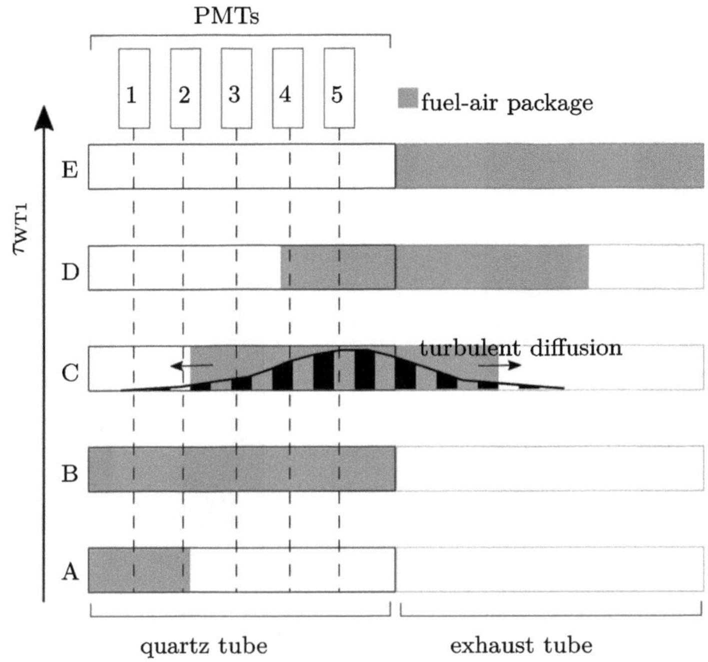

Fig. 10 Position of the fuel-air package in the combustor for an increasing τ_{WT1} value

For an increased τ_{WT2} and a constant τ_{WT1}, no ignition event can be detected at PMTs 3–5. A presumption for this effect is that due to small periods of purging, hot combustion products remain in the combustor due to recirculation areas near the injection geometry and cause contact burning when the next fuel package is injected. Thus, a auto-ignition fails and the combustion is only detected by the first PMTs since there is no time for the fuel-air mixture to travel downstream before ignition. This phenomenon can be observed when τ_{WT2} exceeds a certain threshold that equals the time that is needed to purge all the exhaust gases that have entered the injection geometry and the supply line due to the increased pressure in the combustor after the combustion event.

This threshold for τ_{WT2}, where the success rate changes from 1 to 0 for a constant τ_{WT1}, is shifted upwards by increasing τ_{WT1}. As mentioned earlier, increasing τ_{WT1} causes the burnable mixture to travel further downstream before the auto-ignition takes place. This leads to a smaller amount of hot exhaust gases in the upstream part of the combustor. Thus, a shorter time is needed to purge the line.

By further increasing τ_{WT2}, a second characteristic of the SEC test rig is visible. When the preheated air flow is guided through the bypass, the temperature in the combustor decreases. The more time the air flow is guided through the combustor, the higher the temperature gets. Since the ignition delay time of DME is still sensitive to the mixing temperature as shown in [1, 10], a small value of τ_{WT1} and a large

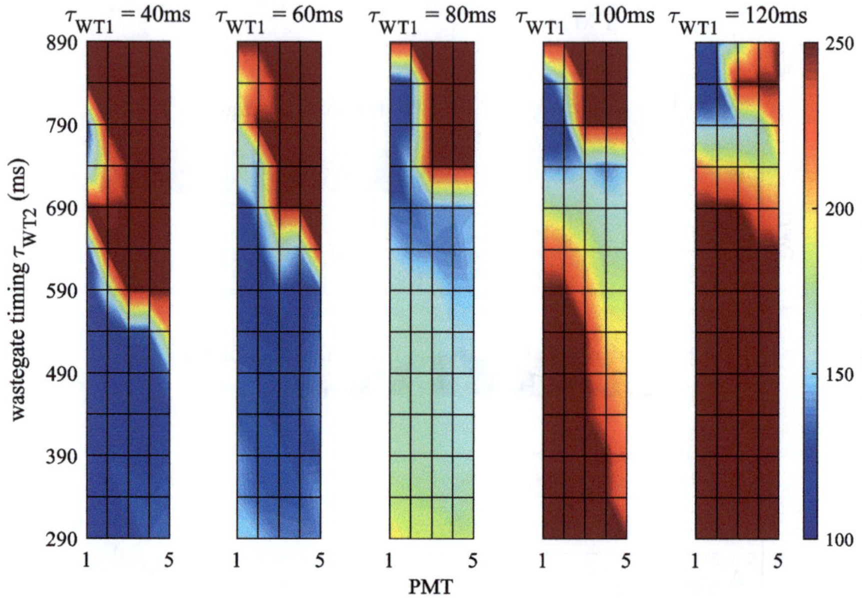

Fig. 11 Investigation of the impact of the purging time on the ignition delay time on every single PMT by varying τ_{WT1} and τ_{WT2}

value of τ_{WT2} causes the ignition delay time to exceed the cycle duration. Therefore, no combustion event is detected by the PMTs. Depending on the purging time the temperature in the combustor varies significantly. A higher purging time causes an increase in temperature but is also the limiting factor for higher frequencies. Low purging times cause burnt gas to remain in the combustor causing an uncontrolled ignition. This proves that the purging time is a decisive parameter in order to achieve an increased homogeneity of the auto-ignition process.

The mean ignition delay time is measured as the time span from the start of injection to the detection until combustion event. The local ignition delay time varies from 100 to 250 ms. When no combustion event is detected, the ignition delay time exceeds the value of 250 ms (see Fig. 11).

The velocity over time shown in Fig. 5 implies that an increase in τ_{WT1} leads to an increased time span in which the mixture is exposed to high turbulence. Turbulent diffusion causes mixing of the flammable mixture with the surrounding air at the bounds of the fuel-air package (see Fig. 10 case C). This leads to a decrease in the gradient of the equivalence ratio with increasing τ_{WT1}. Thus, the observed ignition delay time increases with increasing τ_{WT1}, which is contradicting the wish for a specifically shaped profile of fuel stratification. Increasing τ_{WT2} induces a slight decrease in the ignition delay time.

5 Conclusion and Outlook

The SEC process was investigated under atmospheric pressure conditions using a modified test rig. Wastegates, which replaced the former used fluidic switch and valve heads, were used to obtain the influence of purging time and to find improved conditions for successfully reducing the ignition delay time at atmospheric conditions. It has been shown that actively controlled wategates enable precise control of the fuel distribution. It has been proved that the purging time has a decisive impact on the performance of a shockless explsoion combustor. New parameters were found that can be utilized as input data for closed-loop control in order to optimize the spatial fuel distribution which is essential for achieving a homogeneous auto-ignition.

The obtained data showed that the auto-ignition itself is a complex process with a high number of variables. This work gives a detailed look to the sensitivity of the SEC process to the switching timings of the wastegates as an important control parameter.

The next step is to significantly decrease the ignition delay times. Therefore, experimental investigations in an intermediate pressure test rig are planned. A decrease in ignition delay time would lead to higher firing frequencies that are needed to realize resonant operation. A new stainless steel combustion chamber has been designed, where ionization probes will be used for flame detection at elevated pressures.

Acknowledgements The authors gratefully acknowledge support by the Deutsche Forschungsgemeinschaft (DFG) as part of Collaborative Research Center SFB 1029 “Substantial efficiency increase in gas turbines through direct use of coupled unsteady combustion and flow dynamics” on project A01.

References

1. Bobusch, B.C., Berndt, P., Paschereit, C.O., Klein, R.: Shockless explosion combustion: an innovative way of efficient constant volume combustion in gas turbines. Combust. Sci. Technol. **186**(10–11), 1680–1689 (2014)
2. Stathopoulus, P., Vinkeloe, P., Paschereit, C.O.: Thermodynamic evaluation of constant volume combustion for gas turbine power cycles. In: 11th International Gas Turbine Congress, Tokyo, Japan, pp. 15–20 (2015)
3. Roy, G.D., Frolov, S.M., Borisov, A.A., Netzer, D.W.: Pulse detonation propulsion: challenges, current status, and future perspective. Prog. Energy Combust. Sci. **30**(6), 545–672 (2004)
4. Lu, F.K., Braun, E.M.: Rotating detonation wave propulsion: experimental challenges, modeling, and engine concepts. J. Propul. Power **30**(5), 1125–1142 (2014)
5. Yao, M., Zheng, Z., Liu, H.: Progress and recent trends in homogeneous charge compression ignition (HCCI) engines. Prog. Energy Combust. Sci. **35**(5), 398–437 (2009)
6. Stanglmaier, R.H., Roberts, C.E.: Homogeneous charge compression ignition (HCCI): benefits, compromises, and future engine applications. Technical report, SAE Technical Paper (1999)
7. Meyer, J.W., Cohen, L.M., Oppenheim, A.K.: Study of exothermic processes in shock ignited gases by the use of laser shear interferometry. Combust. Sci. Technol. **8**(4), 185–197 (1973)
8. Cai, L., Pitsch, H.: Tailoring fuels for a shockless explosion combustor. In: Active Flow and Combustion Control 2014, pp. 299–315. Springer (2015)

9. Reichel, T.G., Schäpel, J.-S., Bobusch, B.C., Klein, R., King, R., Paschereit, C.O.: Shockless explosion combustion: experimental investigation of a new approximate constant volume combustion process. J. Eng. Gas Turbines Power **13**(2), 021504 (2017)
10. Berndt, P.: On the use of the hll-scheme for the simulation of the multi-species euler equations. In: Finite Volumes for Complex Applications VII-Elliptic, Parabolic and Hyperbolic Problems, pp. 809–816. Springer (2014)
11. Bobusch, B.C.: Fluidic devices for realizing the shockless explosion combustion process (2015)

Part Load Control for a Shockless Explosion Combustion Cycle

Florian Arnold, Giordana Tornow and Rudibert King

Abstract Since a significant increase in the efficiency of conventional gas turbines is unlikely due to various reasons, new concepts are needed. One option is to redesign the thermodynamic process itself. Replacing the constant pressure combustion with constant volume combustion (CVC) offers such an increase in efficiency. A promising new process that approximates constant volume combustion is the so-called shockless explosion combustion (SEC). SEC utilizes a homogeneous auto-ignition inside a combustion tube to avoid gas expansion during combustion. An acoustic interaction within the tube is exploited to ensure a self-sustained cyclic operation. For this, chemical and acoustic time-scales have to match. As this is impossible under ambient pressure conditions, for which SEC has been tested experimentally, this study focuses on simulations that mimic the situation of elevated pressure to design a controller. Herein, a control system is introduced within the numerical simulation of SEC that is capable of driving the process to different operating points. It expands on an iterative learning control from recent publications, which adjusts ignition time over the length of the tube. The control system proposed here can be used to realize a part load operation within the observed simulation.

Keywords Iterative learning control · Part load · Shockless explosion combustion

F. Arnold (✉) · R. King
Chair of Measurement and Control Department of Process Engineering,
Technische Universität Berlin, Hardenbergstr. 36a, 10623 Berlin, Germany
e-mail: f.arnold@tu-berlin.de

R. King
e-mail: rudibert.king@tu-berlin.de

G. Tornow
Department of Mathematics, Geophysical Fluid Dynamics Freie Universität Berlin,
Arnimallee 6, 14195 Berlin, Germany
e-mail: gtornow@math.fu-berlin.de

R. King (ed.), *Active Flow and Combustion Control 2018*,
Notes on Numerical Fluid Mechanics and Multidisciplinary Design 141,
https://doi.org/10.1007/978-3-319-98177-2_9

Nomenclature

SEC	Shockless explosion combustion
ILC	Iterative learning control
Ψ	Hammerstein compensator
τ	Ignition delay time
$\underline{b}$	Model offset
E_k, E_{max}	Energy
$\underline{e}$	Control error
e_P	Position control error
e_{T_F}	Fuel injection error
$\underline{f}_k$	Fuel injection curve
$\mathbf{G}$	Model matrix
$\mathbf{I}$	Identity matrix
J	Cost function
k	Cycle index
$\mathbf{L}$	Learning matrix
m	Number of ignition sensors
n	Size of the control trajectory
$\underline{r}$	Ignition time reference
r_P	Reference position
$t_{ign,x}$	Ignition time at position x with respect to the beginning of the injection
$\underline{u}_k$	Control trajectory (ignition delay time)
$\underline{\tilde{u}}_k$	Injection trajectory (ignition delay time)
$\Delta\underline{u}$	Change of the control trajectory
$\mathbf{W}_e$, $\mathbf{W}_{\Delta u}$, $\mathbf{W}_r$, $\mathbf{W}_V$	Weighting matrices
w_u, w_r, w_V	Weighting parameters
$\underline{y}_k$	Control variable (ignition time)

1 Introduction

Gas turbines provide safe and reliable energy conversion to produce electric energy or thrust in aircrafts. The efficiency of gas turbines has been improved substantially over recent decades, but currently approaches a plateau. A further increase would require a higher turbine-input temperature, which state-of-the-art turbines cannot handle. For a significant improvement in efficiency, an optimization of the combustion cycle itself offers high potential. Conventional gas turbines rely on the Joule cycle, which assumes combustion with constant pressure. By replacing the combustion with constant volume combustion (CVC), efficiency can be increased. The resulting Humphrey cycle provides a higher amount of usable work for the same initial conditions. An innovative process that approximates a constant volume combustion is the so-called shockless explosion combustion (SEC), which was first introduced by Bobusch et al. [1].

The SEC process consists of several phases that run in a sequence before starting over again. A single cycle with all phases is shown in Fig. 1. At the beginning of a cycle, the combustion tube is filled with a combustible mixture of fuel, an air buffer, and the remaining exhaust gas of the previous combustion. The fuel-air mixture, and by this the ignition delay time, is stratified during the fuel injection process in order to compensate for variations in the residence time of individual fuel particles and finally achieve a homogeneous auto-ignition. Since the ignitable mixture is sensitive to high temperatures, it is separated from the hot exhaust gas by an air buffer. That way, premature ignition is avoided (see phase I in Fig. 1). After the homogeneous auto-ignition of the fuel-air mixture takes place, the pressure and the temperature rise within the related part of the tube. A pressure wave propagates downstream to the outlet on the right (phase II) and is reflected as a suction wave (phase III). When the suction wave reaches the closed inlet on the left side, the pressure drops below the input pressure (phase IV). This allows a refilling of the tube with a fresh air buffer followed by the fuel-air mixture (phase V). Berndt [2] assumes, as an order of magnitude estimate, that 40% of the tube has to be filled with the fuel-air mixture for a full load SEC operation. In this contribution, we will study the effect of lower fillings to realize a part load operation. For a detailed introduction to SEC, see Bobusch et al. [1].

This work assumes that SEC replaces the conventional combustion in a gas turbine. One of the main advantages of gas turbines is their ability to quickly change between different operating points. For most gas turbines, this is done by lowering the fuel effort, which implies a lower heat addition within the combustion cycle. It is thus necessary that a process like SEC be able to be adjusted easily to react to different demands. However, due to the intricate relation between chemical and acoustic time-scales, as proposed in the original idea of SEC [1], it is not obvious how this could be achieved. Moreover, the controller that has been proposed so far [3] has resulted in a homogeneous auto-ignition within a combustion region that did not start at the inlet of the tube. The scope of this work therefore will be the introduction of a control system that provides the capability of driving the SEC process to requested part load points of operation, i.e., set-points holding these states, and achieving ignition from the start of the tube.

Recent investigations [4] have shown that resonant operation of SEC is not possible with affordable equipment under the ambient conditions that have been studied

Fig. 1 SEC cycle

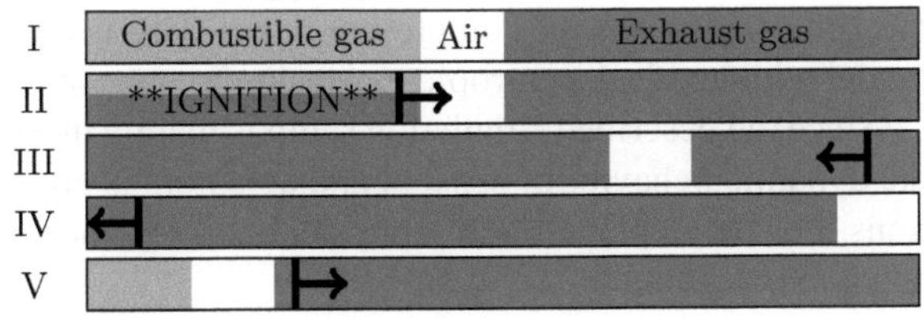

experimentally so far. Since the development of a part load control system requires a stable resonant operation, the approaches in this work are implemented within a numerical simulation of SEC with an elevated pressure. The used 1-D Euler simulator built by Berndt [2].

The remainder of this contribution is organized as follows: Sect. 2 presents a short overview over the numerical SEC simulation followed by a definition of the term part load for the SEC process in Sect. 3. Based on this, a control system is introduced in Sect. 4 that is capable of driving and holding the process from full load to part load and back. At the end, in Sect. 5, representative results are shown to illustrate how the system handles different requirements, before conclusions are drawn in Sect. 6.

2 Simulation of the Shockless Explosion Combustion

The numerical simulator of the SEC process used here was developed by Berndt [2] and has since been maintained by Tornow in Klein's group at FU Berlin. The simulation is restricted to a 1-D regime and is based on the Euler equations extended by chemical kinetics. The fuel used within the simulation is hydrogen. This choice is motivated by similar CVC studies, although hydrogen necessitates unrealistically long combustion tubes. That is a result of the inert ignition behavior of hydrogen. The combustion behavior is approximated by the detailed H_2/O_2 kinetic model introduced by Burke et al. [5]. The reaction mechanism includes 11 substances and more than 20 reaction paths.

Since it is planned that the SEC process works within a gas turbine, it is essential to define appropriate, realistic boundary conditions. To model the inlet of the tube, a pressure-related boundary condition has been implemented. A constant pressure and temperature generated by a compressor is assumed in front of the tube. If the pressure inside the tube is higher than the pressure provided by the compressor, the boundary is considered to be closed. The solid wall is implemented by using ghost cells that reflect the state of the first cell after the inlet. In contrast, if the pressure inside the tube drops below the compressor output pressure, the boundary condition models an isentropic inflow into the tube. The ghost cell before the inlet is then set according to the state reached by adiabatic expansion from the compressor output state.

For simplicity, neglecting a plenum behind the tube and neglecting a detailed turbine model, the open end of the tube at the outlet is modeled by an isentropic expansion to the pressure level of a hypothetical plenum, which is assumed to have the same pressure as the compressor outlet. This is followed by a very simple turbine model with an ideal isentropic expansion to atmospheric conditions.

Since the described simulation setup is mainly used for theoretical investigations, the dynamic behavior of sensors and actuators is not considered within the calculations. For the application at a physical test rig, these additional aspects have to be taken into account.

3 Part Load of SEC

This work proposes a part load control for the SEC process in the context of a gas turbine. It is thus necessary to take a look at the term *part load* operation and clarify it for SEC. The part load operation of the SEC is defined by the amount of energy output E_k within a cycle k. Every state with an energy output below the maximum, $E_k < E_{max}$, can be considered as a part load set-point. The highest amount of energy is available at the outlet if a homogeneous auto-ignition occurs in the largest possible region of the combustion tube. The combustible region as shown in Fig. 1 during phase I is located between the tube inlet and the air buffer. The combustion region is bounded by the size of the air buffer that is necessary to separate the combustible mixture upstream and the hot exhaust gas from the last cycle further downstream. To avoid premature ignition of the fuel-air mixture by the hot exhaust gas, the air buffer size is set to 10% of the length of the tube. This spatial separation leaves enough space between both parts, even when some gas mixes with the air buffer. It is assumed that under full load conditions, at the beginning of every cycle, the tube is filled with the gas of two cycles. The first 50% after the inlet contains the gas of the current cycle, and the other 50% consists of the gas of the previous combustion. This leaves about 40% of the length of the tube for the combustion itself. A full load operation thus requires a homogeneous auto-ignition, which implies a constant volume combustion, in the first 40% of the tube behind the inlet.

The SEC process offers two options to reduce heat release during the combustion, or the energy output of one cycle. The common way, used for conventional gas turbines, would be the adjustment of the equivalence ratio. A lower equivalence ratio would result in a lower temperature and pressure rise. Consequently, a lower energy emission at the outlet would be seen. But since a variation of the equivalence ratio has to be used in the SEC to realize fuel stratification, an additional requirement for a low-averaged equivalence ratio might increase the auto-ignition time scale relative to the shortest ignition delay and therefore disturb the resonant operation. Therefore, the size of the combustion region as a second option is used here to reduce the fuel effort and lower the energy output.

The size of the combustion region can be scaled by the size of the air buffer. A larger air buffer leads to a decreased region of combustible mixture upstream and thus also diminishes energy emission at the outlet. The adjustment of the combustion region is independent of the control for the ignition time. Therefore, the timescale of the ignition delay is not influenced during part load operation. This is an important aspect since the efficiency of the SEC process mainly depends on a reliable homogeneous auto-ignition.

4 Combustion Control

To change the operating point of SEC, the controller has to adjust the size of the combustion region and generate a fuel stratification that ensures homogeneous auto-ignition. As these two aspects are independent from each other, it is possible to consider both parts separately. A short summary of fuel stratification is presented first, as this task was already successfully solved for full load operation. Then, the new control of the combustion region will be included.

4.1 Control of the Fuel Stratification

To achieve homogeneous auto-ignition inside the combustion tube, it is necessary to adjust the ignition delay time as a function of the distance from the tube's inlet, i.e., as a function of the total residence time of a fuel particle before combustion. In experimental investigations, ignition is detected by several photomultipliers [4]. For the theoretical analysis considered here, the ignition time is calculated for simplicity in every grid cell of the numerical scheme. For practical use, the relevant data for ignition detection, like spatial pressure and temperature values, have to be estimated from available measurements. Such a state estimation approach for SEC has been already introduced by Schäpel et al. [6] and will be part of future work concerning part load operation. The ignition delay time depends on the equivalence ratio, the temperature, and the pressure of the fuel-air mixture. Since the last two parameters cannot be adjusted easily, fuel stratification is the only option to command the ignition delay time.

Regarding the cyclic characteristic of SEC, it seems reasonable to use an iterative learning controller (ILC), which improves fuel stratification from cycle to cycle. Such an ILC setup for SEC simulation has been introduced in a former study of full load operation [3]. As the scope of this work is focused on control of the length of the combustion region, the ILC system and the applied model will only be summarized shortly in the next two sections, giving further details not presented previously [3].

4.1.1 Model

ILC relies on a linear model that describes the impact of the control trajectory $\underline{u}_k \in \mathbb{R}^n$ on the control variable $\underline{y}_k \in \mathbb{R}^m$, where $\underline{u}_k$ and $\underline{y}_k$ contain the sampled values of one cycle k. Due to real-time requirements and technical restrictions, the numbers n and m of input and output variables, respectively, had to be limited in the experimental investigation of SEC [4]. This makes sense in the simulation study as well, as tests have shown. However, for the numerical simulation, a finer discretization of the input with a time increment Δt has to be chosen. As will be seen below, the control input to be calculated by the controller will be chosen, rather unconventionally, as commanded ignition delay times τ. To that end, a linear interpolation is applied to obtain $\underline{\tilde{u}}_k = [\tau_1, \tau_2, \dots, \tau_q]^T$ for $t_j = j\,\Delta t < t_{0,k} + T_{F,k}$ and $j = \{1, 2, \dots, q\}$ from the actual control trajectory $\underline{u}_k = [\tau_1, \tau_2, \dots, \tau_n]^T$ The time $t_{0,k}$ specifies the beginning of the injection for cycle k and $T_{F,k}$ is the fuel injection duration. The trajectory $\underline{\tilde{u}}_k$ with an ignition delay time for every time instance of the simulation can then be used to compute the necessary equivalence ratio for this commanded

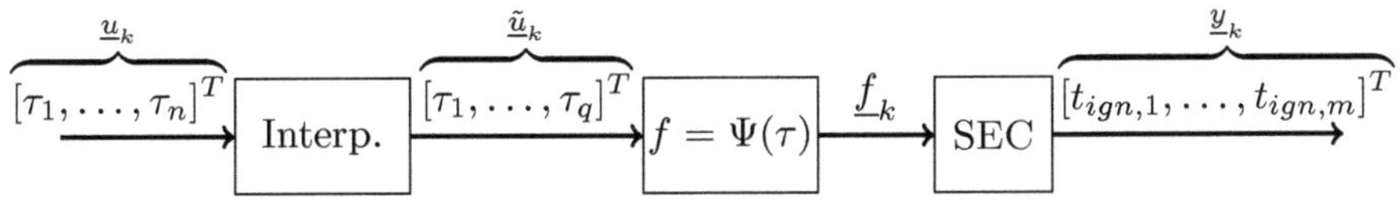

Fig. 2 Hammerstein compensator

value by a steady non-linear map. This non-linear map $f = \Psi(\tau)$ can be interpreted as a Hammerstein compensator, Fig. 2, as an almost linear relation between the commanded ignition delay time $\underline{\tilde{u}}_k = [\tau_1, \tau_2, \ldots, \tau_q]^T$ and the actually measured ignition times $\underline{y}_k = [t_{ign,1}, t_{ign,2}, \ldots, t_{ign,m}]^T$ is obtained. The Hammerstein map is generated from a detailed zero-dimensional combustion model that assumes constant pressure and temperature.

If in cycle k an ignition is detected in m spatial grid cells, the ignition times $\underline{y}_k = [t_{ign,1}, t_{ign,2}, \ldots, t_{ign,m}]^T$ can be calculated with respect to the beginning of the injection in this cycle. The impact of the actual control trajectory $\underline{u}_k = [\tau_1, \tau_2, \ldots, \tau_n]^T$ represented by a number of commanded ignition delay times on the ignition time $\underline{y}_k = [t_{ign,1}, t_{ign,2}, \ldots, t_{ign,m}]^T$ for cycle k can then be approximated by linear model:

$$\underline{y}_k = \mathbf{G}_k \cdot \underline{u}_k + \underline{b}_k(t_I). \tag{1}$$

The additional vector $\underline{b}_k(t_I)$ contains the offset, which results from the time a fuel particle is injected. As the combustion region might vary between different cycles of SEC, the model has to be adapted to the current combustion cycle. This may cause adaptions of the dimensions of $\mathbf{G}_k$, $\underline{b}_k$, and $\underline{y}_k$ as well. For more detailed information about the model in Eq. 1, see Rähse et al. [3].

4.1.2 ILC

An iterative learning control improves the injection trajectory $\underline{u}_k$ from one cycle to the next. The ILC takes the measurement $\underline{y}_k$ of one cycle and calculates an improved trajectory $\underline{u}_{k+1}$ for the next cycle. During one cycle, the controller does not adjust the control trajectory, as the already-injected fuel cannot be corrected. The improvement is based on a model like Eq. (1), which describes the relation between the control trajectory $\underline{u}_k$ and the controlled variable $\underline{y}_k$.

For cycle $k+1$, the control error $\underline{e}_{k+1} = \underline{r} - \underline{y}_{k+1}$ can be calculated from the reference $\underline{r}$ and the measurement $\underline{y}_{k+1}$ for this cycle. In case of SEC, the vector $\underline{r}$ contains identical entries specifying the time of ignition. The controller aims to find a control trajectory $\underline{u}_{k+1}$ that minimizes the control error concerning a given norm $||\underline{e}_{k+1}||_{\mathbf{W}_e}$. Using the assumed model, Eq. (1), the control error of cycle $k+1$ can be calculated as follows:

$$\underline{e}_{k+1} = \underline{r} - \underline{y}_{k+1} = \underline{r} - \underline{b} - \mathbf{G}\,\underline{u}_{k+1} \tag{2}$$

If Eq. (1) described real behavior perfectly, it would be possible to analytically calculate the necessary control trajectory $\underline{u}_{k+1}$ that would minimize the following cost function:

$$J_{k+1} = ||\underline{e}_{k+1}||_{\mathbf{W}_e} = \underline{e}_{k+1}^T\,\mathbf{W}_e\,\underline{e}_{k+1}. \tag{3}$$

But since the model might be inaccurate, the cost function, Eq. (3), has to be expanded by an expression that penalizes the change of the control trajectory $\Delta\underline{u}_{k+1} = \underline{u}_{k+1} - \underline{u}_k$ to increase robustness:

$$J_{k+1} = \underline{e}_{k+1}^T\,\mathbf{W}_e\,\underline{e}_{k+1} + \Delta\underline{u}_{k+1}^T\,\mathbf{W}_{\Delta\underline{u}}\,\Delta\underline{u}_{k+1}. \tag{4}$$

The positive definite, symmetric matrices $\mathbf{W}_e \in \mathbb{R}^{m\times m}$ and $\mathbf{W}_{\Delta\underline{u}} \in \mathbb{R}^{n\times n}$ weight the control error and the change of the control trajectory, respectively. The cost function, Eq. (4), can be used to find the optimal control trajectory in every iteration. By choosing the optimum for every cycle, the algorithm gets closer to the minimal control error iteratively.

To find the minimum of the cost function, the derivative with respect to $\underline{u}_{k+1}$ is computed, set to zero, and solved for $\underline{u}_{k+1}$. As the reference $\underline{r}$ and the disturbances $\underline{b}$ are assumed to be invariant for all cycles, Eq. (2) is also applicable for cycle k:

$$\underline{r} - \underline{b} = \underline{e}_{k+1} + \mathbf{G}\,\underline{u}_{k+1} = \underline{e}_k + \mathbf{G}\,\underline{u}_k \tag{5}$$

Using this expression finally leads to the ILC law:

$$\underline{u}_{k+1} = \underline{u}_k + \mathbf{L}\,\underline{e}_k, \tag{6}$$

$$\mathbf{L} = \left(\mathbf{W}_{\Delta u} + \mathbf{G}^T\mathbf{W}_e\mathbf{G}\right)^{-1}\mathbf{G}^T\mathbf{W}_e \quad . \tag{7}$$

The adjustment of the control trajectory depends on the learning matrix $\mathbf{L}$, which contains all model information.

4.2 Control of the Size of the Combustion Region

The extent of the combustion region is mainly influenced by the size of the air buffer represented by air injection time $T_{Air,k}$ and the fuel injection duration $T_{F,k}$. For part load operation of SEC, the controller mainly has to follow a required reference for the length of the combustion region. When the power requirement is reduced or increased, the air buffer size has to be adjusted accordingly. However, if the air buffer size changes, the impact of the suction wave might vary from one cycle to the next. This results in a variation of the low pressure period at the inlet necessary for sucking in fresh air, which is paraphrased here as valve opening and closing times at the inlet. If a constant fuel injection duration $T_{F,k}$ were applied, one of two different problems are likely to occur. For a valve closing before the fuel injection was completed, a part of the current control trajectory would have to be dropped. This would lead to incorrect information about the influence of the control trajectory, as described by Eq. (1), if not accounted for appropriately. Thus, the results of the other controller, the ILC for the ignition time control, might be impaired.

In contrast, a too-long phase with an open valve could evoke a second, undesired air buffer behind the fuel-air mixture, as fuel injection would end earlier. After the ignition, a second pressure wave travelling to the left would occur. This would be reflected at the closed inlet after passing through the small air buffer between the combustion region and the inlet. This unwanted pressure wave would interact in an undesired fashion with the other pressure/suction waves and might disturb the resonant operation of SEC. For this reason, two controllers are proposed here for the combustion region control. One will be responsible for the position of the air buffer, and the other for the timing between injection and the period of low pressure.

A sketch of the injection process is shown in Fig. 3 in an x/t-diagram. It shows the time from the beginning of the air injection until the ignition on the vertical axis and the axial location inside the tube on the horizontal axis. The error e_P describes the difference between the last spatial element for which an ignition was detected in the last cycle and a reference position r_P. Accordingly, an error e_{T_F} is calculated between the end of the fuel injection duration T_F and the time T_V the valve closes, see Fig. 3. The goal is to adjust both values, $T_{Air,k}$ and $T_{F,k}$, so that, at the moment of the first ignition, an ignitable fuel-air mixture lies exactly between the inlet and a given position of the tube. This was not possible with the previously proposed controller [3]. The desired position r_P is the reference that can be changed to decrease or increase the main control variable for part load control, i.e., the length of the combustion region.

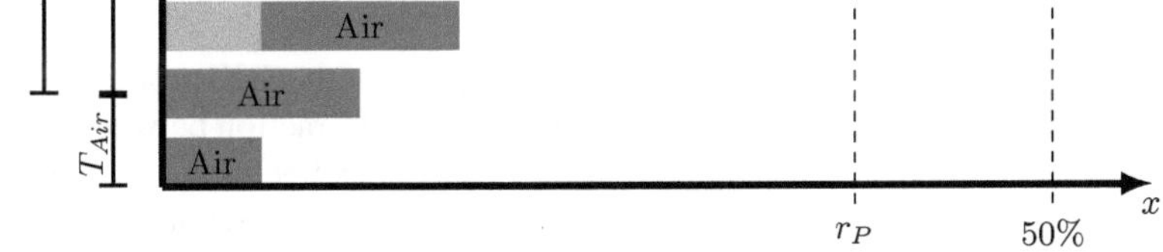

Fig. 3 SEC injection process

The end of the combustion region can be manipulated by the air injection time $T_{Air,k}$. It is thus intuitive that a first controller commands the air buffer size based on the error e_P between the desired and the detected ends of the combustion region in the last cycle. Since this is a pretty basic control task, a PI controller is applied to adjust the air buffer size. To support the controller, an additional feed-forward control $F_{T_{Air,k}}(r_P)$ is implemented to adjust the air buffer to the current set-point. The fresh air inflow duration for cycle k is calculated by Eq. (8). The values for the feed-forward control for a given reference are calculated from a map that contains the steady-state solutions for the set-points. The control law, directly specifying a time, reads as follows:

$$T_{Air,k+1} = F_{T_{Air,k}}(r_P) + P_{T_{Air}} \cdot e_{P,k} + I_{T_{Air}} \sum_{i=1}^{k} e_{P,i} \, , \tag{8}$$

with proportional and integral gain $P_{T_{Air}}$ and $I_{T_{Air}}$, respectively. A second PI controller is set up to command the fuel injection duration $T_{F,k+1}$ based on the observed error $e_{T_F,k}$ between injection duration and the time during which the valve was open in the last cycle. A feed-forward control $F_{T_{F,k}}$ is implemented for the fuel injection duration as well, similar to the air buffer control. The fuel injection duration $T_{F,k+1}$ for cycle $k+1$ is calculated by Eq. (9):

$$T_{F,k} = F_{T_{F,k}} + P_{T_F} \cdot e_{T_F,k} + I_{T_F} \sum_{i=1}^{k} e_{T_F,i} \tag{9}$$

All parts of the control system interact with each other. Figure 4 shows a diagram of the whole system to illustrate the dependencies. These relations have to be considered for the controller tuning that will be described in the following section. A main aspect is the interaction between the control for the size of the combustion region to realize variable load and the ILC responsible for homogeneous auto-ignition. The time-span the control trajectory $\underline{u}_k$ of ILC considers has to be adapted according to the length of the combustion region. A close-to-constant length for the combustion region will improve the results of ILC noticeably.

4.3 Controller Tuning

The complete control system handles individual aspects of the SEC process. Although all parts of the control system are set up separately, the interaction between the single controllers has to be considered. Therefore, each controller of the system has to be tuned properly to obtain a satisfying performance for the whole system. More specifically, the controllers must be designed such that they work on different time scales.

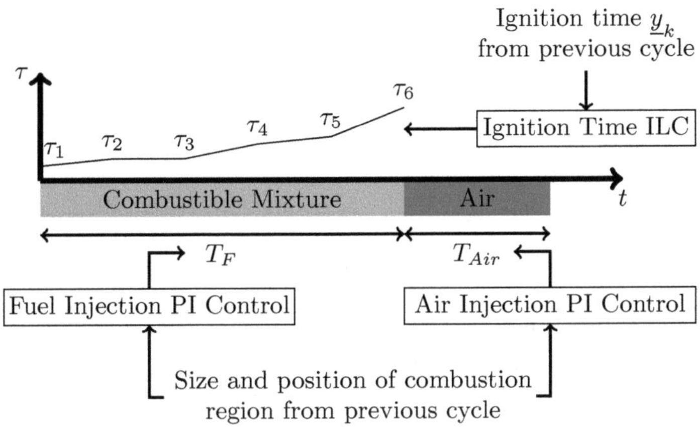

Fig. 4 Diagram of the control system

ILC for the ignition time control is the most important aspect with respect to the SEC process. A proper SEC operation is not possible without a homogeneous auto-ignition. Thus, all parameters of the PI controllers are set with regard to the ignition time controller, which is designed first. Since previously [3] we were able to produce adequate results with an ILC system for full load operation, the same parameter settings are adopted for this work as well. This includes the weighting of the absolute control error $\mathbf{W}_r$ as well as the penalization of its variance with $\mathbf{W}_V$ within the cost function, see Eq. (4) and Rähse et al. [3] for more details:

$$\mathbf{W}_e = \mathbf{W}_r + \mathbf{W}_V \tag{10}$$

$$\mathbf{W}_r = w_r \mathbf{I} \tag{11}$$

$$\mathbf{W}_V = w_V \frac{1}{m-1}\left(\mathbf{I} - \frac{1}{m}\mathbb{1}\right) \tag{12}$$

The parameters are set to $w_r = 1$ and $w_V = 2$ so that the variance between the detected ignition times is considered more important than the absolute control error. This setting relies on the assumption that it is more important for the ignition to be as homogeneous as possible than to achieve an exact ignition time. Additionally, the change of the control trajectory is also penalized as follows:

$$\mathbf{W}_{\Delta u} = w_u \mathbf{I}. \tag{13}$$

This generates a robustification of the ILC algorithm. To avoid overshoots, the weight for the change is set to $w_u = 20$. This tuning ensures that ILC converges within the first 50 cycles.

Since ILC works best for a combustion region with a constant length, the controllers for the air buffer and the fuel injection duration have to work much faster

Table 1 Controller settings

Parameter	P	I	
Fuel injection time T_F	0.7	0.2	
Air buffer size T_{Air}	0.5	0.2	
Parameter	w_r	w_V	w_u
Ignition time	1	2	20

than the ignition time control. A constant size of the combustion region should be reached as soon as possible so that ILC may start improving the combustion itself. The parameters for both PI controllers are set to reach the reference within 5–10 iterations, depending on the point of operation. All parameters for the control system are shown in Table 1. To avoid interference of both PI controllers, the fuel injection PI control works slightly faster than the air buffer control. To avoid unintended behavior when switching between set-points, both PI controllers are reset. They are also turned off for 10 cycles after the set-point change while the feed-forward control drives the process toward the new set-point.

5 Results

The part load control was tested for hydrogen combustion and compressor pressure ratios between 24 and 50. The conditions are similar to the work of Rähse et al. [3]. However, the results shown in this section are focused on a single compressor pressure ratio, as the results for all considered values show a similar trend. The following results are calculated for a compressor pressure ratio of 34 and a temperature of around 919 K. It should be noted again that H_2/O_2 is a poor choice for SEC due to the immense length of the combustion tube needed [7]. At this set-point a tube length of 19.16 m would be required. Other fuels, such as dimethyl-ether, would feature significantly shorter lengths even for much smaller pressure ratios. For this study, the spatial grid resolution was set to 100 cells. The reference ignition time for the ILC is $\underline{r} = 45$ ms, since this is the minimal ignition delay time for these conditions and thus the fastest ignition that is possible.

5.1 *Control of the Size of the Combustion Region*

A sequence of set-points was used that required the system to initially run with a combustion region of $r_P = 40\%$ of the tube's length for 50 cycles and then drive down to $r_P = 20\%$ for 100 cycles before returning to full load, $r_P = 40\%$. It has to be pointed out that all three sub-controllers were running in this simulation, the

ILC and both PI controllers. The detected end of the combustion region is shown in Fig. 5. At the beginning, the system reached the set-point after fewer than 20 cycles. This is a very satisfying result since ILC also had to change the fuel injection substantially because the optimal control trajectory $\underline{u}_k$ was unknown at the beginning. Furthermore, the choice of the initial conditions had a major influence during the first iterations. The responses following subsequent set-point changes showed that the controllers drove the process closer to the new set-points in fewer cycles. The controller for the size of the combustion region was able to adjust new set-points seamlessly and avoid large overshoots. Furthermore, the controller maintained the process at a desired operating point. The remaining variation of 1% is a result of the spatial grid resolution of the simulation.

Figure 6 shows the injection time for the air buffer on the left and for the fuel on the right. It is noticeable that the variation of the air buffer is much smaller than the modulation of the fuel injection time. This is a result of the different response times of the two PI controllers. The PI control of the fuel injection time responded slightly faster. Therefore, the fuel injection time controller already counteracted the disturbances before the air buffer PI controller started to respond.

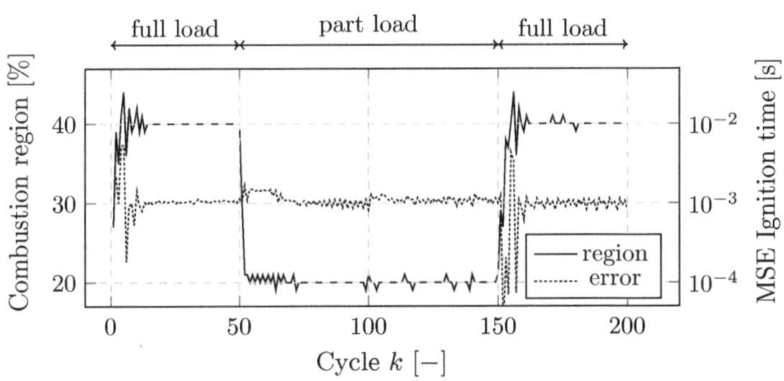

Fig. 5 Combustion region and mean squared error (MSE) of ignition time

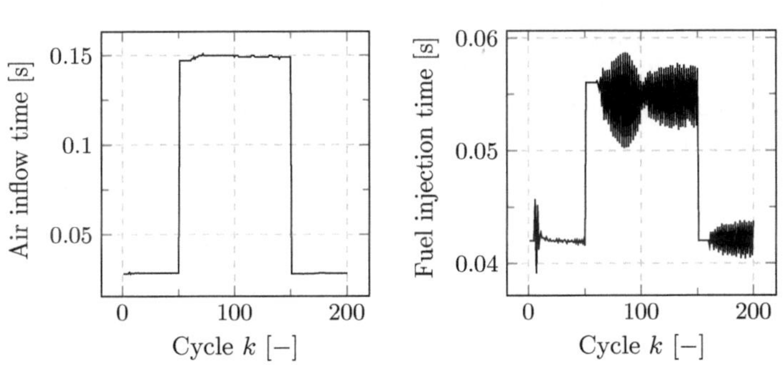

Fig. 6 Air buffer injection duration (left) and fuel injection duration (right)

5.2 Ignition Time Control

When the combustion region decreases in size during a part load operation period, it is essential that ILC still guarantees the homogeneous auto-ignition necessary to approximate a constant volume combustion and achieve higher efficiency compared to constant pressure combustion. Figure 5 also shows the mean squared error for all detected ignition times within one cycle for a range of 200 cycles. Convergence was achieved within the first 20 cycles, and the controller maintained this control quality for each set-point change. Even when the system was switched to part load operation, after cycle $k = 50$, the error between the reference and the detected ignition time remained on a similar level. This is a significant result, since it demonstrates that the control system is capable of running the SEC cycle in part load operation with a homogeneous auto-ignition as well as in full load mode.

5.3 Part Load Aspects

The results presented above show that the control system can operate the SEC process at different set-points and guarantees a seamless transition between them, even when the load is changed in a drastic, step-wise fashion, which would not be done in a real application. As mentioned, a part load operation should offer a lower energy output at the outlet. A common way to investigate the usable work of the process is to add an expansion over a turbine as shown by Stathopoulos et al. [8]. This is a realistic assumption to analyze the whole combustion process. For the investigation in this work, the absolute values are not as relevant as a general statement about the level of energy emission at the outlet. Therefore, a more basic approach was used. The energy change is described as the difference of enthalpy for an isentropic expansion to atmospheric conditions, cumulated over one cycle. The enthalpy difference is shown in Fig. 7 for two cycles during full load operation as well as for part load operation. It is clearly recognizable that the energy output is lower during part load operation, which is a result of a decreased heat release and consequently lower mean temperatures and pressures. In part load operation, the cumulated enthalpy difference of one cycle is about 10^8 J smaller than in full load operation.

Part load operation also influences the firing frequency of the combustion tube. The higher amount of cold air that comes with an increased air buffer size leads to a lower mean sonic speed. Therefore, the pressure and suction waves travel at a lower velocity between the inlet and the outlet, slowing down the whole process. If the combustion region is reduced from 40 to 20% of the length of the tube, the firing frequency drops from around 12.5–4.8 Hz. A lower frequency for the decreased combustion region also influences the fuel and air injection times. It is obvious that a larger air buffer requires a longer period of time for air intake, but Fig. 6 shows that the fuel injection duration also rises. This illustrates the deceleration of the process.

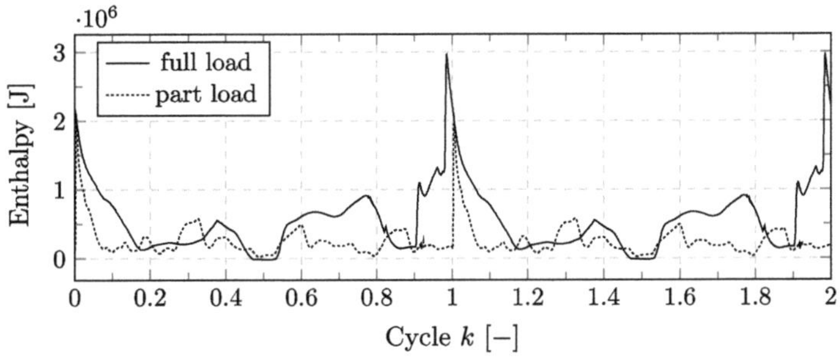

Fig. 7 Enthalpy difference for two normalized cycles during full load and part load operation. Absolute cycle lengths differ

Consequently, it takes longer to fill 20% of the tube during part load than 40% during full load operation.

Combined with the reduced energy emission, the lower frequency leads to a change in the power of the SEC process. By decreasing the size of the combustion region to 20% of the length of the tube, the power drops from 8.3×10^9 to 2.9×10^9 W. These high values are results of the unrealistic tube size mentioned earlier. However, besides the dimension, these values show how well the proposed control system is able to realize a part load operation.

6 Conclusion

This work introduced a control system, consisting of three individual controllers, that is capable of maneuvering the SEC process within a certain regime of set-points. By adjusting the size of the combustion region, it is possible to reduce the energy emission at the outlet to realize a part load operation. Furthermore, the frequency can be influenced by changing the combustion region. Homogeneous auto-ignition is marginally influenced by the set-point adjustments, which is an important insight since homogeneous auto-ignition is essential to achieve the thermodynamic benefit of shockless explosion combustion.

This work provides a control architecture that can be used for further theoretical or numerical investigations of the SEC process in general and specifically for the study of the part load operation. To get closer to the real application, it will be necessary to work with more realistic conditions for the compressor as well as the turbine. One solution, as mentioned above, would be the application of a realistic map similar to the work of Stathopoulos et al. [8]. With this expansion model, it will be possible to discuss a part load operation for an overall gas turbine model based on the SEC process in more detail. Finally, more appropriate fuels for an SEC operation have to be included in the simulator.

Acknowledgements The authors gratefully acknowledge support by the Deutsche Forschungsgemeinschaft (DFG) as part of collaborative research center SFB 1029 "Substantial efficiency increase in gas turbines through direct use of coupled unsteady combustion and flow dynamics" in project A05.

References

1. Bobusch, B., Berndt, P., Paschereit, C.O., Klein, R.: Shockless explosion combustion: an innovative way of efficient constant volume combustion in gas turbines. In: 24th ICDERS, July 28–August 2, 2013, Taiwan (2013)
2. Berndt, P.: Mathematical modeling of the shockless explosion combustion. Ph.D. thesis, Freie Universität Berlin (2016)
3. Rähse, T.S., Stathopoulus, P., Schäpel, J.-S., Arnold, F., King, R.: On the influence of fuel stratification and its control on the efficiency of the shockless explosion combustion. In: Proceedings of ASME Turbo Expo 2018: Turbomachinery Technical Conference and Exposition GT2018, Oslo, Norway, GT2018-76435 (2018)
4. Reichel, T.G., Schäpel, J.-S., Bobusch, B.C., Klein, R., King, R., Paschereit, C.O.: Shockless explosion combustion: experimental investigation of a new approximate constant volume combustion process. J. Eng. Gas Turbines Power **139**(2), 021504 (2016)
5. Burke, M.P., Chaos, M., Ju, Y., Dryer, F.L., Klippenstein, S.J.: Comprehensive H_2/O_2 kinetic model for high-pressure combustion. Int. J. Chem. Kinet. **44**(7), 444–474 (2011)
6. Schäpel, J.-S., Wolff, S., Schulz, P., Berndt, P., Klein, R., King, R.: State estimation for reactive Euler equation by Kalman Filtering. CEAS Aeronaut. J. **8**(2), 261–270 (2017)
7. Rähse, T.S., Paschereit, C.O., Stathopoulos, P., Berndt, P., Klein, R.: Gas dynamic simulation of shockless explosion combustion for gas turbine power cycles. In: Proceedings of ASME Turbomachinery Technical Conference and Exposition, Charlotte, USA, GT2017-63439 (2017)
8. Stathopoulos, P., Vinkeloe, J., Paschereit, C.O.: Thermodynamic evaluation of constant volume combustion for gas turbine power cycles. In: Proceedings of the 11th International Gas Turbine Congress, Tokyo, Japan (2015)

Knock Control in Shockless Explosion Combustion by Extension of Excitation Time

Lisa Zander, Giordana Tornow, Rupert Klein and Neda Djordjevic

Abstract Shockless Explosion Combustion is a novel constant volume combustion concept with an expected efficiency increase compared to conventional gas turbines. However, Shockless Explosion Combustion is prone to knocking because it is based on autoignition. This study investigates the potential of prolonging the excitation time of the combustible mixture by dilution with exhaust gas and steam to suppress detonation formation and mitigate knocking. Analyses of the characteristic chemical time scales by zero-dimensional reactor simulations show that the excitation time can be prolonged by dilution such that it exceeds the ignition delay time perturbation caused by a difference in initial temperature. This may suppress the formation of a detonation because less energy is fed into the pressure wave running ahead of the reaction front. One-dimensional simulations are performed to investigate reaction front propagation from a hot spot with various amounts of dilution. They demonstrate that dilution with exhaust gas or steam suppresses the formation of a detonation compared to the undiluted case, where a detonation ensues from the hot spot.

Keywords Shockless explosion combustion · Dilution · Detonation formation

L. Zander (✉) · N. Djordjevic
Institute of Fluid Dynamics and Technical Acoustics, Combustion Kinetics,
Technische Universität Berlin, Müller-Breslau-Str. 8, D-10623 Berlin, Germany
e-mail: lisa.zander@tu-berlin.de

N. Djordjevic
e-mail: neda.djordjevic@tu-berlin.de

G. Tornow · R. Klein
Department of Mathematics, Geophysical Fluid Dynamics, Freie Universität Berlin,
Arnimallee 6, D-14195 Berlin, Germany
e-mail: gtornow@math.fu-berlin.de

R. Klein
e-mail: rupert.klein@math.fu-berlin.de

R. King (ed.), *Active Flow and Combustion Control 2018*,
Notes on Numerical Fluid Mechanics and Multidisciplinary Design 141,
https://doi.org/10.1007/978-3-319-98177-2_10

1 Introduction

A substantial efficiency increase is expected for constant volume combustion systems compared to conventional gas turbines utilizing constant pressure combustion. A novel system utilizing constant volume combustion is Shockless Explosion Combustion (SEC). It was first described by Bobusch et al. [1]. A qualitative analysis indicates an efficiency gain of the SEC process compared to the Joule cycle [2].

SEC is based on quasi-homogeneous autoignition of the gas in the combustion tube and utilizes pressure waves for filling and purging. To achieve quasi-homogeneous autoignition the fresh gas mixture's equivalence ratio is stratified such that the ignition delay time of each discrete gas volume is correlated to its residence time in the combustion tube. As a result, the total volume of the combustion mixture auto-ignites simultaneously after filling is completed.

Because SEC relies on homogeneous autoignition it has several advantages compared to constant volume combustion systems that use detonation waves. Sharp pressure rises which can be harmful to the machine are avoided. Smaller exergy losses are expected for SEC because the kinetic energy in detonation waves cannot be used entirely. Furthermore, losses due to the deflagration to detonation transition which appear in detonative combustion processes are not present in SEC.

Nonetheless, a process that relies on autoignition is strongly dependent on the chemical-kinetic properties of the combustible mixture. Under certain circumstances, a detonation wave can ensue from a more reactive spot caused by premature ignition and lead to knocking. Consequently, the SEC process is very sensitive to deviations in temperature or mixture composition from the ideal distributions that guarantee nearly homogeneous autoignition after completion of the charging process. The formation of detonations in SEC can be explained with the SWACER (Shock Wave Amplification by Coherent Energy Release) mechanisms proposed by Lee et al. [3]. Premature ignition in a more reactive spot, e.g. a hot spot, generates a pressure wave which moves into the not yet ignited gas. Additionally, the gradient in ignition delay time around this reactive spot leads to an autoignition wave. If the autoignition wave propagates at a speed, such that the heat release is in phase with the pressure wave running ahead of the autoignition wave, the pressure wave is amplified and a detonation may be formed.

Multiple researchers investigated the conditions for the occurrence of detonation waves in general. Zeldovic et al. [4] identified that the slope of the temperature gradient influences whether a detonation is formed for a combustible mixture with an inhomogeneous initial temperature distribution. Later, Zeldovich [5] distinguished four modes of reaction front propagation for mixtures with inhomogeneous initial temperature distributions theoretically: weak detonation (also referred to as supersonic deflagration), with propagation governed by autoignition at a velocity higher than Chapman–Jouget (C–J) velocity and without the formation of a shock wave; developing detonation and detonation; subsonic flame propagation controlled by autoignition and flame propagation which is governed by heat conduction and diffusion. Zeldovich regards adiabatic explosion at constant volume (also referred to as thermal explosion) as a limiting case of weak detonation, because chemical conversion happens simultaneously in the complete combustion volume and the propagation speed reaches infinity. Gu et al. [6] identified all of the modes described in [5] in simulations with hot spots of different radii and temperature gradients. They underlined the importance of the rapidness of the heat release into the shock for the successful formation of a detonation. Based on this idea, they proposed a regime diagram for the occurrence of detonations defined by two non-dimensional quantities, the normalized temperature gradient of the hot spot and the acoustic time normalized by the excitation time. The range of values for which detonations can occur in this regime diagram is commonly referred to as detonation peninsula.

Sources of perturbations in SEC can be heat convection or radiation from the combustion tube's walls, equivalence ratio perturbations caused by the filling process or residual gas that remains in the combustion tube from the previous cycle. These fluctuations are difficult to predict and control and the process has to be robust against them. Sources of perturbations in SEC that may be predicted up to a certain level are temperature changes due to the filling process. Their order of magnitude amounts to $\mathcal{O}(10)$ K [7]. When using fuels with two-stage ignition the heat release of the first stage can cause inhomogeneous ignition and possibly the formation of a detonation wave [7].

Different strategies were investigated to alleviate the sensitivity of the SEC process to inhomogeneities. The effects of equivalence ratio perturbations on homogeneous ignition can be mitigated by reducing the range of equivalence ratios used for the fuel stratification [8]. The negative temperature coefficient (NTC) behavior of most hydrocarbons, which leads to an increase in ignition delay time with increased initial temperature over a certain range of temperatures, can be utilized to tailor the combustible mixture for SEC. Mixing fuels with and without NTC behavior yields a fuel blend with temperature independent ignition delay time over a range of initial temperatures, which ideally eliminates the effect of temperature perturbations in the SEC process [9].

Berndt et al. investigated the SEC process by means of simulations with a simplified reaction mechanism [10]. They showed that even when the temperature dependency was reduced for a range of initial temperatures, the heat release of the first ignition stage still led to inhomogeneous ignition and concluded that the temperature dependency of the ignition delay time needs to be reduced further. Additionally, Berndt et al. determined the detonation peninsula in the regime diagram qualitatively [7]. They proved that the lower bound for detonation development in SEC is linked to the C–J speed and found that long excitation times can prevent detonation formation, even when there are fair perturbations in the ignition delay time.

To avoid the formation of a detonation caused by inhomogeneous ignition not only the temperature dependency of the mixture has to be decreased but an increase in excitation time would substantially reduce the demands on the accuracy of mixture stratification and temperature homogeneity. However, it was not possible to increase the excitation time by blending different fuels [9] because relevant fuels have similar excitation times.

In order to prolong the excitation time and mitigate knocking we consider the possibility of diluting the fresh gas mixture with exhaust gas and steam in the present study. To evaluate the proposed methods we conduct simulations of a combustible mixture subjected to a temperature inhomogeneity in form of a hot spot with various amounts of dilution.

2 Numerical Setup and Methods

One-dimensional numerical simulations are conducted to assess the effect of dilution on ignition and wave propagation in a setup with an initial temperature perturbation. A section of a cylindrical tube with a hot spot is considered. The gas mixture is at rest initially. The hot spot is modeled as a linear temperature increase at the left part of the domain. The initial temperature distribution $T(x, t = 0)$ containing a hot spot is described by the following equation:

$$T(x,t=0) = \begin{cases} (x - x_{hs})\left(\frac{\partial T_0}{\partial x}\right) + T_0, & x \leq x_{hs} \\ T_0, & x > x_{hs} \end{cases},$$

where T_0 is the temperature of the gas outside of the hot spot, x_{hs} is the hot spot's spacial extension and $\frac{\partial T_0}{\partial x}$ is the temperature gradient in the hot spot (it is negative to obtain a temperature elevation within the hot spot). Figure 1 sketches the initial temperature profile defined by the equation above. The hot spot's extension is chosen to be 0.5 mm and the maximum temperature elevation (at $T(x = 0, t = 0)$) amounts to 10 K. The gas in the tube is assumed to be radially homogeneous. Thus, the problem reduces to one dimension along the tube axis. The section that is modeled has a length of 5 cm. Because the domain represents merely a section of a tube there are no reflections at the boundaries. Hence, continuous boundaries are used which force the gradients to be zero. The gas mixture in the tube section is a homogeneous dimethyl ether (DME)/air mixture with stoichiometric composition. Various amounts of steam and exhaust gas are added to the reactants. The exhaust gas is assumed to be composed according to the global products composition of stoichiometric DME/air combustion. The initial thermodynamic state in the tube is calculated assuming the gas is compressed with a pressure ratio of 20 from ambient conditions before entering the tube with an isentropic efficiency of 90%. The pressure ratio is chosen such that the ignition delay time is of the order of 1 ms to ensure feasible tube lengths (compare e.g. [11] for the relation between tube length and ignition delay time). From this reasoning, the conditions inside the tube result to 20 bar and 755 K.

The processes inside the tube are simulated using a code developed for simulating the SEC process [12]. The reactive Euler equations with chemical source terms in the energy equation and species mass fraction conservation equations are solved using the finite volume method with a HLL (Harten, Lax, van Leer) solver with Einfeldt correction. It was demonstrated in [13] that the choice of the solver is appropriate for the problem to be modeled. A second order integration scheme is used. Chemical

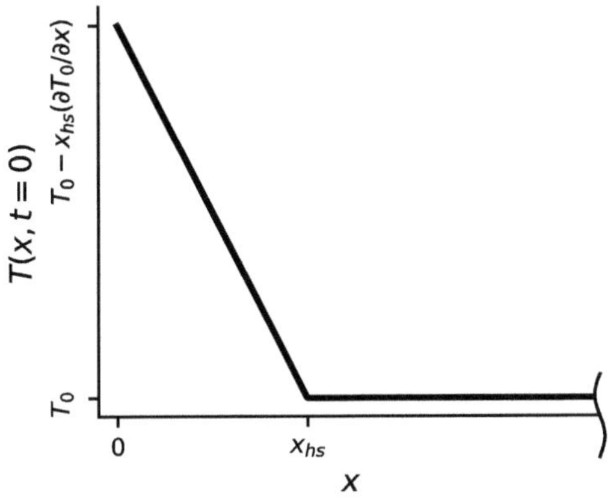

Fig. 1 Sketch of initial temperature distribution with a hot spot

kinetics is integrated into the scheme by Strang splitting. The chemical kinetics of DME is modeled using a detailed mechanism developed by Zhao et al. [14] which was validated for a range of temperatures and pressures relevant to the present study. However, none of the available mechanisms for DME was validated for dilution with exhaust gas and steam and experimental data of ignition delay times in such mixtures are necessary to quantitatively prove the effects of dilution on the chemistry. Nevertheless, the physical influence of dilution on ignition and the characteristic time scales is reproduced.

All simulation results provided in this work were obtained with a spacial resolution of 5×10^{-5} m and a CFL (Courant, Friedrichs, Lewy) number of 0.5.

3 Results

As described in the introduction, the two time scales of interest are the ignition delay time τ_i and the excitation time τ_e. To investigate how these time scales change with dilution, zero-dimensional isochoric reactor simulations are performed with the software package Cantera [15]. The ignition delay time is determined through the time it takes the gas mixture to reach the maximum temporal change in temperature. The excitation time is defined as in [12] as the time needed for the temperature to rise from

$$0.85T(t=0) + 0.15T(t=t_\infty)$$

to

$$0.15T(t=0) + 0.85T(t=t_\infty).$$

In the equation above $T(t=t_\infty)$ denotes the temperature the gas mixture attains in equilibrium after ignition. Other researchers define the excitation time from the instant when a given fraction of the maximum heat release rate (e.g. 5% in [6] and 1%

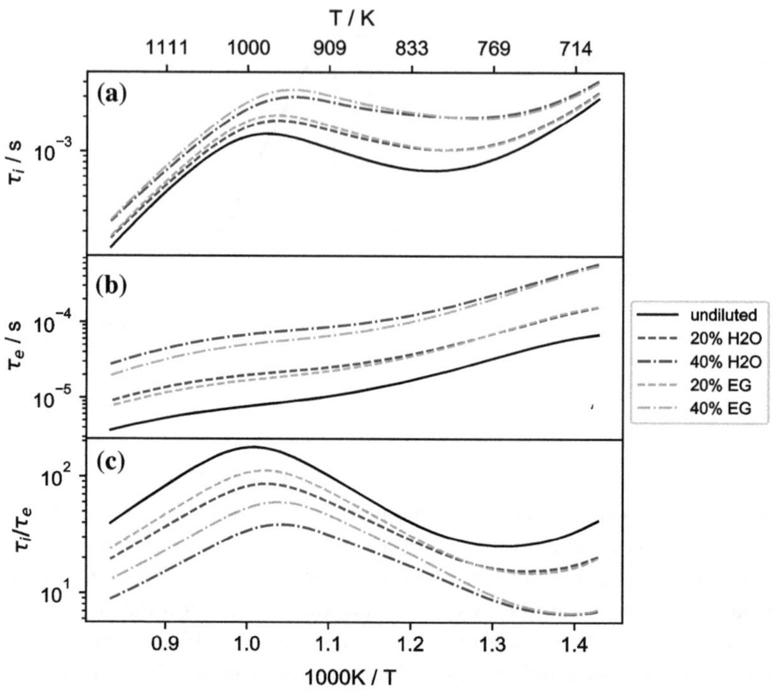

Fig. 2 Ignition delay time **a**, excitation time **b** and their ratio **c** over initial mixture temperature for 0%, 20 vol% and 40 vol% dilution with steam (H2O) and exhaust gas (EG) for a pressure of 20 bar

in [9]) is achieved until the instant of maximum heat release. For the considered mixtures these definitions are ambiguous due to the presence of multiple ignition stages for certain conditions. The definition from [12] was chosen because the monotonous increase of the temperature yields an unambiguous definition of the excitation time. The predominant trends in the following results are similar regardless of the chosen excitation time definition.

The variation of excitation times and ignition delay times with initial temperature is depicted in Fig. 2a, b for a pressure of 20 bar as in the one-dimensional simulations. The excitation time increases with increasing amount of dilution (Fig. 2b), and can thus be used to mitigate knocking as proposed. The prolongation of excitation times is higher for steam dilution, except for lower temperatures and lower dilutions rates, where excitation times are similar for both diluents.

However, dilution with steam or exhaust gas also increases the ignition delay time (Fig. 2a), which influences the combustion process of SEC including filling and purging. For intermediate to high temperatures the ignition delay time is prolonged more with exhaust gas dilution compared to the same amount of steam dilution, while ignition delay times are similar for both diluents at low temperatures. Furthermore, the ignition delay times in Fig. 2a show that DME exhibits a pronounced NTC behavior in the temperature range between 820 and 980 K for the undiluted case. The range of temperatures where NTC behavior is prevailing shifts to lower temperatures with increasing amount of dilution.

In order to evaluate the effect of initial temperature on both chemical time scales the ratio of ignition delay time to excitation time is depicted in Fig. 2c. Dilution enables the desired prolongation of the excitation time (Fig. 2b), but it also increases the ignition delay time (Fig. 2a), thus influencing the combustion process and operation of SEC. It is desired that the effect of dilution on the excitation time is stronger than on the ignition delay time. Hence, a low ratio of ignition delay time to excitation time τ_i/τ_e is desired. With increasing dilution this ratio is decreased (Fig. 2c). For intermediate to high temperatures the ratio of ignition delay to excitation time is smaller with steam dilution, while it is similar at low temperatures for both diluents.

When the temperature in a combustion volume is perturbed at a specific spot, ignition can appear there earlier compared to the surrounding mixture. Generally, this is the case for hot spots, or for cool spots within the NTC region. In the following analysis we refer to the temperature of the perturbation as T_p and define the ignition delay time deviation $\Delta\tau_i$ from the surrounding mixture's ignition delay time caused by a temperature perturbation as

$$\Delta\tau_i = \tau_i(T_p) - \tau_i(755\ K).$$

In the equation above, the reference temperature which represents the surrounding mixture's temperature is chosen to be 755 K as in the one-dimensional simulations.

Depending on the temperature gradient of the perturbation, the perturbation size and mixture properties, a detonation can ensue from an ignition spot. In the following, the influence of the rapidness of heat release on detonation formation is investigated. If the temperature perturbation in a hot spot leads to premature ignition, the local heat release results in a local increase in pressure. The pressure then propagates into the gas surrounding the hot spot, which is already close to autoignition. The compression due to the pressure propagation from the hot spot may accelerate the autoignition of the surrounding gas and trigger a detonation. However, if the time scale of the pressure rise in the hot spot is longer than the difference in ignition delay time between the hot spot and the surrounding gas, the surrounding gas will have enough time to autoignite when its actual ignition delay time expired without being affected by the hot spot. Thus, the hot spot would pose no risk with respect to detonation formation [12]. Since the time scale of the pressure rise is determined by the excitation time τ_e, this is qualitatively fulfilled if

$$|\Delta\tau_i| < \tau_e. \tag{1}$$

In order to examine the fulfillment of condition (1), Fig. 3 shows the ratio of the ignition delay time deviation $\Delta\tau_i$ to the excitation time of the temperature perturbation $\tau_e(T_p)$ over the perturbation's temperature T_p.

When the ratio $\Delta\tau_i/\tau_e$ depicted in Fig. 3 is negative, the ignition delay time of the perturbation is smaller than the surrounding mixture's ignition delay time. In

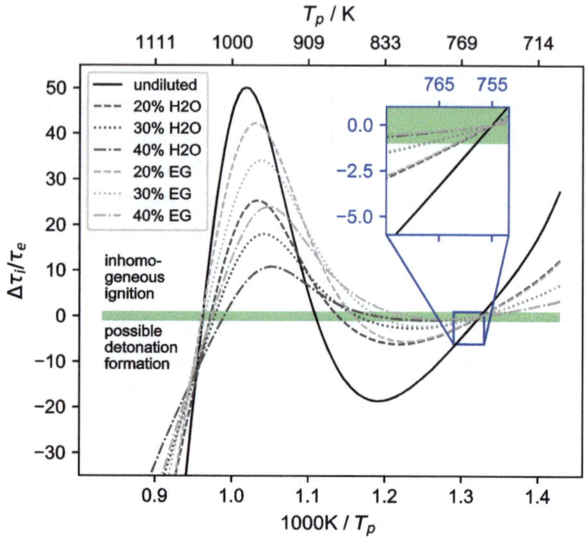

Fig. 3 Ratio of ignition delay time deviation $\Delta\tau_i$ caused by a temperature perturbation to the excitation time of the mixture with perturbed temperature τ_e over the perturbation temperature T_p with various amounts of steam (H2O) and exhaust gas (EG) dilution. The green shaded area denotes condition (1)

the opposite case, where the ratio $\Delta\tau_i/\tau_e$ is positive, the ignition delay time of the surrounding mixture has expired before the hot spot ignites. In Fig. 3 the ignition delay time of the hot spot is higher than the ignition delay time of the surrounding mixture in the range of perturbation temperatures between 900 and 1040 K for the undiluted mixture because of the NTC behavior of DME.

As explained above, only negative ignition delay time deviations which describe a decrease in ignition delay of a hot spot compared to the surrounding mixture's ignition delay time can lead to premature ignition and the formation of a detonation wave. Nonetheless, positive ignition delay time deviations can lead to inhomogeneous ignition. In order to investigate the possibility of knock control by dilution, only negative ignition delay time deviations $\Delta\tau_i$ (respectively negative ratios $\Delta\tau_i/\tau_e$) will be considered because they can possibly result in a detonation.

The condition in Eq. (1) can be transformed to

$$\left|\frac{\Delta\tau_i}{\tau_e}\right| < 1. \tag{2}$$

Condition (2) is represented by the green shaded area in Fig. 3.

The magnitude of the ratio of ignition delay deviation to excitation time $|\Delta\tau_i/\tau_e|$ is decreased by dilution (Fig. 3). Exhaust gas dilution is decreasing it more for intermediate temperatures, while steam dilution does for high temperatures. For dilution with 40 vol% steam the ratio of ignition delay time deviations to the excitation time remains within the green shaded area for temperature perturbations of up to 860 K, indicating that condition (1) is fulfilled and knocking may effectively be mitigated for these temperatures. For perturbation temperatures above 1040 K dilution is less effective, because the magnitude of the ignition delay time deviation increases much more with increasing perturbation temperature than the excitation time. The insert in Fig. 3 shows the diagram magnified to a perturbation temperature of 765 K, which is the amplitude of the hot spot present in the one-dimensional simulations. Dilution with more than 30 vol% exhaust gas or steam decreases the ratio $|\Delta\tau_i/\tau_e|$ such that it fulfills condition (1), which indicates that the prolongation of excitation times may become effective for reducing knocking in the one-dimensional simulations at about this amount of dilution.

To conclude, dilution alters the characteristic time scales of the mixture in such a way that condition (1) is met for a wider range of perturbation temperatures compared to the undiluted case. This indicates that the increase in excitation time caused by dilution may in fact prevent detonation formation. However, the aforementioned zero-dimensional analysis does not consider the influence of the properties of the temperature perturbation, such as its radius and temperature gradient.

While the previous analysis shows the influence of dilution on the magnitude of the relevant time scales, the following considers the effect of dilution on the gradient in ignition delay time which influences the autoignition wave propagation mode. The propagation speed u of an autoignition wave equals the inverse of the ignition delay time gradient [5, 6] and is related to the temperature gradient in a mixture as follows:

$$u = \left(\frac{\partial\tau_i}{\partial x}\right)^{-1} = \left(\frac{\partial\tau_i}{\partial T_0}\frac{\partial T_0}{\partial x}\right)^{-1} \tag{3}$$

When the temperature gradient in a mixture close to autoignition reaches a certain critical value, such that the autoignition wave generated by this temperature gradient propagates at approximately the speed of sound a into the unburned gas a detonation can be initiated through coupling of the pressure wave with the reaction front [5]. This critical temperature gradient can be expressed as [6]

$$\left(\frac{\partial T_0}{\partial x}\right)_c = a^{-1}\left(\frac{\partial\tau_i}{\partial T_0}\right)^{-1}. \tag{4}$$

It is usually defined from the initial temperature distribution in a combustion volume. However, the ignition delay time gradient in a hot spot will be altered by species diffusion and heat conduction. Therefore, a detonation develops not exclusively for the critical temperature gradient, but for a range of temperature gradients [6].

Equation (4) is evaluated to determine the critical temperature gradient over a range of temperatures (Fig. 4). Due to NTC behavior cold spots (with a positive temperature gradient) can initiate detonations in the intermediate temperature range. At the transition temperatures from NTC to the non-NTC region there are two singular points, because a change in temperature results solely in a small change in ignition delay time. In general, dilution of the fresh gas alters the critical temperature gradient.

After identifying the main parameters influencing the occurrence of knocking qualitatively with zero-dimensional calculations, the effect of dilution is studied for the case of a hot spot in a tube by means of one-dimensional simulations. Figure 5b, c

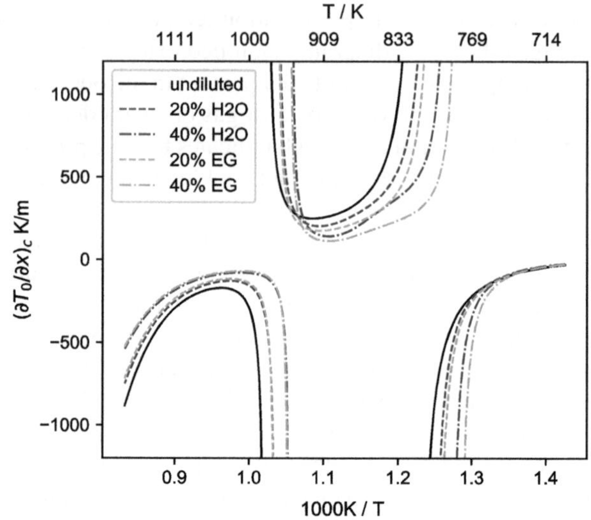

Fig. 4 Critical temperature gradient over initial temperature for 0%, 20 vol% and 40 vol% dilution with steam (H2O) and exhaust gas (EG)

shows temperature and pressure distributions at different times inside the combustion tube for the undiluted case of a stoichiometric DME/air mixture. The reaction front and shock wave are traced by the maximum change in OH mass fraction and pressure respectively. Their trajectories are represented by a white and dashed blue curve in the space-time diagram of the temperature in Fig. 5a and their propagation speeds are plotted over the axial coordinate in Fig. 5d. In all of the performed one-dimensional simulations temperature, pressure and OH mass fraction are sampled every μs. At $t = 0.95$ ms the temperature has risen to a value of approximately 1000 K throughout the domain, which indicates that some heat was already released (Fig. 5b). Later, at $t = 0.96$ ms, a shock forms, which is indicated by the sharp pressure rise in Fig. 5c and the letter **s**. It is propagating ahead of the reaction front (Fig. 5a). The shock and reaction front couple at $x \approx 0.015$ m where the propagation speeds of both waves coalesce in Fig. 5d and the detonation wave is fully developed. Eventually the propagation speed of the detonation wave approaches the speed of a C–J detonation (Fig. 5d) and the C–J temperature and pressure are distinguishable in the profiles in Fig. 5b, c (C–J conditions are computed according to [16, 17]). This wave propagation mode corresponds to the shock-detonation mode observed by Dai et al. [18].

In the following, dilution is added to the reactants mixture aiming at mitigation of knock behavior. Figures 6, 7 and 8 show the ignition processes inside the tube when adding different amounts of exhaust gas and steam dilution. Compared to the undiluted case the shock propagates a longer distance in front of the reaction front before they form a detonation wave in the 20 vol% steam or exhaust gas diluted mixture. This can again be observed by the trace of the reaction front and pressure wave in Fig. 6a, e or by the distance it takes until their propagation velocities are equal in Fig. 6d, h. At an axial location of $x \approx 0.03$ m for exhaust gas dilution, respectively $x \approx 0.035$ m for steam dilution, the reaction front reaches a propagation speed close

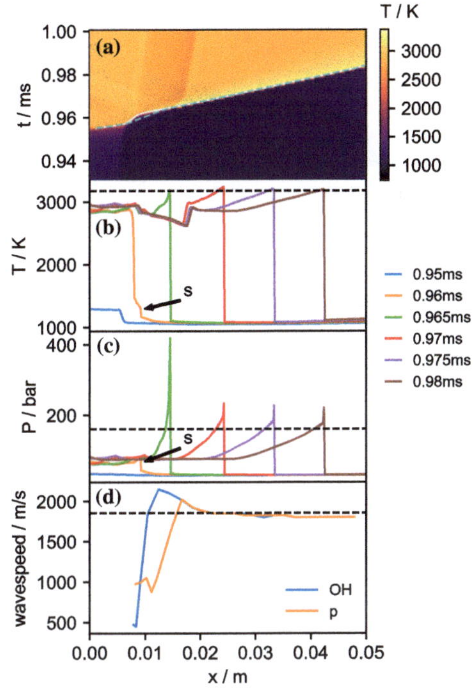

Fig. 5 **a** Space-time diagram of the temperature, white curve—reaction front, dashed blue curve—pressure wave. **b, c** Temperature and pressure distributions at different times. **d** Velocities of reaction front (OH) and leading pressure wave (p). Dashed lines in **b–d** C–J values of temperature, pressure and C–J detonation velocity

to a C–J detonation. The distributions in Fig. 6b, c, f, and g exhibit pressures and temperatures close to the C–J state. Dilution with 30 vol% exhaust gas delays the formation of a fully developed detonation even more (Fig. 7a, d).

When diluting with 30 vol% steam the formation of a detonation fails, as indicated by the temperature distributions in Fig. 7f and the pressure distributions in Fig. 7g, which reach values below their respective C–J values and do not exhibit steep profiles as in the undiluted case in Fig. 5. Ignition and wave propagation for dilution with 40 vol% exhaust gas or steam show a similar behavior (Fig. 8). The heat release of the reaction front creates a pressure wave which runs ahead of it. It compresses the gas and raises the temperature. However, the pressure wave does not steepen into a shock wave (Fig. 7g). Eventually the reaction front propagates at a speed greater than C–J detonation velocity, which indicates that the wave is driven by autoignition (Fig. 7h). This propagation mode is described in [5, 6] as supersonic autoignitive deflagration. It is an approximation to constant volume combustion [12] and indicates that successful SEC can be achieved without knocking even with the presence of a hot spot.

The simulation results show that dilution decreases the detonation propensity of the mixture in fact and indicates that knocking can effectively be reduced by diluting the reactants mixture with steam or exhaust gas. Although the Euler equations do not completely represent all processes relevant to combustion, they do cover the mechanism of detonation formation. Diffusion processes are generally unimportant in detonation formation except in the course of the initial formation of a detonation.

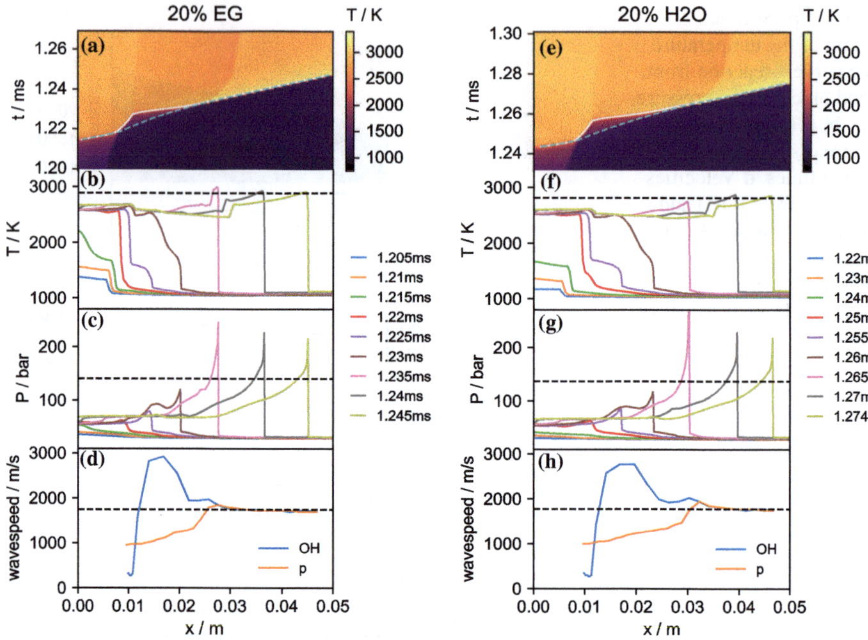

Fig. 6 **a/e** Space-time diagram of the temperature, white curve—reaction front, dashed blue curve—pressure wave. **b/f**, **c/g** Temperature and pressure distributions at different times. **d/h** Velocities of reaction front (OH) and leading pressure wave (p). Dashed lines in **b–d/f–h** C–J values of temperature, pressure and C–J detonation velocity. For 20% exhaust gas (EG) and steam (H2O) dilution

Once autoignition is taking place in a regime prone to detonation formation, the gas dynamic and reactive time scales are far shorter than those of diffusive processes. As a consequence, molecular transport does not have enough time to sizeably interfere with the ignition event. Of course, Euler simulations are limited in that they cannot accurately capture the course of events when temperature gradients are rather steep and diffusion controlled deflagrations develop, but this regime is not of interest here.

In general, dilution alters multiple gas properties that support avoiding detonation formation and knocking:

1. When diluting the reactants mixture, the volumetric energy density decreases. Less amount of energy is transferred from the reaction front into the pressure wave. For a sufficient amount of dilution only a pressure wave is produced, that does not develop into a shock. A similar observation was made in [19] by changing the initial temperature of the gas. Rudloff et al. [20] pointed out that the energy in the gas determines how severe knocking can get. This indicates that even when knocking appears in SEC, it might be less harmful for diluted fresh gas mixtures.

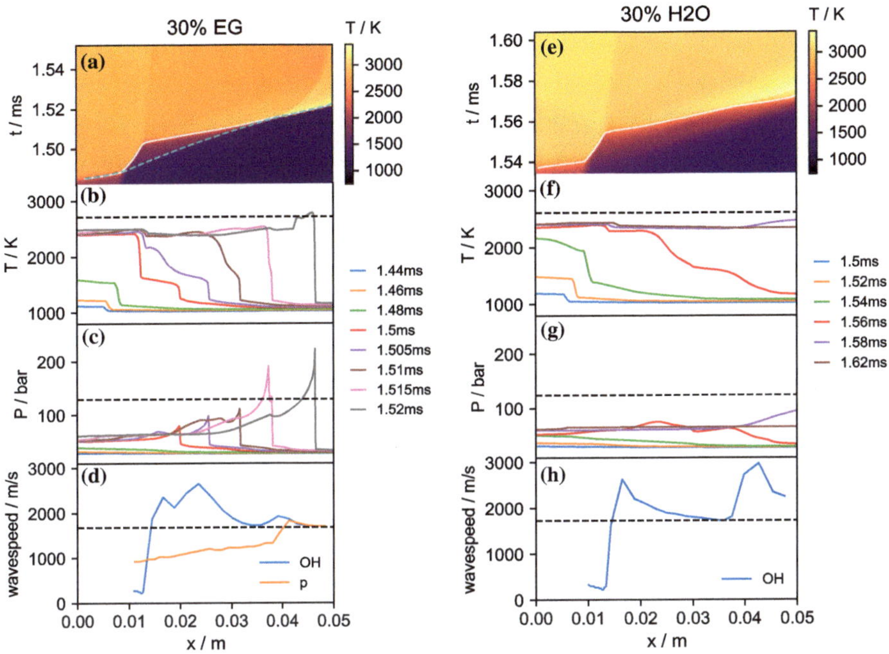

Fig. 7 **a/e** Space-time diagram of the temperature, white curve—reaction front, dashed blue curve—pressure wave. **b/f**, **c/g** Temperature and pressure distributions at different times. **d/h** Velocities of reaction front (OH) and leading pressure wave (p). Dashed lines in **b–d/f–h** C–J values of temperature, pressure and C–J detonation velocity. For 30% exhaust gas (EG) and steam (H2O) dilution

2. Dilution alters both the speed of sound in the unburned gas as well as the gradient in ignition delay $\partial\tau_i/\partial x$ (by changing $\partial\tau_i/\partial T$). Hence, the propagation speed of the autoignition and pressure wave emanating from the hot spot are different such that they may not couple. This is expressed in the critical temperature gradient (Fig. 4).
3. Dilution increases the excitation time (Fig. 2b), which decreases the rapidness of heat release into the shock. Figure 3 shows that the heat release caused by premature ignition is slow compared to the perturbation in ignition delay time when the combustible mixture is diluted. This can mitigate detonation formation.

The performed one-dimensional simulations show, that dilution of the combustible mixture with exhaust gas or steam alters the aforementioned gas properties such that detonation formation is suppressed.

Dilution changes both non-dimensional parameters in the regime diagram proposed by Gu et al. [6]. Dai et al. [19] underlined that a decrease in volumetric energy density narrows the detonation peninsula. This indicates that dilution shifts the location of the detonation peninsula and narrows it, which can benefit engine operation because operation points that are prone to detonation formation may be decimated.

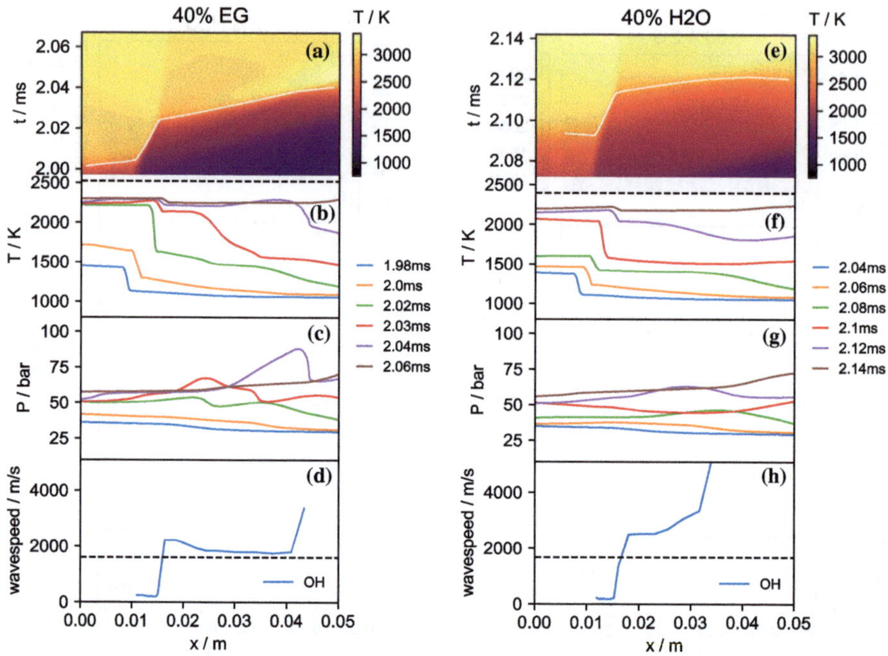

Fig. 8 **a/e** Space-time diagram of the temperature, white curve—reaction front. **b/f**, **c/g** Temperature and pressure distributions at different times. **d/h** Velocities of reaction front (OH) and leading pressure wave (p). Dashed lines in **b–d/f–h** C–J values of temperature, pressure and C–J detonation velocity. For 40% exhaust gas (EG) and steam (H2O) dilution

4 Conclusions

We investigated the influence of dilution of the combustible mixture with exhaust gas and steam on knocking in SEC. The main influences were identified by analysis of the change of mixture properties with dilution. It showed that the excitation time is increased while the volumetric energy density is decreased and the critical temperature gradient is altered, which are beneficial for the prevention of detonations in a combustion system. Simulations with the Euler equations with different amounts of dilution proved that the propensity of the mixture to detonate is decreased with increasing amount of dilution. For a hot spot with a temperature elevation of 10 K diluting with 30% steam or 40% exhaust gas is sufficient to prevent the formation of a detonation.

The results show that the approach is promising to prevent knocking. In order to support the development and implementation of SEC further, the location of the detonation peninsula in the regime diagram needs to be determined for relevant mixtures with and without dilution.

Furthermore, the influence of dilution onto the whole SEC process needs to be assessed. The SEC design has to consider that dilution increases the ignition delay time and therefore the combustion tube's length. Moreover, the impact of dilution on the efficiency needs to be determined.

A challenge for further research is that most kinetic models are not validated for dilution with exhaust gas or steam. In order to increase the confidence in the determination of the detonation peninsula as well as for the process design and control of SEC with diluted gas mixtures the experimental database needs to be extended for ignition delay times of mixtures with steam and exhaust gas dilution. If necessary, kinetic models need to be adjusted.

Nevertheless, the study shows that by influencing the excitation time through dilution it is possible to prevent knocking. Exhaust gas or steam are well suited as potential diluents because their integration into a conventional, respectively wet, gas turbine cycle is feasible.

Acknowledgements The authors gratefully acknowledge support by the Deutsche Forschungsgemeinschaft (DFG) as part of collaborative research center SFB 1029 "Substantial efficiency increase in gas turbines through direct use of coupled unsteady combustion and flow dynamics" on project A08.

References

1. Bobusch, B.C., Berndt, P., Paschereit, C.O., Klein, R.: Shockless explosion combustion: an innovative way of efficient constant volume combustion in gas turbines. Combust. Sci. Technol. **186**(10–11), 1680–1689 (2014)
2. Rähse, T.S., Paschereit, C.O., Stathopoulos, P., Berndt, P., Klein, R.: Gas dynamic simulation of shockless explosion combustion for gas turbine power cycles. In: ASME Turbo Expo 2017: Turbomachinery Technical Conference and Exposition, pp. V003T06A005–V003T06A005. American Society of Mechanical Engineers (2017)
3. Lee, J.H., Knystautas, R., Yoshikawa, N.: Photochemical initiation of gaseous detonations. Acta Astronaut. **5**(11–12), 971–982 (1978)
4. Zeldovich, Y.B., Librovich, V.B., Makhviladze, G.M., Sivashinsky, G.I.: On the development of detonation in a non-uniformly preheated gas **15**, 313–321 (1970)
5. Zeldovich, Y.B.: Regime classification of an exothermic reaction with nonuniform initial conditions. Combust. Flame **39**(2), 211–214 (1980)
6. Gu, X.J., Emerson, D.R., Bradley, D.: Modes of reaction front propagation from hot spots. Combust. Flame **133**(1), 63–74 (2003)
7. Berndt, P., Klein, R.: Modeling the kinetics of the shockless explosion combustion. Combust. Flame **175**, 16–26 (2017)
8. Bobusch, B.C., Berndt, P., Paschereit, C.O., Klein, R.: Investigation of fluidic devices for mixing enhancement for the shockless explosion combustion process. In: Active Flow and Combustion Control 2014. Springer (2015)
9. Cai, L., Pitsch, H.: Tailoring fuels for a shockless explosion combustor. In: Active Flow and Combustion Control 2014. Springer (2015)
10. Berndt, P., Klein, R., Oliver Paschereit, C.: A kinetics model for the shockless explosion combustion. In: Proceedings of the ASME Turbo Expo (2016)
11. Bobusch, B.C.: Fluidic devices for realizing the shockless explosion combustion process. Ph.D. thesis, Technische Universitt Berlin (2015)
12. Berndt, P.: Mathematical modeling of the shockless explosion combustion. Ph.D. thesis, Freie Universität Berlin (2016)

13. Berndt, P.: On the use of the hll-scheme for the simulation of the multi-species euler equations. In: Finite Volumes for Complex Applications VII-Elliptic, Parabolic and Hyperbolic Problems, pp. 809–816. Springer (2014)
14. Zhao, Z., Chaos, M., Kazakov, A., Dryer, F.L.: Thermal decomposition reaction and a comprehensive kinetic model of dimethyl ether. Int. J. Chem. Kinet. **40**(1), 1–18 (2008)
15. Goodwin, D.G., Moffat, H.K., Speth, R.L.: Cantera: an object-oriented software toolkit for chemical kinetics, thermodynamics, and transport processes (2017). http://www.cantera.org, Version 2.3.0
16. Gordon, S., McBride, B.J.: Computer program for calculation of complex chemical equilibrium compositions and applications. I: analysis. Technical report, NASA (1994). NASA Reference Publication 1311
17. Lee, J.H.S.: The Detonation Phenomenon. Cambridge University Press, The Detonation Phenomenon (2008)
18. Dai, P., Chen, Z., Chen, S., Ju, Y.: Numerical experiments on reaction front propagation in n-heptane/air mixture with temperature gradient, vol. 35, pp. 3045–3052. Elsevier (2015)
19. Dai, P., Qi, C., Chen, Z.: Effects of initial temperature on autoignition and detonation development in dimethyl ether/air mixtures with temperature gradient, vol. 36, pp. 3643–3650. Elsevier (2017)
20. Rudloff, J., Zaccardi, J.-M., Richard, S., Anderlohr, J.M.: Analysis of pre-ignition in highly charged SI engines: emphasis on the auto-ignition mode, vol. 34, pp. 2959–2967. Elsevier (2013)

Reduced Order Modeling for Multi-scale Control of Low Temperature Combustion Engines

Eugen Nuss, Dennis Ritter, Maximilian Wick, Jakob Andert, Dirk Abel and Thivaharan Albin

Abstract Internal combustion engines face tightening limits on pollutant and greenhouse gas emissions. Therefore, new solutions for clean combustion have to be found. Low Temperature Combustion is a promising technology in this regard, as it is able to reduce pollutant emissions while increasing the engine's efficiency. Recent research has shown that closed-loop control manages to stabilize the process. Nevertheless, sensitivity to varying boundary conditions and a narrow operating range remain unfavorable. To investigate new control concepts such as in-cycle feedback, computationally feasible cycle-resolved models become necessary. This work presents a low order model for Gasoline Controlled Auto Ignition (GCAI) that is continuous in time and computes the pressure trace over the entire combustion cycle. A comparison between simulation and measurement supports the suitability of the modeling approach. Furthermore, the model captures the characteristic transition of system dynamics in case GCAI during late combustion.

E. Nuss · D. Ritter · D. Abel
Institute of Automatic Control, RWTH Aachen University,
Campus-Boulevard 30, 52074 Aachen, Germany
e-mail: e.nuss@irt.rwth-aachen.de

D. Ritter
e-mail: d.ritter@irt.rwth-aachen.de

D. Abel
e-mail: d.abel@irt.rwth-aachen.de

M. Wick · J. Andert
Institute for Combustion Engines, RWTH Aachen University,
Forckenbeckstraße 4, 52074 Aachen, Germany
e-mail: Wick_m@vka.rwth-aachen.de

J. Andert
e-mail: andert@mechatronics.rwth-aachen.de

T. Albin (✉)
Institute for Dynamic Systems and Control, ETH Zürich, Sonneggstrasse 3,
8092 Zürich, Switzerland
e-mail: albinrat@ethz.ch

R. King (ed.), *Active Flow and Combustion Control 2018*,
Notes on Numerical Fluid Mechanics and Multidisciplinary Design 141,
https://doi.org/10.1007/978-3-319-98177-2_11

Keywords Gasoline controlled autoignition · Low temperature combustion
Low order HCCI modeling

1 Introduction

Combustion engines are a crucial mobile drive for transportation purposes due to the characteristics of good availability, high energy density and easy storage capability of liquid fuels. An important goal for the future is the reduction of combustion related pollutant emissions and greenhouse gases.

A promising technology to improve the combustion process is Low Temperature Combustion (LTC). LTC promises to fulfill existing requirements on power and high efficiency while simultaneously reducing pollutant emissions. Additionally, it is applicable to both diesel and gasoline engines. The latter is called Gasoline Controlled Auto Ignition (GCAI) in this paper. A common property of different LTC applications is the high degree of homogenization of the air-fuel mixture and the need of self-ignition to trigger the combustion.

While the LTC concept has significant advantages, challenges arise with respect to process stabilization. The process of LTC is substantially determined by chemical reaction kinetics which leads to a lack of the stabilizing mechanism of the mixture-controlled high-temperature combustion. As result, LTC has a high sensitivity with respect to the global and local thermodynamic state in the combustion chamber. The thermodynamic state is characterized by a high number of parameters such as temperature, chemical composition and stratification. To reduce the sensitivity to these parameters, the application of closed-loop control methods has been established.

Very roughly speaking, the general stabilization of the process has been shown in different research groups. However, there are still two aspects which cause the necessity for further development in this field. On one side, the limited operating range remains unfavourable. On the other side, stabilization of the process under all possible boundary conditions still needs to be ensured.

Conventional control techniques operate on a cycle-to-cycle basis. For each combustion cycle, a surrogate parameter is calculated which represents the combustion behaviour. This parameter is used as a controlled variable within the closed-loop controller. Typical values are the indicated mean effective pressure (IMEP), the crank-angle of 50% burned fuel mass (CA50) or the maximum pressure rise gradient (DPMAX). The cycle basis also results in updating the actuated values only once per cycle. In this way, disturbances which act on the current combustion cycle cannot be rejected.

A possibility to consider the disturbances which arise on a smaller time scale are control strategies which can be classified as multi-scale control algorithms. These algorithms allow for in-cycle cylinder control where the feedback is based on the measurements of the current cycle. First experiments to show the feasibility of this approach can be found in [1]. An important factor for the successful application are new reduced order models for the purpose of control design. In contrast to mean

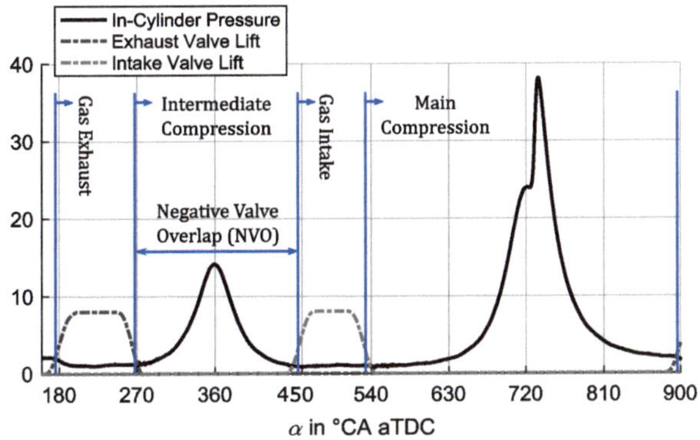

Fig. 1 Pressure trace of GCAI with exhaust gas recompression—different sections of one cycle and the Negative Valve Overlap (NVO)

value models, these need to capture the dynamics of the entire cylinder pressure and also have low order to be suitable for control design.

In this paper, an innovative modeling approach for the purpose of multi-scale control for GCAI engines is investigated. The main requirement lies in a model with low complexity that is capable of capturing the entire cylinder pressure trace. The particular process studied uses a fully variable valve train in combination with exhaust gas recompression for process control. The flexible choice of valve timings enables a symmetric collocation of exhaust valve closing and inlet valve opening around the gas exchange Top Death Center (TDC) (see Fig. 1). The resulting Negative Valve Overlap (NVO) retains an amount of the residual gas inside the cylinder and thus influences the dilution and temperature of the charge. This in turn affects the auto ignition, which is used to control the combustion timing [2].

Section 2 presents the control-oriented, time-continous GCAI model for the specified GCAI process. In Sect. 3, simulation results are compared to experimental data of a GCAI test bed. Finally, Sect. 4 concludes the paper with the summary and drawn conclusions.

2 Gasoline Controlled Auto-Ignition Model

Recent research has shown that a cycle-to-cycle control approach generally enables GCAI operation [3, 4]. Multi-scale control methods hold promise for further improvement in terms of process stabilization and extension of the operating range. But these methods require control oriented system models that reproduce the in-cycle behavior. Despite the underlying nonlinear process characteristics, the model's complexity must be limited to be suitable for control synthesis. This section introduces a first principle based zero-dimensional modeling approach for GCAI.

Events that only last few degrees of crank-angle (CA) relative to the entire engine cycle have time constants that are small compared to other processes within the engine cycle. Therefore, the exhaust valve opening (EVO), exhaust valve closing (EVC),

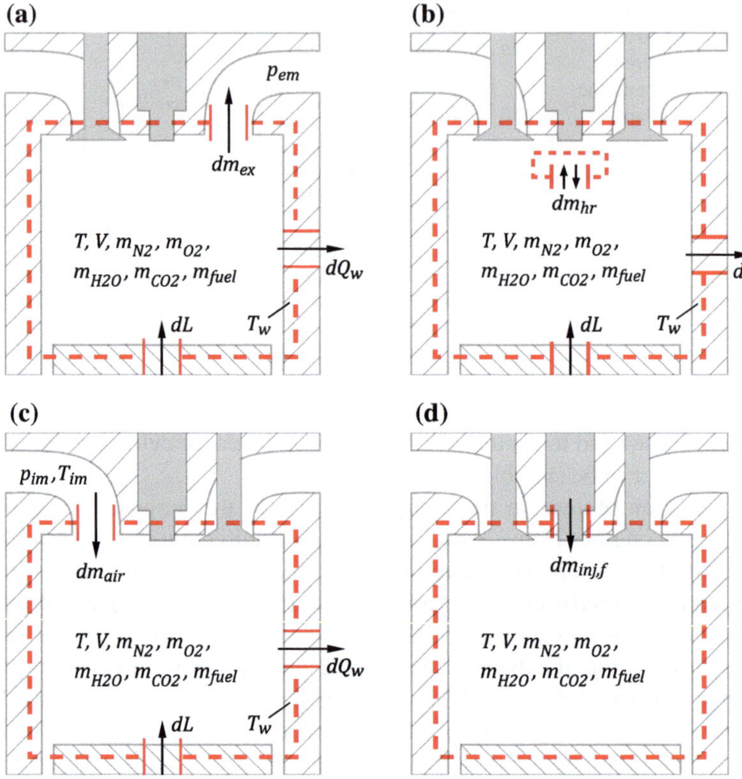

Fig. 2 GCAI stages. **a** Gas exhaust (ex). **b** Intermediate or main compression (ic/mc). **c** Gas intake (in). **d** Fuel injection (inj, f)

inlet valve opening (IVO), inlet valve closing (IVC) as well as the fuel injection (inj, f) are modeled as zero-time events. In case of the valve events, opening or closing is assumed when 10% of the maximum valve lift is reached. The resulting discrete valve timings divide the engine cycle into the four seperate stages gas exhaust (ex), intermediate compression (ic), gas intake (in) and main compression (mc). Figure 2a–c show the distinct system dynamics of each stage according to the current valve positions. The fuel's heat release can take place during both, intermediate and main compression. Despite being included in schematic (b), the combustion event is also modeled as a discrete time event and addressed specifically later in this section. Contrary to the valve timings, the fuel injection does not separate two different system dynamics, but causes a discontinuous jump in the system states. The underlying physically motivated equations of this impulsive event are based on Fig. 2d.

The resulting multi-stage model with eight states and six control inputs is:

$$
\begin{aligned}
\frac{dx}{dt} &= f_i(x) \qquad\qquad i = ex, ic, in, mc \\
x(t_{0,i}) &= x(t_{end,i-1}) \\
y &= h(x) \\
x &= \left(T,\ \alpha,\ m_{N_2},\ m_{O_2},\ m_{H_2O},\ m_{CO_2},\ m_{fuel},\ T_w\right)^T \\
u &= \left(\alpha_{evo},\ \alpha_{evc},\ \alpha_{ivo},\ \alpha_{ivc},\ \alpha_{soi},\ m_{inj,f}\right)^T \\
y &= p
\end{aligned}
\tag{1}
$$

where x is the state, u the input and y the output vector.

The engine model repetitively develops through the continuous state spaces $f_i(x)$ of each introduced stage in the defined order. The valve timings actuate as external switches between the stages. When an external switch is triggered, the state of the current stage serves as the initial condition for following one.

According to Fig. 2, the state vector x of all functions $f_i(x)$ consists of the in-cylinder temperature T, the crank angle α, the masses of the five species $m_{N_2}, m_{O_2}, m_{H_2O}, m_{CO_2}$ and m_{fuel} as well as the temperature T_w of the walls surrounding the charge. Throughout the work, iso-octane C_8H_{18} is assumed as fuel. The valve timings are expressed through the crank-angle at which the valve opening or closing takes place, stated in ° CA after Top Death Center (aTDC). Further control inputs are the crank-angle at start of fuel injection α_{soi} and the injected fuel mass $m_{inj,f}$. Due to modeling the valve timings as external switches and the fuel injection as a discrete jump, none of the control inputs directly appears in $f_i(x)$. The only model output is the in-cylinder pressure p. The following subsections present the physical and phenomenological equations the model is based on.

2.1 Energy Balance

The general energy balance covering all cases of Fig. 2 is used in the form:

$$
\mathrm{d}U = \mathrm{d}Q_w + \mathrm{d}L + h_{in} \cdot \mathrm{d}m_{in} + h \cdot \mathrm{d}m_{out} \tag{2}
$$

First, Eq. 2 is further specified by considering only compression work and assuming ideal gases. Second, the terms including the differential of specific internal energy $\mathrm{d}u$ and species $\mathrm{d}Y_k$ are combined in order to achieve a direct dependence on the species' masses and eliminate the differential of the internal energy.

$$
\mathrm{d}U = \mathrm{d}Q_w - \frac{m \cdot R \cdot T}{V}\mathrm{d}V + h_{in} \cdot \mathrm{d}m_{in} + h \cdot \mathrm{d}m_{out} \tag{3}
$$

$$\mathrm{d}U = m \cdot \mathrm{d}u + m \cdot c_v \cdot \mathrm{d}T + \sum_k u_k \cdot \mathrm{d}Y_k \tag{4}$$

$$= m\frac{R}{\kappa - 1}\mathrm{d}T + \sum_k u_k \cdot \mathrm{d}m_k \tag{5}$$

Solving for $\mathrm{d}T$ and deriving with respect to time yields the general differential equation for the in-cylinder temperature:

$$\Rightarrow \dot{T} = \frac{\kappa - 1}{m \cdot R}\left(\dot{Q}_w - \frac{m \cdot R \cdot T}{V} \cdot \dot{V} + h_{in} \cdot \dot{m}_{in} + h \cdot \dot{m}_{out} - \sum_k u_k \cdot \dot{m}_k\right) \tag{6}$$

Applying the simplifications for the stages according to Fig. 2 results in:

$$\dot{T}_{ex} = \frac{\kappa - 1}{m \cdot R}\left(\dot{Q}_w - \frac{m \cdot R \cdot T}{V} \cdot \dot{V} + h \cdot \dot{m}_{ex} - \sum_k u_k \cdot \dot{m}_k\right) \tag{7}$$

$$\dot{T}_{ic} = \dot{T}_{mc} = \frac{\kappa - 1}{m \cdot R}\left(\dot{Q}_w - \frac{m \cdot R \cdot T}{V} \cdot \dot{V}\right) \tag{8}$$

$$\dot{T}_{in} = \frac{\kappa - 1}{m \cdot R}\left(\dot{Q}_w - \frac{m \cdot R \cdot T}{V} \cdot \dot{V} + h_{air} \cdot \dot{m}_{air} - \sum_k u_k \cdot \dot{m}_k\right) \tag{9}$$

As the combustion event is treated seperately, Eq. 6 reduces to Eq. 8 for the intermediate and main compression.

2.2 *Mass Balance*

The change in the total charge mass during the exhaust and intake stroke, respectively, is evaluated by the valve flow equations [5]:

$$\dot{m}_{ex} = A_{s,ex} \cdot \sqrt{\frac{2 \cdot p}{R \cdot T}} \cdot \sqrt{\frac{\kappa}{\kappa - 1}\left(\left(\frac{p_{em}}{p}\right)^{\frac{2}{\kappa}} - \left(\frac{p_{em}}{p}\right)^{\frac{\kappa+1}{\kappa}}\right)} \tag{10}$$

$$\dot{m}_{in} = A_{s,in} \cdot \sqrt{\frac{2 \cdot p_{im}}{R_{air} \cdot T_{im}}} \cdot \sqrt{\frac{\kappa_{air}}{\kappa_{air} - 1} \cdot \left(\left(\frac{p}{p_{im}}\right)^{\frac{2}{\kappa_{air}}} - \left(\frac{p}{p_{im}}\right)^{\frac{\kappa_{air}+1}{\kappa_{air}}}\right)} \tag{11}$$

where A_s represents the valve specific isentropic flow cross-sectional area and the in-cylinder pressure p is determined through the ideal gas law. Depending on the stage, either fresh air is inducted into the cylinder or the in-cylinder gases flow into the exhaust manifold.

2.3 Species Balance

During the exhaust stroke, the mass fractions remain constant as the charge is assumed to be a homogeneous mixture. Using Eq. 10, the following relation holds for all species during the exhaust stroke:

$$\dot{m}_{k,ex} = Y_k \cdot \dot{m}_{ex}, \qquad k = N_2,\ O_2,\ H_2O,\ CO_2,\ fuel \tag{12}$$

During the gas intake phase, only air with a constant mass distribution is inducted. Consequently, only the masses of N_2 and O_2 change according to:

$$\dot{m}_{N2,in} = Y_{N2,air} \cdot \dot{m}_{air} \tag{13}$$

$$\dot{m}_{O2,in} = Y_{O2,air} \cdot \dot{m}_{air} \tag{14}$$

2.4 Heat Release and Fuel Injection

Heat release during main and intermediate combustion as well as the fuel injection are modeled as discrete jumps in the state vector x. For these zero-time events, the piston position does not change and wall heat losses are neglected. Starting from Eq. 6, the discrete step in the in-cylinder temperature due to heat release is:

$$\mathrm{d}T_{hr} = \frac{1-\kappa}{m \cdot R} \cdot \sum_k u_k \cdot \mathrm{d}m_k \tag{15}$$

The global combustion reaction for iso-octane establishes a relation between amount of burned fuel $m_{f,hr}$ and the modeled species. In order to match the models units, the reaction equation is transformed with the respective mole masses.

$$C_8H_{18} + O_2 \rightarrow CO_2 + H_2O \tag{16}$$

$$\Rightarrow c_{fuel} \cdot [C_8H_{18}]_{kg} + c_{O2} \cdot [O_2]_{kg} \rightarrow c_{CO2} \cdot [CO_2]_{kg} + c_{H2O} \cdot [H_2O]_{kg} \tag{17}$$

The introduced constants are listed in Table 1. The change in the species can now be stated in dependence of the burned fuel mass:

$$\mathrm{d}m_k = c_k \cdot m_{f,hr}, \qquad k = O_2,\ H_2O,\ CO_2,\ fuel \tag{18}$$

A difference between intermediate and main combustion lies in determining the amount of burned fuel. As for the intermediate compression, the amount of burned fuel is assumed proportional to the remaining fuel in the cylinder. The introduced

Table 1 Model parameters

Param.	Value	Unit	Eq	Param.	Value	Unit	Eq
h_{air}	2.5313e + 04	$J.kg^{-1}$	9	β_3	−2.5139e + 02	–	19
$A_{s,ex}$	1.5000e − 04	m^2	10	β_4	4.0000e − 01	–	20
p_{em}	1.0000e + 05	Pa	10	γ_0	5.4790e − 01	rad	21
$A_{s,in}$	2.0000e − 04	m^2	11	γ_1	1.9696e + 05	$Pa^{-\gamma_2}$	21
p_{im}	1.0000e + 05	Pa	11	γ_2	7.0670e −01	–	21
T_{im}	3.2315e + 02	K	11	γ_3	6.5131e + 03	K	21
κ_{air}	1.3980e + 00	–	11	γ_4	1.3270e + 00	–	22
R_{air}	2.8818e + 02	$J.kg^{-1}K^{-1}$	11	γ_5	−4.0457e + 00	rad	22
$Y_{N_2,air}$	7.6700e − 01	–	13	$h_{evap,f}$	2.6954e + 05	$J\,kg^{-1}$	23
$Y_{O_2,air}$	2.3300e − 01	–	14	$h_{inj,f}$	−1.9142e + 06	$J.kg^{-1}$	23
c_{fuel}	−1.0000e + 00	–	17	D	8.4000e − 02	m	24
c_{O2}	−3.5088e + 00	–	17	T_{w1}	1.2000e − 01	s	25
c_{CO2}	3.0877e + 00	–	17	δ_0	3.5000e + 02	K	25
c_{H2O}	1.4211e + 00	–	17	V_c	5.2501e − 05	m^3	32
β_0	9.5000e − 01	–	19	r_1	4.5000e − 02	m	32
β_1	1.9757e + 01	rad^{-1}	19	r_2	1.5900e − 01	m	32
β_2	−3.9793e + 05	kg^{-1}	19	*fuel*	C_8H_{18}	–	–

proportional constant β_4 is eventually tuned to fit the model to measurement data. For the main compression on the other hand, a varying combustion efficiency η_{hr} is introduced, which is initially presented in [6]. These incomplete combustions lead to a dynamic coupling between consecutive cycles.

$$m_{f,hr,mc} = \beta_0 \cdot \left(1 + e^{\beta_1 \cdot CA50 + \beta_2 \cdot m_{inj,f} + \beta_3}\right)^{-1} \cdot m_{fuel} \tag{19}$$

$$m_{f,hr,ic} = \beta_4 \cdot m_{fuel} \tag{20}$$

Furthermore, it is necessary to determine the instant at which the combustion occurs. During main compression, the center of combustion *CA50* is used and identified by applying a semi-empirical Arrhenius rate threshold approach [7]. The integral

$$\gamma_0 = \int_{\alpha_{ivc}}^{\alpha_{soc}} \gamma_1 \cdot p^{\gamma_2} \cdot e^{-\frac{\gamma_3}{T}} d\alpha \tag{21}$$

is numerically integrated alongside the state space model until it reaches the prescribed threshold γ_0. The crank-angle at this instance represents the start of combustion. Linear extrapolation yields

$$CA50 = \gamma_4 \cdot \alpha_{soc} + \gamma_5 \tag{22}$$

Analogous to [6], the heat release during the intermediate compression is modeled in the beginning of the stage.

The only change in the species' masses caused by fuel injection trivially is $\mathrm{d}m_{fuel} = m_{inj,f}$. The corresponding change in the in-cylinder temperature is:

$$\mathrm{d}T_{inj,f} = \frac{\kappa - 1}{m \cdot R} \left(h_{inj,f} - h_{evap,f} - u_{inj,f} \right) \cdot m_{inj,f} \tag{23}$$

2.5 Wall Heat Loss

The wall heat losses $\dot{Q}_w$ are calculated by using the Hohenberg correlation [8] for estimating the heat transfer coefficient α_w and the ideal gas law for the in-cylinder pressure p.

$$\begin{aligned} \dot{Q}_{wall} &= \alpha_w \cdot A_w \cdot (T_w - T) \\ \alpha_w &= 130 \cdot V^{-0.06} \cdot \left(\frac{p}{10^5} \right)^{0.8} \cdot T^{-0.4} \cdot \left(\frac{4 \cdot \dot{V}}{\pi \cdot D^2} + 1.4 \right)^{0.8} \\ A_w &= \frac{4 \cdot V}{D} + \frac{\pi \cdot D^2}{2} \end{aligned} \tag{24}$$

The dynamics of the wall temperature T_w are approximated by a first order lag element with the in-cylinder temperature T as the input:

$$\dot{T}_w = \frac{1}{T_{w,1}} (T - T_w - \delta_0) \tag{25}$$

The time constant $T_{w,1}$ is initially chosen so that a step response in reaches a stationary level after three cycles. Additionally a constant cooling factor δ_0 is introduced and determined by tuning the paramater to fit measurement data.

2.6 Auxilary Equations

The derived set of equations still depends on a set of parameters that can partly be expressed through functional relations. The component's specific heat capacity $c_{p,k}$ and enthalpy h_k are expressed through the in-cylinder temperature T by using NASA polynomials, as:

$$
\begin{aligned}
c_{p,k} &= \left(a_1 + a_2 \cdot T + a_3 \cdot T^2 + a_4 \cdot T^3 + a_5 \cdot T^4\right) \frac{\overline{R}}{\overline{M}_k} \\
h_k &= \left(a_1 \cdot T + a_2 \cdot T^2 + a_3 \cdot T^3 + a_4 \cdot T^4 + a_5 \cdot T^5 + a_6\right) \frac{\overline{R}}{\overline{M}_k}
\end{aligned}
\tag{26}
$$

Given $c_{p,k}$ and h_k, the specific enthalpy h, the isentropic exponent κ, the specific gas constant R and the internal energy u_k of each component k are determined with:

$$h = \sum_k Y_k \cdot h_k \tag{27}$$

$$c_p = \sum_k Y_k \cdot c_{p,k} \tag{28}$$

$$R = \sum_k Y_k \cdot \frac{\overline{R}}{\overline{M_k}} \tag{29}$$

$$\kappa = \frac{c_p}{c_p - R} \tag{30}$$

$$u_k = h_k - \frac{\overline{R}}{\overline{M}_k} T \tag{31}$$

As the state vector contains the mass of each modeled component, Y_k can always be directly computed.

The current cylinder volume V and its time derivative $\dot{V}$ are gained through the slider-crank formula [5] with the bore diameter D, the volume at top death center V_c and the two rod lengths r_1 and r_2 ($\lambda = \frac{r_2}{r_1}$):

$$
\begin{aligned}
V &= V_c + \frac{\pi \cdot D^2}{4} \left(r_1 \cdot \left(1 - \sqrt{1 - \lambda^2 \cdot \sin(\alpha)^2}\right) + r_2 \cdot (1 - \cos(\alpha)) \right) \\
\dot{V} &= \frac{\pi \cdot D^2}{4} \left(r_1 \cdot \frac{\lambda^2 \cdot \cos(\alpha)}{\sqrt{1 - \lambda^2 \cdot sin(a)^2}} + r_2 \right) \cdot \omega \cdot \sin(\alpha)
\end{aligned}
\tag{32}
$$

2.7 Model Summary

The complete set of state space functions is as follows:

$$
\begin{aligned}
f_{ex}(x) = \big(&\dot{T}_{ex}\,,\ \omega,\ \dot{m}_{ex} \cdot Y_{N_2},\ \dot{m}_{ex} \cdot Y_{O_2},\ \dot{m}_{ex} \cdot Y_{H_2O},\ \ldots \\
&\dot{m}_{ex} \cdot Y_{CO_2},\ \tfrac{1}{T_{w,1}} \left(T - T_w - \delta_0\right) \big)^T
\end{aligned}
\tag{33}
$$

$$f_{ic}(x) = \left(\dot{T}_{ic},\ \omega,\ 0,\ 0,\ 0,\ 0,\ 0,\ \tfrac{1}{T_{w,1}} \left(T - T_w - \delta_0\right) \right)^T \tag{34}$$

$$f_{in}(x) = \Big(\dot{T}_{in},\ \omega,\ \dot{m}_{in} \cdot Y_{N_2,air},\ \dot{m}_{in} \cdot Y_{O_2,air},\ \ldots \\ 0,\ 0,\ 0,\ \tfrac{1}{T_{w,1}}\left(T - T_w - \delta_0\right)^T\Big) \tag{35}$$

$$f_{mc}(x) = \left(\dot{T}_{mc},\ \omega,\ 0,\ 0,\ 0,\ 0,\ 0,\ \tfrac{1}{T_{w,1}}\left(T - T_w - \delta_0\right)\right)^T \tag{36}$$

The engine speed n is substituted by ω in order to emphasize the use of SI-units in the model. The discrete state jumps due to fuel injection and combustion are:

$$\Delta x_{hr} = \big(\mathrm{d}T_{hr},\ 0,\ c_{N2} \cdot m_{f,hr,i},\ c_{O2} \cdot m_{f,hr,i},\ c_{H2O} \cdot m_{f,hr,i},\ \ldots \\ c_{CO2} \cdot m_{f,hr,i},\ c_{C8H18} \cdot m_{f,hr,i},\ 0\big)^T \qquad i = ic,\ mc \tag{37}$$

$$\Delta x_{inj,f} = \left(\mathrm{d}T_{inj,f},\ 0,\ 0,\ 0,\ 0,\ 0,\ m_{inj,f},\ 0\right)^T \tag{38}$$

Figure 3 illustrates the implementation of the hybrid multi-stage model. Once the crank-angle α crosses a control input in form of a valve opening or closing, the subsequent state space function becomes active and uses the current state vector as an initial condition. The conversion of remaining fuel during intermediate compression is modeled through a discrete combustion event in the beginning of the phase. With regard to implementation, this results in a manipulation of the respective initial condition. Fuel injection also takes place during the intermediate compression and adds an offset to the state vector at the start of injection α_{soi}. The integrator is reset with the updated state during this step. The main combustion is triggered through the Arrhenius ignition delay model and produces a step in the state vector at the center of combustion *CA50*. In contrast to the intermediate combustion the amount of burned fuel is not constant but varies according to the empirical combustion efficiency. All parameters of the model are listed in Table 1.

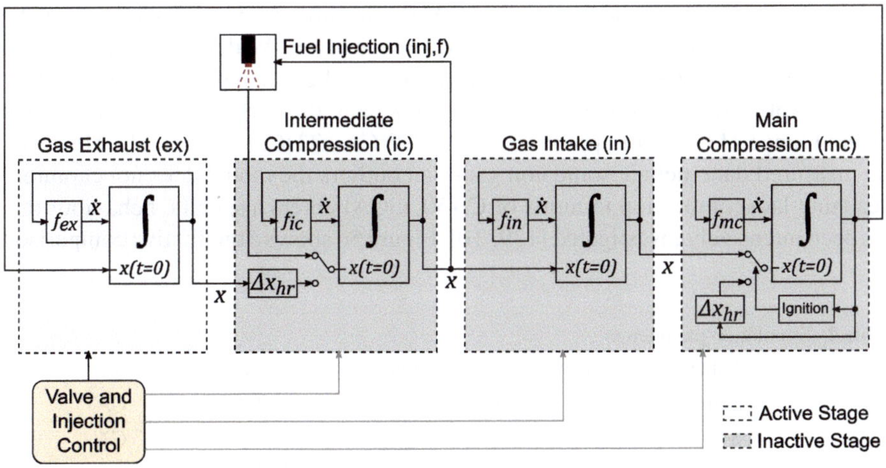

Fig. 3 Schematic of the continuous-cycle GCAI model—shown during the exhaust stroke

3 Simulation Results

In this section, the simulated output of the model presented in Sect. 2 is compared to measurement data. In the experimental setup, 14 different NVO settings are recorded at an engine speed of 1500 rpm. Each of the 14 measurement points contains 200 consecutive cycles, excluding the transition from one measurement point to the other. Regarding the system inputs, EVC and IVO are set symmetrically to the gas exchange TDC and are adapted to achieve an NVO sweep from 190° − 183° CA. EVO, IVC and the injected fuel mass remain constant. For reasons of better homogenization, the start of fuel injection α_{soi} is referenced to the intake valve opening event.

The discrete valve timings necessary for simulation are extracted from the valve lifts of each measured cycle. Also in the simulation, each measurement point is held for 200 cycles. The simulation step size is set to achieve a crank-angle discretization $\Delta\alpha_{sim}$ of 0.1° CA. Table 2 lists the set points for the valve timings and the remaining parameters useds for simulation.

Figure 4 compares the simulated and the measured pressure trace of an arbitrary cycle for three different measurement points. As the experimental raw data is recorded with a continuous advancement in the crank-angle, the simulation results are adapted and plotted over steadily increasing ° CA instead of ° CA *a*TDC which is reset each 720°.

In general, the model achieves good resemblance with regard to the measured pressure trace. The intermediate compression is fit well in all three cases. Considering the main compression, the decrease starting from the point of maximum pressure is approximated well also. The maximum pressure peak, however, is overestimated in case (a), while the simulated CA50 is inaccurate in all three cases, varying up to 5.5° CA.

As it is unpractical to compare the large number of cycles in detail, the center of combustion CA50 of all measured and simulated cycles is shown in the return map in Fig. 5a. In case of the model, CA50 can directly be extracted while the measurement data was processed by heat release analysis. With reducing NVO, also the temperature of the cylinder charge decreases and consequently the combustion center CA50 shifts towards later combustion. Again, the overall fit, particularly the covered range of CA50 of approximately 4–12° CA *a*TDC, is acceptable. However, the measured data include some non gaussian outliers the model does not capture. Reaching late combustion in terms of CA50, GCAI enters a specific behaviour that has been intensively investigated [1, 9, 10]. Figure 5b shows a qualitative comparison

Table 2 Simulation parameters

Param.	Value	Unit	Param.	Value	Unit
α_{evo}	160	° CA *a*TDC	NVO	190–183	° CA
α_{ivc}	545	° CA *a*TDC	$\Delta\alpha_{sim}$	0.1	° CA
α_{soi}	455–448.5	° CA *a*TDC	n_{engine}	15000	rpm

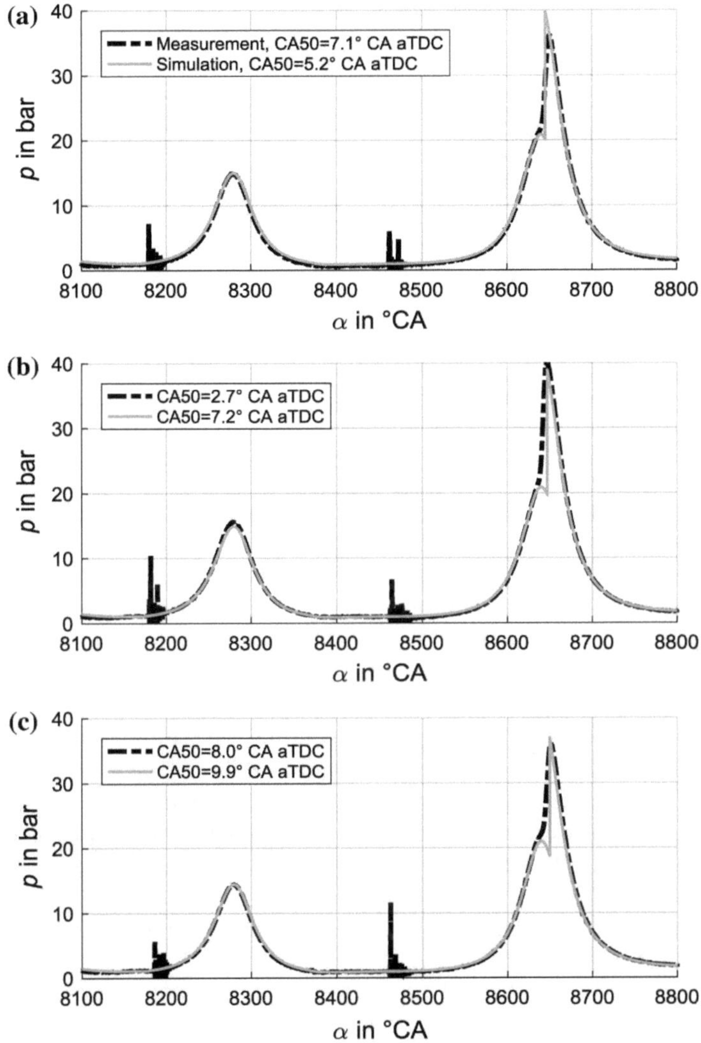

Fig. 4 Simulated versus measured pressure trace. **a** Measurement point 1. **b** Measurement point 2. **c** Measurement point 3

between an open loop simulation of the presented model and another experimental GCAI setup. In the simulation, NVO is further decreased to 170° CA and a disturbance in form of white noise is added to the initial condition of the in-cylinder temperature at inlet valve closing. Again, α_{soi} is referenced to IVO and the remaining control inputs stay as before. The simulation outcome of this setup shows a clear correspondence to the experimental data. In general, the characteristic change in the dynamic behaviour can thus be reproduced by the derived model. Further research is

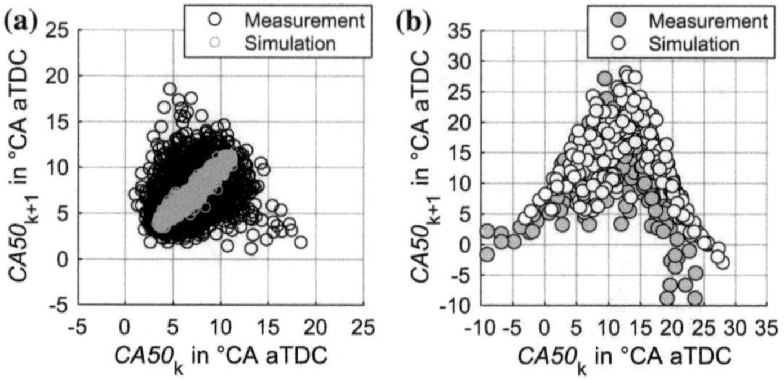

Fig. 5 Return map of center of combustion CA50. **a** Simulated against measured data with measured input. **b** Qualitative comparison between simulation and measurement for small NVO and late combustion

necessary to adapt the moment of transition to the same point as for the experimental data in Fig. 5a.

The averaged computation time for simulation of one cycle on an Intel i7 processor using Simulink's ode3 (Bogacki-Shampine) is 250 ms. Consequently, admissible simplifications must be found for the model to be suitable for control synthesis. In case of model predictive control, which has shown to have favourable properties for the present control problem [11], the model is used in context of optimization and consequently needs to be as simple as possible. Furthermore, the prediction of the ignition delay has to be improved, as there is a discrepancy with regard to the values determined by heat release analysis. As the overall pressure trace shows good correspondence, however, the continuous-cycle model allows using other metrics than CA50 for controlling the process. If the onset of the cyclic variations seen in Fig. 5b is to be avoided by the controller, a proper tuning of the parameters needs to be conducted, so the predicted change in system dynamics occurs at the same conditions as at the test bed.

4 Conclusions

The paper presents a modeling approach for GCAI with the purpose of multi-scale control synthesis. In this particular case, it is required to model the process time-resolved, instead of the cycle-to-cycle resolution of state-of-the-art mean-value models. The paper provides a physically motivated, control-oriented GCAI model which is continuous in time. A comparison between the simulated model output and the continuous pressure traces of a selected set of measurement data verifies the general assumptions made. The in-cycle resolution of the combustion cycle allows for new control approaches such as in-cycle control and therefore motivates further research. In particular, the ignition delay model must be improved as the prediction of CA50 is not accurate. In addition, further work needs to be conducted concerning the reduction of complexity. Despite the comparably low order, the simulation time of 250 ms for an engine cycle is excessively high. Finally, the model parameters need to be identified and the covered operating range of model assessed.

Acknowledgements The research was performed as part of the Research Unit (Forschergruppe) FOR2401 "Optimization based Multiscale Control for Low Temperature Combustion Engines", which is funded by the German Research Association (Deutsche Forschungsgemeinschaft, DFG). The support is gratefully acknowledged.

References

1. Lehrheuer, B., Pischinger, S., Wick, M., Andert, J., Berneck, D., Ritter, D., Albin, T., Thewes, M.: A study on in-cycle combustion control for gasoline controlled autoignition. In: SAE Technical Paper 2016-01-0754 (2016)
2. Yao, M., Zheng, Z., Liu, H.: Progress and recent trends in homogeneous charge compression ignition (HCCI) engines. Progress Energy Combust. Sci. **35** (2009)
3. Ebrahimi, K., Koch, C.: Model predictive control for combustion timing and load control in HCCI engines. In: SAE Technical Paper 2015-01-0822 (2015)
4. Jade, S., Hellstöm, E., Larimore, J., Stefanopoulou, A.G., Jiang, L.: Reference governor for load control in a multicylinder recompression HCCI engine. IEEE Trans. Control Systems Technol. **22**(4) (2014)
5. Shaver, G.M., Gerdes, J.C., Jain, P., Caton, P.A., Edwards, C.F.: Modeling for control of HCCI engines. In: Proceedings of the American Control Conference (2003)
6. Jade, S., Larimore, J., Hellstöm, E., Stefanopoulou, A.G.: Controlled load and speed transitions in a multicylinder recompression HCCI engine. IEEE Trans. Control Syst. Technol. **23**(3) (2015)
7. Jade, S., Larimore, J., Hellstöm, E., Jiang, L., Stefanopoulou, A. G.: Enabling large load transitions on multicylinder recompression HCCI engines using fuel governors. In: American Control Conference (ACC), Washington, USA (2013)
8. Hohenberg, G.F.: Advanced approaches for heat transfer calculations. In: 1979 SAE International Off-Highway and Powerplant Congress and Exposition (1979)
9. Shahbakhti, M., Koch, C.R.: Characterizing the cyclic variability of ignition timing in a homogeneous charge compression ignition engine fuelled with n-heptane/iso-octane blend fuels. J. Int. Engine Res. **9** (2008)
10. Ritter, D., Andert, J., Abel, D., Albin, T.: Model-based control of gasoline-controlled autoignition. Int. J. Engine Res. **19** (2018)
11. Albin, A., Zweigel, R., Hesseler, F.: A 2-stage MPC approach for the cycle-to-cycle dynamics of GCAI (Gasoline Controlled Auto Ignition). In: 7th IFAC Symposium on Advances in Automotive Control (2013)

Part III
Constant Volume Combustion

The Influence of the Initial Temperature on DDT Characteristics in a Valveless PDC

Fabian E. Völzke, Fatma C. Yücel, Joshua A. T. Gray, Niclas Hanraths, Christian O. Paschereit and Jonas P. Moeck

Abstract The detonation velocity and the detonation cell width are determined experimentally as a function of the initial mixing temperature in a valveless pulse detonation combustor (PDC). The initial temperature was varied from 290 K up to 650 K. To shorten the run–up distance to the deflagration–to–detonation transition (DDT), the detonation tube was equipped with six orifice plates which support the flame acceleration. Ionization probes are used to record the combustion event at several axial positions. Sooted foils inside the downstream section of the detonation tube are used to record the imprint of the detonation front and to determine the detonation cell width. It was found that the propagation speed of the detonation front decreases with increasing mixing temperature, which agrees with the theoretical temperature dependence of the CJ–velocity. The detonation cell width decreases linearly for elevated initial temperatures.

Keywords PDC · Detonation tube · Detonation cell width · Detonation velocity
Temperature variation · Flame acceleration

F. E. Völzke (✉) · F. C. Yücel · J. A. T. Gray · N. Hanraths · C. O. Paschereit · J. P. Moeck
Technische Universität Berlin, Institut für Strömungsmechanik und Technische Akustik,
Müller-Breslau-Street 8, 10623 Berlin, Germany
e-mail: fabian.voelzke@tu-berlin.de

F. C. Yücel
e-mail: f.yuecel@tu-berlin.de

J. A. T. Gray
e-mail: joshua.gray@tu-berlin.de

N. Hanraths
e-mail: niclas.hanraths@tu-berlin.de

C. O. Paschereit
e-mail: oliver.paschereit@tu-berlin.de

J. P. Moeck
e-mail: jonas.moeck@tu-berlin.de

R. King (ed.), *Active Flow and Combustion Control 2018*,
Notes on Numerical Fluid Mechanics and Multidisciplinary Design 141,
https://doi.org/10.1007/978-3-319-98177-2_12

1 Introduction

Pressure gain combustion is a promising way of achieving a remarkable increase in the thermal efficiency of gas turbines [1]. Although there are several approaches of realizing an approximate constant volume combustion (aCVC), e.g., wave rotor [2], shockless explosion combustor (SEC) [3], and rotational detonation combustor (RDC) [4], the pulse detonation combustor (PDC) [5] is the most investigated design. Nevertheless, the design of a reliable and efficient PDC is not a straightforward task. Many parameters have a notable influence on the behavior of a PDC, among them are the detonability of the mixture, the detonation cell width, the initial conditions, and the geometry.

The effect of the initial temperature on the deflagration-to-detonation transition (DDT) is one of the least studied aspects. Only few and partially contradictory experimental and numerical data related to this topic have been published. As a characteristic length scale of a detonation wave, the detonation cell width λ is a meaningful parameter when predicting the operation of a PDC. However, the influence of the temperature of hydrogen–air mixtures on the cell width is only rarely investigated. Ciccarelli et al. [6] found that the cell width for a stoichiometric mixture decreases from $\lambda \approx 9\,\text{mm}$ to $\lambda \approx 5\,\text{mm}$ by increasing the initial temperature from $T = 300\,\text{K}$ to $T = 650\,\text{K}$ at a constant initial pressure of 1 bar. However, no data was found for initial temperatures in between the named temperature range. A one-dimensional numerical study, performed by Djordjevic et al. [7] suggests a non-linear dependence in which detonation cell width takes on a minimum value for an initial mixing temperature of about 320 K. For a further increase in temperature, the cell width calculated by Djordjevic et al. increases. As shown in [7], these contradicting results regarding the dependence of the cell width on the initial temperature are caused by two competing effects of increased initial temperature. The post-shock temperature rises while the von-Neumann pressure decreases due to the increased speed of sound in the unburned mixture causing a decrease in the Mach number of the detonation wave. The influence of these state variables on the cell width is still not fully understood and so the influence of the temperature on the resulting detonation cell width cannot be predicted with certainty.

Due to the few available data and the difficulty of prediction, the detonation velocity and the detonation cell width are determined in this work for an initial temperature from 290 K to 760 K. The measurements are carried out in a valveless PDC test rig that is run in single-cycle operation, using it effectively as a single shot detonation tube. Thermocouples and ionization probes are used to identify the initial temperature and the propagation speed of the detonation front, respectively. Sooted foils are used to record the imprint of the triple points of the detonation wave and to determine the detonation cell width.

2 Experimental Setup

2.1 PDC Test Rig

A valveless PDC (Fig. 1) has been designed to enable a wide spectrum of investigations on pulse detonation combustion.

The detonation tube has a total length of $L = 2.2\,\text{m}$ and an inner diameter of $D = 40\,\text{mm}$. The main air flow is sent through the preheater, where it can be heated up to 1100 K. The air mass flow is set to a constant value of 80 kg/h using a closed–loop control including a Coriolis mass flow meter (Endress+Hauser Proline Promass 83 F) and an electronic control valve (Bürkert 8630). Due to disturbances in the supply and the inertance of the control, there is a fluctuation of the air mass flow of ± 5 kg/h. Four solenoid valves (Bosch 0 280 158 827) are used for the control of the fuel mass flow, which was measured by a Coriolis mass flow meter (Endress + Hauser Cubemass DCI). The pressure upstream of the solenoid valves is set manually in order to match the mass flow of 2.34 kg/h when all valves were open, which corresponds to an equivalence ratio of $\Phi = 1$. The measurement error of the fuel mass flow due to the accuracy of the Coriolis mass flow meter and pressure fluctuations in the supply line is assumed to be ± 0.05 kg/h. These uncertainties in the mass flow of air and fuel result in an actual equivalence ratio of $0.91 < \Phi < 1.09$. Ciccarelli et al. [6] and Djordjevic et al. [7] independently found that the deviation of the cell width over this range of Φ is negligible. Due to this, the results are expected to represent the dependence of the detonation cell width on the initial temperature reliably. The applied mass flow rates result in a filling of the entire detonation tube in about 100 ms. Nevertheless, the solenoid valves are opened for 2 s before the mixture is ignited. When the solenoid valves are opened, the supply pressure drops until it stabilizes at a certain level. This pressure level is a function of the geometry of the supply line and the installed components. The variation of the upstream pressure induces a change in the pressure drop along the valves, resulting in an unsteady fuel mass flow. The filling time of 2 s is chosen to minimize the transient effects during the initial phase after opening the valves on the mixture in the detonation tube right before ignition.

Two pressure transmitters (FESTO SPTW) were used to record the static pressure in the fuel line upstream and downstream of the solenoid valves. The measured data confirm a constant pressure gradient during the last 500 ms of the filling process, which ensures a constant fuel mass flow over this time span and so induces

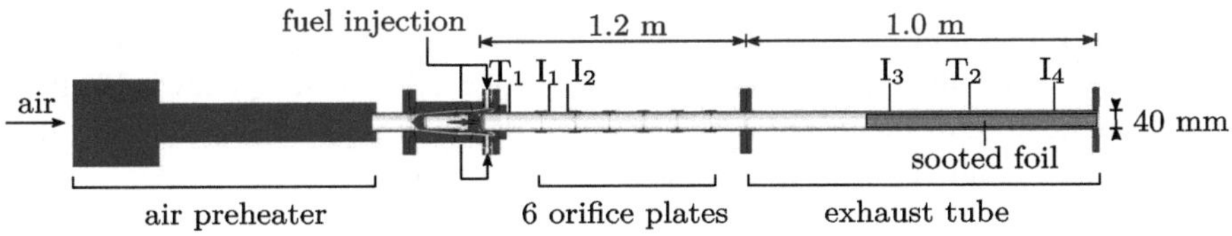

Fig. 1 Sketch of the PDC test rig with thermocouples T_1 and T_2 and ionization probes I_1–I_4

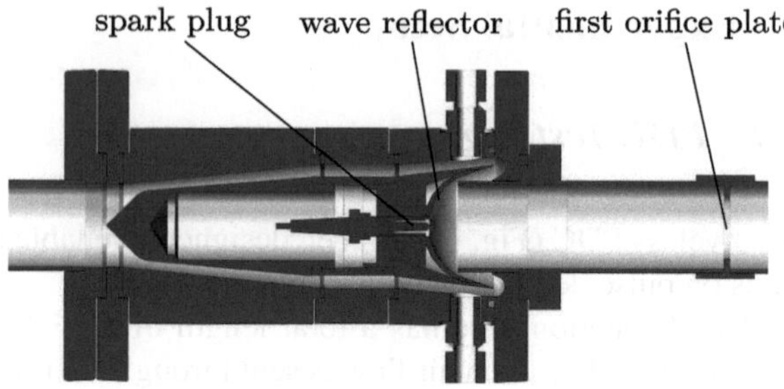

Fig. 2 Detailed view of the inlet geometry of the PDC test rig

a homogeneous equivalence ratio along the tube. The mixture is ignited using an automotive spark plug that initiates a deflagration that travels downstream. Orifice plates with a blockage ratio of 0.43 lead to an acceleration of the flame and enhance DDT. The chosen obstacles and separation distances were found to be a favorable geometry for efficient flame acceleration by Gray et al. [9].

A detailed view of the inlet section of the PDC test rig is shown in Fig. 2. The inlet of the combustion tube is formed by a hemispherical wave reflector that was designed as a fluidic diode. During the filling process, the gas enters the combustion tube with a small pressure loss. After the ignition event, the wave reflector impeeds the hot gas from traveling upstream and supports the flame acceleration due to its high pressure loss coefficient in the upstream direction. The spark plug is located at the center of the hemisphere.

2.2 *Measurement Methods*

2.2.1 Temperature Measurement

Thermocouples are used to measure the temperature downstream of the wave reflector (T_1) and inside of the exhaust tube (T_2) next to the sooted foil (Fig. 1). Sheathed thermocouples of type K with a diameter of 3 mm where chosen to ensure the robustness of the sensors when exposed to detonation waves. The applied setup includes a measurement error of ± 3 K. The recorded output signals allow for determining the mixing temperature immediately before the ignition event. One could expect a large decrease in the measured temperature due to the injection of hydrogen at ambient temperature. For example, the Riemann mixing rule supposes a temperature drop of about 90 K for an initial temperature of 600 K. However, the measured temperature only decreases by about 5 K when injecting hydrogen. This discrepancy is likely to arise from heating of the mixture by hot walls, which is supported by the fact that the associated temperature drop at T_2 is lower that at T_1.

2.2.2 Ionization Probes

Ionization probes are used to detect the flame front at four axial positions in the combustion tube: two probes inside the flame acceleration section and two probes in the exhaust tube (Fig. 1). The axial positions of the probes relative to the spark plug are listed in Table 1.

Figure 3 shows a typical output signal of the four ionization probes. The variation from the voltage output is plotted over time: When the flame front passes an ionization probe, its voltage output decreases as a result of the increased conductivity of the gas. Since the ionization probes I_1 and I_2 detect a deflagration while the probes I_3 and I_4 detect a detonation wave, the output signal differs remarkably between the two pairs of probes. The deflagration wave causes a small decrease in the voltage that lasts for more than 1 ms. In the exhaust tube, the detonation wave provokes a sharp decrease in the voltage of the probes I_3 and I_4. Additionally, the voltage recovers already after about 0.2 ms.

2.2.3 Sooted Foils

The cellular structure of the detonation wave is recorded by a sooted foil, which is inserted into the rear part of the combustion tube. For this purpose, precision spring steel is cut to pieces of 100 mm × 600 mm, which are sooted by a fuel-rich diffusion flame of acetylene in air. Subsequently, the foil is rolled up lengthwise and inserted

Table 1 Axial positions of ionization probes

Ionization probe I_i	I_1	I_2	I_3	I_4
Axial position x_i (mm)	180	220	1050	1450

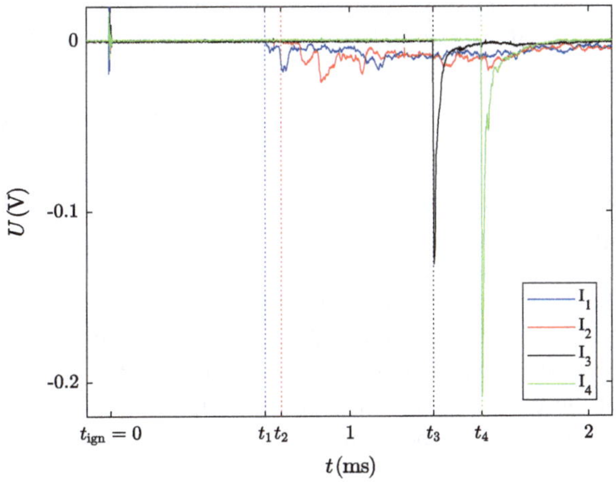

Fig. 3 Exemplary signal of the output voltage U of the ionization probes over time

into the downstream section of the combustion tube (Fig. 1). After a single detonation event in the PDC, the foil is removed from the test rig and the imprint is digitized. Image processing is used to optimize visibility of the detonation cell structure.

Hébral and Shepherd [10] developed a tool, which was applied in this work to determine the detonation cell width from the optimized photographs of the sooted foils. The output of the 2D–correlation tool consists of the size of the most dominant structures of the analyzed picture and the associated relative energies. The most dominant structure (i.e., the structure with the largest relative energy) is assumed to be the detonation cell width. Multiple pictures are extracted from each sooted foil, which are then evaluated separately. The dominant cell width for one foil is determined by averaging the results from the individual pictures weighted by their corresponding relative energy. This weight takes the quality of the picture into account, since a large relative energy suggests a reliable result. This correlates with the quality of the picture which is mainly defined by the contrast and the sharpness of the cellular structure.

3 Results and Discussion

3.1 Detonation Velocity

The time resolved output signals of the ionization probes are used to determine the time at which the detonation wave passes the probe. Considering the axial distance of the two probes in the exhaust tube of $x_4 - x_3 = 400\,\text{mm}$, the propagation velocity of the detonation front v_{det} can be calculated using time–of–flight substracting the flow velocity:

$$v_{\text{det}} = \frac{x_4 - x_3}{t_4 - t_3} - v_{\text{flow}}. \tag{1}$$

The flow velocity of the mixture can be expressed as a function of the mass flow rates $\dot{m}$ of air and fuel and the mixing temperature T_{mix}:

$$v_{\text{flow}} = \frac{T_{\text{mix}}}{pA} \left(\dot{m}_{\text{air}} R_{\text{s,air}} + \dot{m}_{\text{fuel}} R_{\text{s,fuel}} \right), \tag{2}$$

with the initial pressure p, the cross-section area of the detonation tube A and the specific gas constant R_s. This velocity is not negligible and can be as high as $v_{\text{flow}} = 55\,\text{m/s}$ is case of $T_{\text{mix}} = 760\,\text{K}$. Figure 4 shows the measured propagation velocity of the detonation front over the initial temperature.

Since the velocity is calculated from the time instances t_3 and t_4, at which the detonation wave passes the ionization probes 3 and 4 respectively, the accuracy of these time instances is of high importance for the error estimation. The voltage output of the probes is recorded with a sampling frequency of $f_s = 1\,\text{MHz}$. Hence, the maximal accuracy for the determination of t_3 and t_4 is $\pm 1\,\mu\text{s}$. Due to the detection

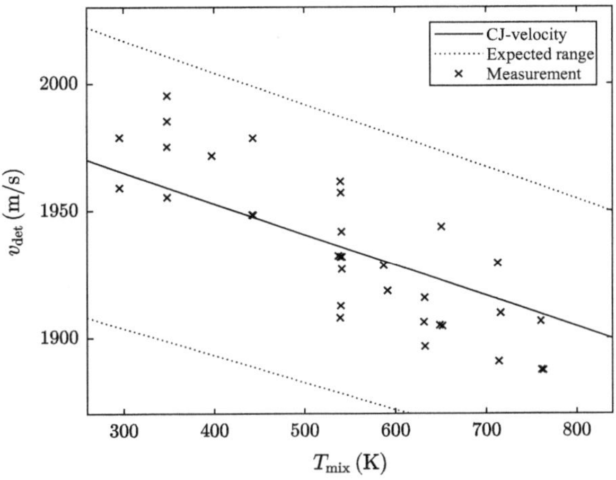

Fig. 4 Measured propagation velocity v_{det} of the detonation front over the initial mixing temperature T_{mix} compared to the CJ-velocity and the expected range of deviation due to uncertainties in the equivalence ratio at a constant initial pressure of 1 bar

of the time at which the reaction front passes the sensors, a measurement error of $\pm 2\,\mu s$ results for $\Delta t = t_4 - t_3$. For an axial distance of $x_4 - x_3 = 0.4\,m$, this leads to an error of $\pm 20\,m/s$ in the range of the measured velocities.

Beside the measured propagation velocities, Fig. 4 shows the CJ-velocity for a perfect mixed fuel–air mixture with an equivalence ratio of $\Phi = 1$ and the expected range of measured data points taking the variation of the actual equivalence ratio and the measurement error of the velocity into account.

It can be seen that all the measured velocities are located in the expected range. The variance of the propagation at a certain temperature can be explained by the measurement error of the velocity by the ionization probes of $\pm 20\,m/s$. For temperatures between 400 K and 500 K the mean detonation velocity exceeds the CJ-velocity, which can be explained by a fuel mass flow that has been set slightly too high for these conditions. Nevertheless, every measured propagation velocity is located in the expected range, which implies that the mass flow rates have been adjusted correctly with the given precision.

3.2 Detonation Cell Width

Figure 5 shows the photographs of three different sooted foils. These foils have been generated at an initial temperature of 290 K, 500 K and 650 K respectively.

Three different types of appearance were found, each visible in one photograph in Fig. 5. For a temperature range of 290–380 K, the foils show a homogeneous distribution of the soot, which results in low contrast in the photographs. The detonation cells are indicated by thin, dark lines (Fig. 5a). The second type, which has been observed for a temperature between 450 and 550 K, is characterized by dark cells separated by thick lines where the soot has been removed by the movement of the triple points. An initial temperature of 600 K and more led to a fish-scale-like appearance of the detonation cell structure. While there is an abrupt change between the first and the

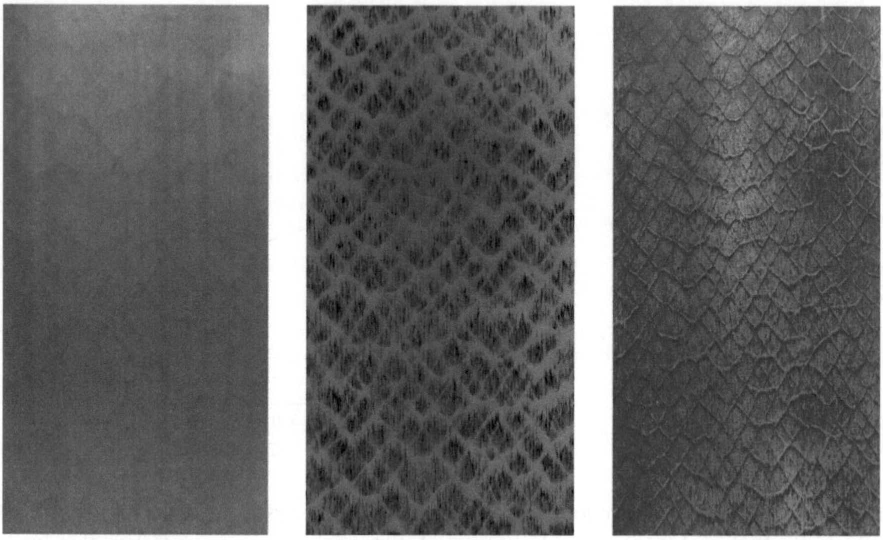

Fig. 5 Processed photographs of sooted foils at three different initial temperatures show various types of appearance of the cellular structure

second type, there is a smooth transition between the second and the third type, which causes a simultaneous appearance of the latter two types for a mixing temperature of 550 and 600 K.

The detonation cell width is determined from photographs of the sooted foils using a 2D–correlation tool developed by Hébral and Shepherd [10] as mentioned in Sect. 2.2. Since this tool has to be operated manually at some points, a random error appears. Due to the small contrast, the photographs of type 1 are the hardest to evaluate which increases the error of the calculated cell width considerably for the respective temperature range. To estimate the human error, the evaluation of the photographs have been performed by two different persons independently. It was found that the individual values might deviate up to 3 mm but the averaged cell width deviates only by a small margin. Figure 6 shows the measured detonation cell width λ over the initial temperature.

As described by Djordjevic et al. [7] the detonation cell width is affected by two opposing effects when the mixing temperature is increased. The speed of sound increases resulting in a lower Mach number of the detonation wave. Therefore, the von-Neumann pressure decreases while the post-shock temperature is increased. The resulting effect on the cell width is dependent on the mixture and the initial conditions. The performed experimental study of the detonation cell width in a stoichiometric hydrogen–air mixture as a function of the initial temperature at a constant initial pressure of 1 bar shows that the cell width decreases over the temperature.

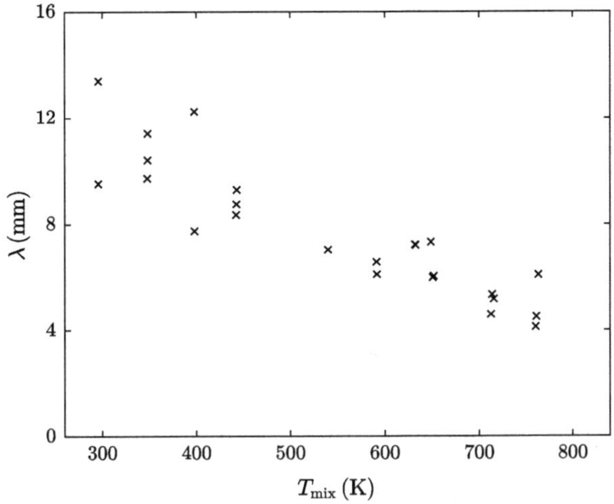

Fig. 6 Detonation cell width λ over the initial mixing temperature T_{mix}, measured every 50 K at a constant initial pressure of 1 bar

The results from Djordjevic et al. are matched only for ambient temperature. For higher temperatures, the calculated cell widths are too large compared to the experimental results. This is most likely due to the fact that the calculations are based on the one-dimensional ZND detonation model, and thus, cannot cover the entirety of the complex shock interactions that take place in a real, multidimensional detonation front. As suggested by Gavrikov et al. [11], these interactions might play a decisive role in defining the initial conditions of the reactive mixture before the onset of combustion under certain circumstances: Due to additional compression by transverse shock waves and resulting shock collisions, the actual strength of the shock that causes the mixture to auto-ignite may exceed the one predicted by the von-Neumann state based on the ZND detonation model. As described before, since several post-shock parameters such as temperature, pressure and relative velocity all contribute to the formation of detonation cells, partly through opposing effects, the influence of multidimensionality should be considered to improve numerical agreement with experimental data.

The experimental results agree well with the investigations by Ciccarelli et al. [6], who observed a decrease from $\lambda(300\,\mathrm{K}) \approx 9\,\mathrm{mm}$ to $\lambda(650\,\mathrm{K}) \approx 5\,\mathrm{mm}$ for atmospheric initial pressure conditions. The measurement results suggest slightly larger values for both temperatures which can be explained by uncertainties in the determination of the cell width. Furthermore, a linear dependency of λ and T was observed in this work for the entire temperature range which exceeds the previously investigated temperature spectrum.

3.3 Initial Flame Acceleration

The ionization probes I_1 and I_2 are used to record the flame velocity between the first and the second orifice plate. As described by Smirnov and Nikitin [8], the detonation run–up distance is strongly dependent on the initial flame acceleration. Hence, the velocity of the deflagration front in the beginning of the combustion tube is a relevant parameter for characterizing the DDT in the valveless PDC used in this work. The coupling criterion implies that the flame propagation speed has to exceed the speed of sound of the unburned mixture before DDT can take place. Thus, the Mach number of the deflagration wave M_{defl} is a good indicator for the magnitude of the DDT run–up distane. Figure 7 shows the Mach number of the flame between the first two ionization probes relative to the unburned mixture. The speed of sound for the stoichiometric mixture of hydrogen and air has been calculated as proposed by Vrkljan [12].

Up to a temperature of $T \approx 400\,\mathrm{K}$, the Mach number decreases significantly. In this regime, the decrease in density apparently has a larger effect on the reactivity than the increase in temperature. When further increasing the temperature, the Mach number of the flame front stays at a constant level of $M \approx 0.3$ which implies the further decrease in the density of the mixture is compensated by the rise in temperature. This relation between mixing temperature and initial flame acceleration implies an increase of the DDT length for a rise in temperature up to 400 K for the used geometry setup. For further increase in temperature, the DDT-length is assumed to be independent of the mixing temperature which is a useful characteristic when designing a PDC for integration in a gas turbine.

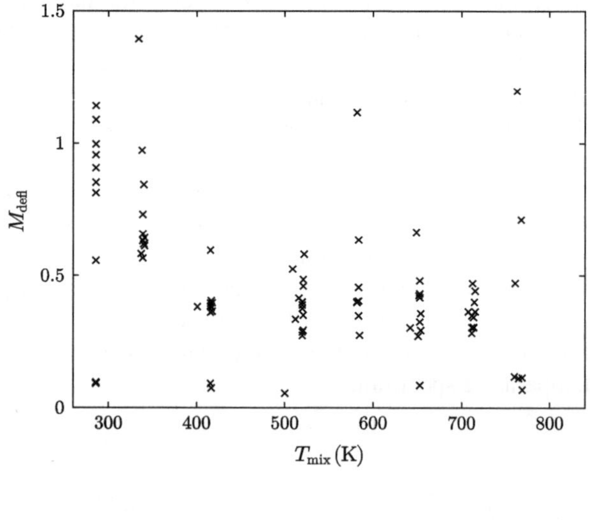

Fig. 7 Mach number of the deflagration wave relative to the unburned mixture M_{defl} between ionization probes I_1 and I_2 over the initial mixing temperature T_{mix} at a constant initial pressure of 1 bar

4 Conclusion

Detonation properties are strongly dependent on the mixture and the test rig geometry. A valveless PDC test rig has been used to record data for stoichiometric hydrogen–air mixtures for a temperature variation from 290 K to 620 K. The detonation cell width has been determined experimentally using sooted foils and digital data post–processing.

It was found that the cell width decreases linearly by increasing the temperature. The measured cell widths for ambient temperature and for $T = 650\,\mathrm{K}$ are in good agreement with the available literature. Additionally, the initial flame velocity and the detonation velocity have been determined from the time–resolved output signal of four ionization probes at different axial positions. For temperatures less than 400 K, the flame acceleration is influenced mainly by the density decrease, leading to a decreasing Mach number of the flame with increasing mixing temperature. For higher temperatures, this effect is nullified by the increase in the initial temperature which leads to an almost constant Mach number of the flame relative to the unburned gas. As mentioned above, the DDT run-up distance is assumed to follow the dependency between the mixing temperature and the initial Mach number, which must be verified in future investigations. When designing a PDC for application in a gas turbine, the knowledge of the influence of the mixing temperature on the detonation cell width and, thus, the detonability of the mixture is crucial to optimize the robustness as well as the efficiency of the system.

The accurate control of the equivalence ratio is a major task when realizing reproducible series of measurement in a detonation tube. When running the PDC at multi-cycle operation, this becomes even more challenging due to unsteady states in the air and fuel supply lines. To ensure reproducibility, it is planned to develop a concept to measure and to control the fuel mass flow at multi-cycle operation.

Acknowledgements The authors gratefully acknowledge support by the Deutsche Forschungsgemeinschaft (DFG) as part of Collaborative Research Center SFB 1029 "Substantial efficiency increase in gas turbines through direct use of coupled unsteady combustion and flow dynamics" on project A01.

References

1. Stathopoulus, P., Vinkeloe, J., Paschereit, C.O.: Thermodynamic evaluation of constant volume combustion for gas turbine power cycles. In: Proceedings of the 11th International Gas Turbine Congress, Tokyo, Japan (2015)
2. Akbari, P., Nalim, R., Mueller, N.: A review of wave rotor technology and its applications. J. Eng. Gas Turbines Power **128**(4), 717–735 (2006)
3. Bobusch, B.C., Berndt, P., Paschereit, C.O., Klein, R.: Shockless explosion combustion: an innovative way of efficient constant volume combustion in gas turbines. Combust. Sci. Technol. **186**(10–11), 1680–1689 (2014)
4. Lu, F.K., Braun, E.M.: Rotating detonation wave propulsion: experimental challenges, modeling, and engine concepts. J. Propuls. Power **30**(5), 1125–1142 (2014)

5. Roy, G.D., Frolov, S.M., Borisov, A.A., Netzer, D.W.: Pulse detonation propulsion: challenges, current status, and future perspective. Progress Energy Combust. Sci. **30**(6), 545–672 (2004)
6. Ciccarelli, G., Ginsberg, T., Boccio, J.L.: Detonation cell size measurements in high-temperature hydrogen-air-steam mixtures at the BNL high-temperature combustion facility. Technical report, Nuclear Regulatory Commission, Washington, DC, United States. Div. of Systems Technology; Brookhaven National Lab., Upton, NY, United States; Nuclear Power Engineering Corp., Tokyo, Japan (1997)
7. Djordjevic, N., Hanraths, N., Gray, J.A.T., Berndt, P., Moeck, J.P.: Numerical study on the reduction of NOx emissions from pulse detonation combustion. J. Eng. Gas Turbines Power **140**(4), 041504 (2018)
8. Smirnov, N.N., Nikitin, V.F.: Effect of channel geometry and mixture temperature on detonation-to-deflagration transition in gases. Combust. Explos. Shock Waves **40**(2), 186–199 (2004)
9. Gray, J.A.T., Paschereit, C.O., Moeck, J.P.: An experimental study of different obstacle types for flame acceleration and DDT. In: Active Flow and Combustion Control 2014, vol. 127, pp. 265–279. Springer (2015)
10. Hébral, J.P., Shepherd, J.E.: Detonation cell size measurement (2000)
11. Gavrikov, A.I., Efimenko, A.A., Dorofeev, S.B.: A model for detonation cell size prediction from chemical kinetics. Combust. Flame **120**(1–2), 19–33 (2000)
12. Vrkljan, V.S.: Über die Berechnung der Schallgeschwindigkeit in Gasmischungen. Il Nuovo Cimento (1955–1965), **17**(6), 845–849 (1960)

Types of Low Frequency Instabilities in Rotating Detonation Combustors

Vijay Anand and Ephraim Gutmark

Abstract Rotating detonation combustors (RDC) offer a significant prospective increase in stagnation pressure across it owing to the presence of one or more rotating detonation waves spinning inside the combustor at the kilohertz regime. Naturally, considerable research impetus has been directed towards this technology in recent years to understand the driving mechanics to harness the associated potential of pressure gain combustion (PGC). One such area of focus has been the off-design operating modes of these devices which cause a myriad of instabilities. The current paper is focused towards the discussion of one such instability regime—low frequency instabilities (LFI)—in RDCs. We review three types of LFIs in RDCs based on prior findings, and propose mechanisms for the same.

Keywords Detonation · Pressure gain combustion · Combustion instability
Low frequency instability · RDE

1 Introduction

The supersonic combustion phenomenon of detonation produces a pressure gain of 13–55 in gases [1] across the wave due to the shock wave linked to the combustion front. This detonative mode of combustion is theorized to provide the highly sought after pressure gain (or more specifically, the gain in stagnation pressure, according the AIAA Pressure Gain Combustion Technical Committee) across the component. While pulsed detonation combustors (PDCs) were the widely investigated type of pressure gain combustion (PGC) systems, the majority of recent research has migrated to rotating detonation combustors (RDCs). The higher power density [2], the lack of a need to regulate periodic ignition and fuel/oxidizer injection,

V. Anand (✉) · E. Gutmark
Department of Aerospace Engineering and Engineering Mechanics,
University of Cincinnati, Cincinnati, OH 45221, USA
e-mail: ganeshvn@mail.uc.edu
URL: http://www.uc.edu

R. King (ed.), *Active Flow and Combustion Control 2018*,
Notes on Numerical Fluid Mechanics and Multidisciplinary Design 141,
https://doi.org/10.1007/978-3-319-98177-2_13

as opposed to a PDC, and the steadier exit flow profile [3] circumvents the notable issues besetting PDCs. Despite the considerable progress made till date on the different facets of RDCs, substantial research is still warranted to ascertain the physics and apply RDCs as a real-world, power-generation device. Until recently, the probable efficiency increases afforded by RDCs, due to detonative burning [4], remained a figment of numerical or analytical solutions, with studies claiming: a notable increased in total impulse over pulsed detonation combustors [3], an increase of up to 9% in fuel efficiency [5], an increase of up to 15% in the total pressure in the combustor due to detonation [6], an increase of 5% in thermal efficiency [7], an increase in thermal efficiency of 1.6% [8], and finally up to 14% increase in power plant efficiency over conventional J class turbines [9].

Multiple factors like the reactants type, combustor geometry, thrust, oxidizer and fuel flow rates, and the rotating detonation wave speed at these conditions have been studied. The next obvious step is to identify and understand the various instabilities in RDCs. The field of combustion instabilities in RDCs is extremely nascent in comparison to the monumental work done in addressing the instabilities in gas-turbine combustors and rocket engines. Though the frequency of the instabilities is more often than not a function of the geometry of the combustor, historically speaking, it has been beneficial to group the instabilities this way since the underlying mechanism is generally relatable for a given frequency range. Traditionally, low frequency instabilities having a frequency range of 1–500 Hz are termed LFI, while those between 500–1000 Hz are termed intermediate frequency instabilities (IFI, or "buzzing" in rocket engines), and those greater than 1000 Hz are named high frequency instabilities (HFI) [10]. Our prior publications present the notable parallels between HFI in rocket engines, and rotating detonation and longitudinal pulsed detonations (LPD) in RDC [11, 12]. However, few studies exist at present that have addressed RDC instabilities to any appreciable degree [13–17].The authors have addressed the four prominently occurring instabilities in an RDC of which LFI are a part of, and subsequently speculated on their mechanism in Ref [13]. We use this abbreviation (of LFI) to be in continuation with the terminology used in gas-turbine combustors and rocket engines.

LFI in an RDC seems to be almost ubiquitous. A brief analysis of the pressure-time traces published by the different RDC facilities worldwide gives concrete evidence of the overarching existence of this instability [13, 15, 18–27]. However, most studies have not made an effort to address either the existence, or the mechanism behind LFI. Considering the crippling effects of LFI in rocket engines, supersonic inlets and hypersonic vehicles owing to their tendency to couple with the natural resonant frequency of the structure [28, 29] and the supply feeds [10, 30], it is imperative to acknowledge and treat LFI as we move forward with RDC research. The current study presents a review of the three types of LFI observed in RDCs, from our prior publications [11, 14, 22, 31]. It is emphasized here that though the first two types are only observed so far in annular RDCs, whereas the third one has been seen only in a hollow RDC, the driving mechanisms do not limit their occurrence to a specific geometric variant of the combustor, and should be construed to be possible in either combustor geometry at amenable conditions.

2 Experimental Methodology

The air and fuel flow rates through the RDC are controlled by a closed-loop system of nitrogen-driven pilot regulators and a set of Flowmaxx sonic nozzles. The equivalence ratios discussed henceforth are global values estimated from the stagnation pressure and temperature upstream of the choked sonic nozzle. Norgren VP50 proportional control valves (pilot) are linked to Norgren pilot-operated regulators to isolate electrical components from the primary fuel supply. GE Unik 5000 sensors are linked to the choked-flow nozzle assemblies to monitor air and fuel flow rates. Fuel flow is administered to the rig through a pneumatically-actuated Bi-Torq isolation ball valve located just upstream of the fuel plenum, which allows fuel flow rates to stabilize within 2 s of fuel introduction. The static pressure evolution inside the air plenum, fuel plenum and combustor before and after ignition can be seen in Ref [22]. The systematic error in flow measurements is only from the scaling error of the instruments. The relative errors in the static pressure sensor and thermocouple are known from the associated instrument specifications. This is used to attain the uncertainty in the pressure and thermocouple sensors used in the reactants delivery, and is found to be ± 0.069 bar and ± 1 K (at the maximum output), respectively, which in turn causes negligible errors in the measured flow rates. For further information about the facility, please consult Refs [31–33].

For the purposes of the current paper, two geometric variations running under air flow rates ($\dot{m}_a$) of 0.2, 0.3 and 0.4 kg/s are to be discussed: annular RDC (running on H2-air [33]) and hollow RDC (C2H4-air [11]). A side and front-view schematic of the two RDCs and the associated instrumentation is shown in Fig. 1. Air and fuel are supplied from two separate plenums (blue and green, respectively) to attain non-premixed mixing. Air is injected radially inwards through a circumferential slot and fuel is injected axially through a circumferentially distributed array of fuel orifices. The mixing scheme is generically visualized in Fig. 1. Dimensions of the parts of interest for both the RDCs are presented in Table 1. The reactants mix through this orthogonal slot-orifice injection scheme and are ignited by an initiator tube (see figure). This tube is fed with opposing jets of ethylene and oxygen that are supplied at the headend through Parker solenoid valves. The presence of a detonation wave inside the pre-detonator tube is verified using two ionization probes (not shown here) that accurately capture the speed of the supersonic wave. For a detailed description of the pre-detonator system, we direct the attention of the readers to our previous work [34]. This initial blast wave from the initiator tube is theorized to cause a complex deflagration-to-detonation transition (DDT) mechanism [34] that finally results in the formation of a detonation wave into the RDC channel (red area).

Both the annular and hollow RDC have four rows of instrumentation ports arranged axially in the combustor. This four row instrumentation scheme is present in three sectors of the combustor (see figure) to spatially resolve the rotating detonation wave dynamics as it propagates azimuthally. Note that the color scheme used to represent the three stations in the two RDCs are used later to denote the pressure/ionization traces acquired from the respective sectors. There are also three

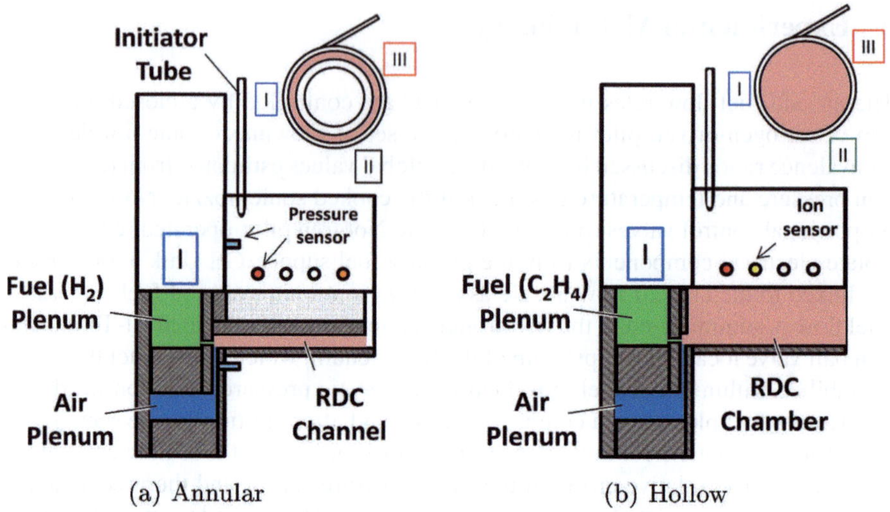

Fig. 1 Schematic with instrumentation of **a** 6" annular RDC, **b** 6" hollow RDC

Table 1 Dimensions of interest—annular and hollow RDC

Part	Geometry measured	Annular RDC dimensions	Hollow RDC dimensions
Fuel injection	Length/diameter of each hole	17	17
Pattern	Circumferential rows	3	1
Air injection slot	Slot width	1.02 mm	1.02 mm
	Total slot area	490 mm^2	90 mm^2
Combustor channel	Width	7.5 mm	–
	Inner diameter	139 mm	–
	Outer diameter	154 mm	154 mm
	Annulus area	760 mm^2	–
	Length	125 mm	125 mm

ports arranged azimuthally in the air inlet (flush-mounted as well) that allow high speed sensing of the air supply plenum dynamics. When there is no sensor in a particular port, a bolt of the proper sizing is used to cover the holes, and is represented as hollow circles in the schematic. In this way, multiple sensor arrangements are possible to analyze RDC dynamics. For the annular RDC, a total of six high speed PCB piezoelectric sensors are used, with three sensors in the combustor in row 1 (red circle in station I in the side view) and three sensors in the air inlet (blue tabs in the side view). Note that this similar arrangement is implemented for the sectors II and III as well. A piezoresistive Kulite sensor in infinite tube pressure (ITP) configuration [35] is integrated to row 4 of the annular RDC, as well. In the hollow RDC, there are three PCB sensors in row 1 (all three sectors) of the combustor. In addition to this, there are also three high speed ion probes (orange circle) in row 2 of the same three stations. This allows us to record the coupling between the shock wave and the combustion wave that together form a detonation wave. The first row of instrumentation ports is 1.9 cm away from the headwall, whereas the other rows are spaced 2.54 cm from each other. The air inlet ports are 2.5 cm away from the annulus, radially. Both the RDCs are also instrumented with capillary tube averaged (CTAP) low speed static pressure sensors [35] that monitor the nominal average static pressures before and during RDC hot-fire operation. Data acquisition is at 1 MHz for the high speed sensors and 1 kHz for the low speed ones.

3 Results and Discussion

3.1 Acoustics-Induced LFI: Spatially Homogenous Oscillations

As mentioned earlier, most operating conditions in an atmospheric annular RDC is prone to a low frequency instability characterized by amplitude modulation in the combustor that is ensconced onto the detonation wave propagation frequency (carrier frequency). This is clearly seen in the piezoresistive pressure trace provided in Fig. 2, which shows the sinusoidal pressure evolution characterized by "waxing and waning" of the peak detonation wave strength. However, two different mechanisms, pertaining to the supply plenum, appear to cause this amplitude modulation (AM) in the combustor. The first mechanism—acoustic oscillation in the form of frequency modulation in the air inlet—is discussed in this section, whereas the next section deals with the other mechanism—detonation wave interaction with the supply plenum. The former manifests as a spatially homogenous LFI, i.e. the sinusoidal oscillation are continuous across all three sectors of the RDC. Remarkably, this type of simultaneous

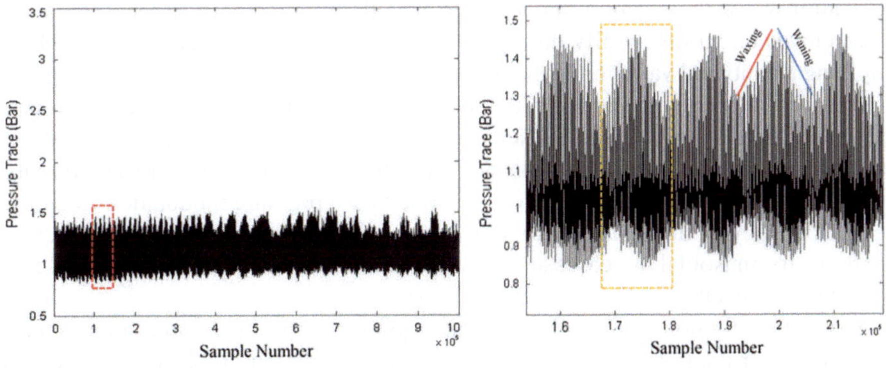

Fig. 2 Pressure-time trace from ITP-mounted piezoresistive sensor at 0.2 kg/s at Φ = 1.8 (left), and a magnified section from the same test point (right)

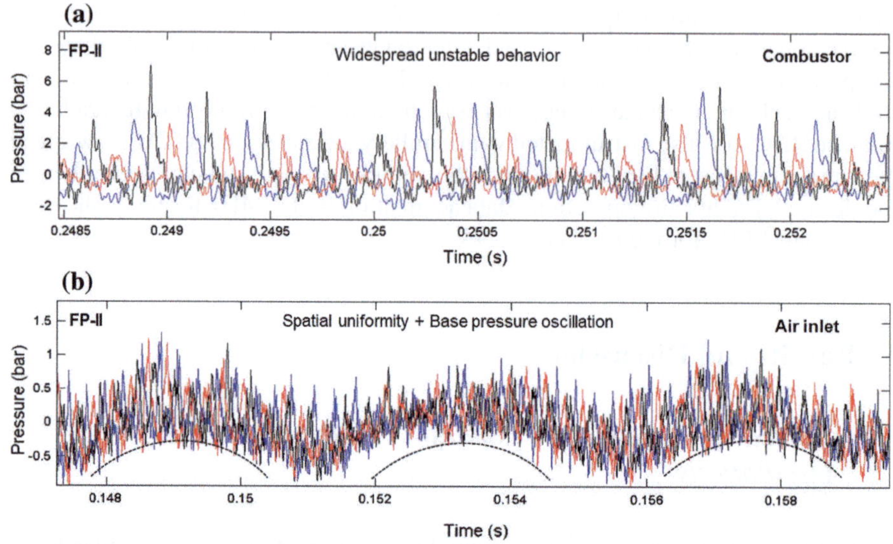

Fig. 3 **a** Pressure series from the combustor, **b** pressure series from the air inlet (0.4 kg/s) showing FM LFI

oscillation is observed in rocket engines at unstable conditions of operation when there is an onset of high frequency combustion instabilities, and is referred to as "chugging" [36], to refer to the changes in flow rates caused by this type of acoustic oscillation linked to the plenum [10]. The latter results in a spatially non-homogenous LFI that is discontinuous across the different RDC sectors, i.e. this LFI appears to be revolving about the combustor annulus. This type of phase-lagged low frequency rotating instability has been observed in rocket engines, as well (Fig. 9.7.1i in [10]). There, it is called a "precessing tangential mode" and is noted to have anywhere between five to hundred tangential wave laps per low frequency cycle.

Figure 3a is an arbitrary pressure trace from the combustor that shows unstable detonation wave propagation that varies sinusoidally in strength as the detonation wave moves circumferentially through the three instrumented sectors of the RDC, implying a continual variation in the rotating detonation wave strength. Fig. 3b gives a pressure time series in the air inlet that shows an oscillation in the base pressure (tagged by dotted curves). The carrier signal in the air inlet is the leaked shock wave from the detonation propagation through the three sectors. This carrier pressure signal, however, can be seen to have a sustained base pressure modulation (thereby, being frequency modulated) that is azimuthally simultaneous. That is, all three sectors are prone to sinusoidal increases and decreases in the base pressure. We will call this spatial homogeneity.

Since spatially homogenous LFI is linked to the base pressure oscillation (chugging) in the air inlet, it is desirable to analyze the oscillation in the air inlet. Fig. 4a shows the Fast Fourier Transform (FFT) plot for three stoichiometric test cases at different air flow rates. While the ff (the RDC operation frequency which is equal to the

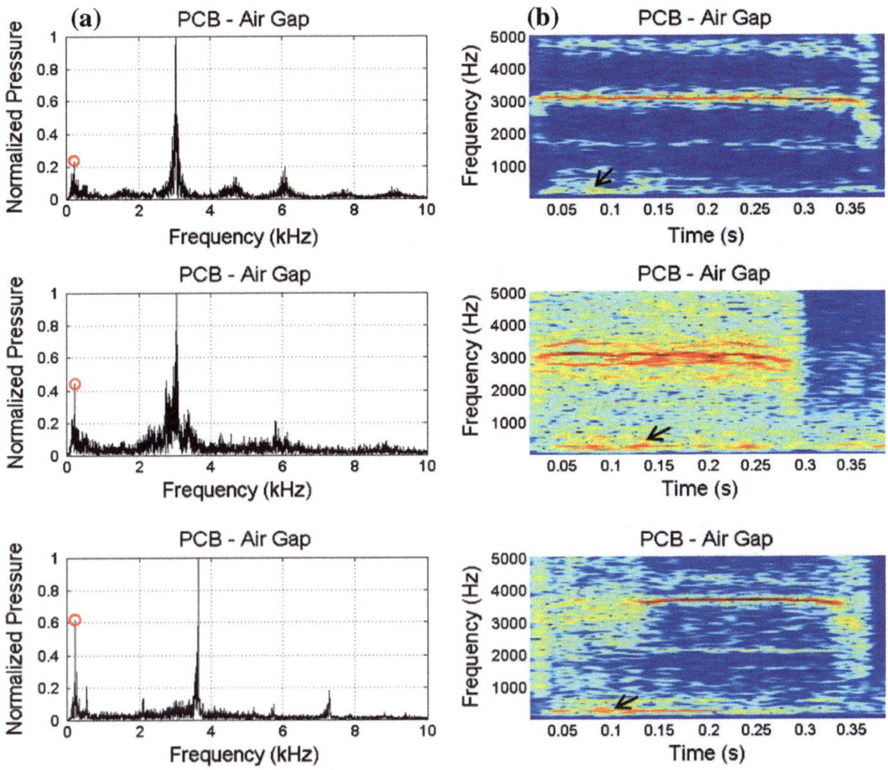

Fig. 4 FFT plot **a** and Spectrogram **b** showing the spatially homogenous LFI frequency in the air inlet for 0.2 kg/s (top), 0.3 kg/s (middle) and 0.4 kg/s (bottom)—all flow rates at an equivalence ratio of 1.0

detonation wave frequency) is easily observable from the three plots at f ≈ 3 kHz, f ≈ 3 kHz and f ≈ 3.8 kHz respectively, there is also considerable LFI activity in all three test cases at f ≤ 0.5 kHz(the frequencies surrounding the red circle in the figure). The secondary dominant frequency is identified to be f ≈ 235 Hz (shown by red circle) and is the same across 0.2, 0.3 and 0.4 kg/s across three different equivalence ratios at the three air flow rates [22]. This is an interesting discovery since it implies that irrespective of the air flow rate (or in other words, the air plenum pressure) or the equivalence ratio of the operating point, the inlet of the RDC in use for the current study always oscillates at f ≈ 235 Hz. The f ≈ 235 Hz oscillation (black arrow) does not extend throughout the test, but is rather concentrated for a few seconds after ignition, as shown by the spectrogram, in Fig. 4b respectively. To confirm that the oscillation is induced by the detonation wave and not due to an inherent oscillation in the air supply, the RDC was operated under cold-flow conditions without ignition of the reactants, which revealed the lack of any activity in the 235 Hz region. Since the cold flow RDC testing did not produce any oscillation at the same frequency, it can be understood that the oscillation in the air gap is linked to detonation wave propagation in the annulus. The Helmholtz frequency for the air plenum under study is estimated to be ≈355 Hz for all flow rates since the speed of sound is relatively unaltered due to an almost constant supply temperature. Here, the length of the radially oriented slot (see Fig. 1) is taken to be the length of the resonator neck, with an added end correction of 0.6 times the equivalent diameter (obtained from the total air injection area). The 0.6 coefficient is recommended for geometries with two open ends, as is the case with the air plenum. But, the complex geometry of the plenum would skew the resonance frequency obtained from the basic Helmholtz equation considerably [37]. For instance, a ≈15% error is incurred between the calculated and the experimentally obtained frequency when the basic Helmholtz equation is used to calculate the resonance frequency for a cylindrical prism with a long neck [37]. Hence, to get an accurate resonance frequency value, geometry-specific equations need to be developed, even for simple geometries. Despite these unknown variables, since the approximately estimated resonance frequency is different from the oscillation frequency in the inlet by only 33%, it is a strong indication of the air plenum's functioning as a Helmholtz resonator due to the excitation produced by the high-frequency detonation wave in the combustor. This is not unlike the chugging phenomenon observed in the F-1 engines due to the quarter-wave mode of acoustic excitation produced in the engine's reactants supply lines [10]. It is hypothesized that the detonation wave excites the air inlet to a resonance frequency which manifests as the spatially homogenous LFI in the combustor, due to the fluctuating air supply, which in turn appears to cause a similar fluctuation in the detonation wave's strength.

3.2 Detonation-Induced LFI: Spatially Non-homogenous Oscillations

Annular RDC supply plenums (studied here) also interact with the combustor dynamics in a spatially non-homogenous way. Here, spatial non-homogeneity' is defined as the phenomenon where for a given arbitrary temporal window, a particular sector of the RDC exhibits notably higher peak pressures (of subsequent detonation laps) than the other sectors. In this case, while there is still a sinusoidal low frequency oscillation, it is segregated to a given sector of the combustor and air inlet; or in other words the low frequency amplitude modulation revolves about the device. The alternating sectoral strength is depicted in Fig. 5, which shows stronger subsequent waves in the air inlet sector I (blue), followed by sector III (red), and finally sector II (black). This is an amplitude modulation (AM) of the carrier shock wave pressure signal. The facet to observe here is the direction of rotation of the detonation wave (trackable through the leaked shock wave into the air inlet). The rotating detonation wave moves in the counter-clockwise direction (black→red→blue/sector II →sector III →sector I) in the represented pressure traces. The spatially varying AM oscillation, however, exhibits the reverse order (blue→red→black / station II→station III→station I). To clarify, the sinusoidal amplitude modulated variation is exhibiting a clockwise rotating motion. This kind of behavior– the spatially varying sinusoidal AM instability

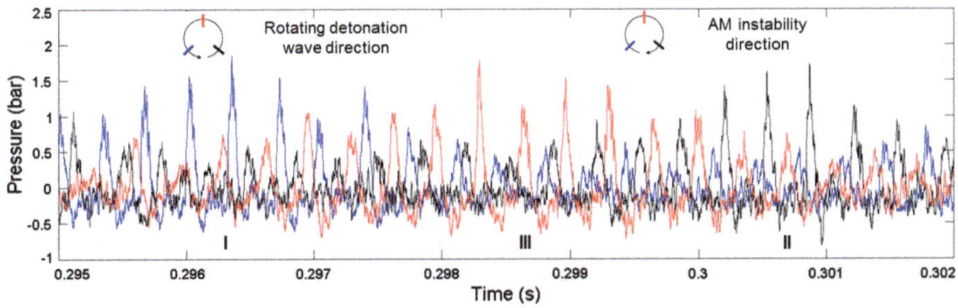

Fig. 5 AM LFI pressure series from air inlet with FP–I at $\dot{m}_a$ = 0.2 kg/s, Φ = 1.0

in the air inlet moving in the opposite direction to the rotating detonation wave– is witnessed when there is spatially non-homogenous LFI. It is thus apparent that AM LFI is characterized by the spatially varying detonation strength is not caused due to a simultaneous stationary acoustic phenomenon in the air inlet unlike the previous case, but rather, a low-speed rotatory event in the air inlet and combustor.

Details of the origins of this rotating instability emerge when an operating point exhibiting the same is studied in entirety. Fig. 6 shows the pressure traces from the three sectors in the air inlet at $\dot{m}_a$= 0.2 kg/s and Φ = 1.0. From Fig. 6a it can be seen that the rotating detonation wave is spinning counterclockwise and AM LFI (the sinusoidal overarching component) is clockwise. Fig. 6b shows the highly dynamic nature of this instability. Initially, there is incoherence and there is no marked sinusoidal spatially varying oscillation. However, from t = 0.14 s, one can observe easily delineated sinusoidal variation existing until the end of the RDC hot-fire run. Additionally, it could also be seen that the sinusoidal variation gradually "thins" with time, occurring faster as time progresses. This kind of dynamic variation in the frequency of occurrence of a phenomenon is usually termed "bootstrapping", a non-linear phenomenon that has been seen to occur in rocket engines [38], albeit for different reasons. The normalized pressure envelope of the low frequency spatially instability is shown in Fig. 6c for all three sensors. The spatial variation and the faster occurrence of the instability throughout the test can be easily noticed. In fact, the plot of the AM LFI velocity vs. time (Fig. 6d) shows that initially the induced velocity is around 40 m/s, and continuously increases from t $\approx$ 0.135 s to t $\approx$ 0.275 s, after which time it plateaus to a speed of 100 m/s. The gradual increase in AM LFI velocity until a terminal condition can be better understood by analyzing the detonation wave speed and directionality plot shown in Fig. 6e. The rotating detonation wave establishes in the clockwise direction initially and rotates in this direction until t $\approx$ 0.135 s. However, at t $\approx$ 0.135 s there is a sudden stochastic flip in the direction to counterclockwise rotation, and from t $\approx$ 0.135 s till t $\approx$ 0.34 s, the rotating detonation wave exhibits stable directionality in the counterclockwise direction. It is thus apparent that the steady increase in AM LFI to a terminal value from t $\approx$ 0.135 s is closely linked to the rotating detonation direction, i.e. as long as the detonation wave maintains its direction of rotation, the AM LFI in the air plenum maintains its direction. The amplitude modulation in the air inlet seems to exhibit a considerable transient operation followed by an apparent steady-state period where the rotary velocity of AM LFI is plateaued. This terminal rotary speed is approximately 100 m/s, for the present test condition. This plateau in the rotary speed is not observed when the detonation wave flips direction frequently [31].

From the above observations, we have enough information to propose a mechanism behind the amplitude modulated spatially non-homogenous instability in an RDC. Fotia et al. [39], in their two-dimensional experimental RDC study, have noted that the trailing shock wave (attached to the bottom of the detonation wave, as established by Schwer and Kailasanath [40]) moves at about 60% of the detonation wave speed. Owing to this relative velocity between the detonation wave in the combustor and the trailing shock wave in the reactants plenum, they postulated a "pressure beating" event when the next detonation wave lap interacts with the trailing shock wave in the plenum from the prior lap. However, their study was a one detonation wave event, and hence the "pressure beating", or more accurately, constructive/destructive interference was not observed. On the other hand, Schwer and Kailasanth [40], in their numerical simulation, have noted that this trailing shock wave (incident wave) travels to the base of the reactants plenum and gets reflected as another relatively strong wave. They tested two different reactant plenum depths (to vary the strength of the reflected wave) and concluded that this reflected wave did not impact the reactants conditions upstream of the next lap of the detonation wave. Additionally, multiple reflected waves are produced at discrete times and locations along the concave surface, depending on curvature and other effects [41]. Evidence for this in an RDC can be seen in the Schlieren images of an air plenum (with curved base) exposed to a detonation wave, obtained by Bedick et al. [42]. Hence, one should expect a similar phenomenon of production of multiple reflected waves moving in the opposite direction of the trailing shock wave when it is incident on the concave curved surface of the base of the air plenum of the RDC. We hypothesize here that the current LFI in an RDC is probably due to this interaction of incident and reflected waves. This is an extension of the acoustic interference hypothesis that Fotia et al. [39] that is expanded to include the effects of the reflected wave as well, in an actual RDC. Naturally, one should expect this complex system of waves to develop much slower in comparison to the detonation wave. This could explain the lower speed of the rotation and the stochasticity in AM LFI when the detonation wave randomly switches direction. In light of this theory, one could contend that the speed of this rotary amplitude modulated motion is inherently a function of the detonation wave speed and the fluidic impedance of the injector element. Increase of the former and decrease of the latter is expected to produce faster moving wave packets. Visualization of the reactants plenum is required to further test the proposed hypothesis.

3.3 Combustion-Induced LFI: Spatially Homogenous

As opposed to the previous two cases where LFI is caused in the combustor due to interaction with the supply plenum, in the current case LFI is dictated seemingly by the combustion dynamics alone. Shown in Fig. 7a are the pressure traces (from the

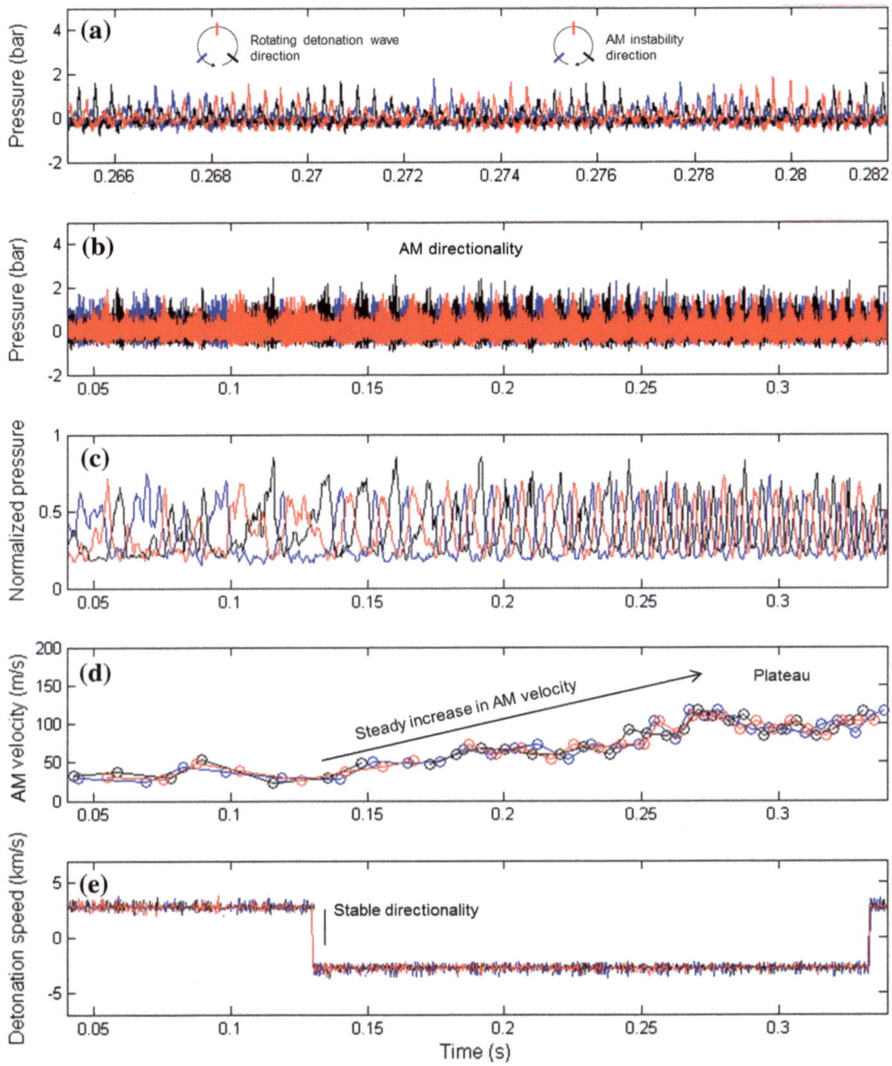

Fig. 6 Variables of interest acquired from air inlet sensors at $\dot{m}_a = 0.2$ kg/s, $\Phi = 1$

pressure sensor in row 1) and the ionization traces (from the ion sensor in row 2), for one such test point in a hollow RDC, running on ethylene-air mixtures. Note that both sensors are at the same sector/azimuth, which gives us the ability to track the relative changes between pressure and ionization. Upon ignition (t $\approx$ 0 s), it is seen from Fig. 7a, that initially there is no pressure activity; just ionization activity. At t $\approx$ 0.02 s, rotating pressure waves with considerable peak pressure magnitudes (and hence rotating detonations) appear. However, at t $\approx$ 0.05 s, the rotating detonation wave seemingly descends into highly unstable propagations that are characterized

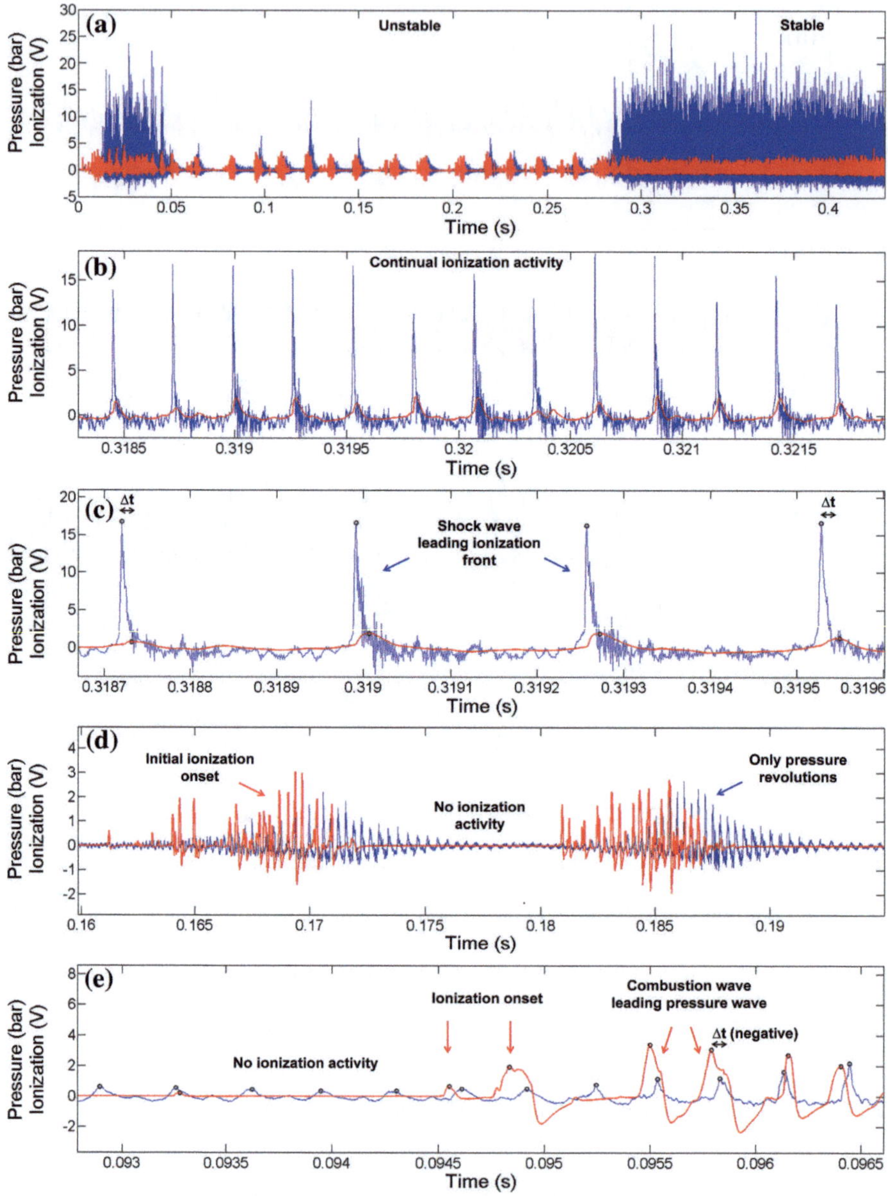

Fig. 7 Pressure and ionization traces at 0.4 kg/s and $\Phi = 1.6$: **a** complete traces, **b** stable propagation, **c** magnified traces during stable propagation, **d** unstable propagation, and **e** magnified traces during unstable propagation

by "packets" of subsequent laps that have the AM sinusoidal component. This AM fluctuation is characteristic of all the unstable rotating pressure wave propagations observed in hollow RDCs [11]. This unstable behavior extends till t $\approx$ 0.28 s, after which there is, once again, sustained periodic rotating detonations without the packets of instability. Of interest is the fact that, for a given packet of instability, ionization (red) is recorded for only about roughly the first half of the sinusoidal packet, whereas the second half is composed of just pressure activity (Fig. 7a). Figure 7b shows pressure and ionization data from an arbitrary duration from the same test point when there is stable rotating detonation wave propagation. It can be seen that the shock wave precedes the ionization (combustion) peak, as is to be expected in a detonation wave. A magnified image of this snippet (Fig. 7c) shows this distinction—pressure peak preceding the ionization peak—better. The black circles denote the peak pressure and ionization values, and are acquired by a time-of-flight algorithm that captures peak values, for a given lap. Fig. 7d contains pressure and ionization data during two subsequent packets of unstable operation, from the same test case. For a given amplitude modulated packet of subsequent rotating pressure waves, ionization activity exists for roughly the first half of the packets. Periods between the two packets do not exhibit any ionization, but do exhibit very weak rotating pressure waves that do not exceed 0.5 bar. It is once again emphasized here that even during this very low amplitude unstable, AM behavior, the pressure waves are confirmed to be rotating (since we have three circumferentially distributed sensors in row 1). A striking difference from the pressure-ionization coupling behavior seen during stable operation (Fig. 7c) is evident in Fig. 7e, during unstable behavior. During unstable propagation, most laps in a given cycle have the peak pressure succeeding the ionization/combustion peak. This is, in fact, observed for all unstable propagations. Hence, it is imperative to quantify the time lag, Δt, between the peak pressure event and the peak ionization event to ascertain the difference between the two propagation mechanisms. We use the system of positive Δt if the peak pressure precedes the peak ionization and negative Δt if it succeeds it.

Figure 8 gives the pressure trace (3 sensors in row 1), ionization trace (3 sensors in row 2), wave speed (from a time-of-flight algorithm) and time lag between pressure peak and ionization peak (negative value if ionization peak precedes pressure peak) for two different test points that exhibit unstable propagation. Note that all plots presented are for the whole duration of testing. The following observations can be readily made. During unstable operation, there are packets of rotating pressure waves with very low peak pressure magnitudes. The corresponding ionization traces reveal a similar packet-type behavior (Fig. 8). The sinusoidal oscillation evident in the peak pressure magnitudes of subsequent laps also results in sinusoidal oscillations in the lap-to-lap wave speeds, when the propagation is unstable. This is suggestive of a periodically strengthening and weakening rotating wave. It is imperative to note that during stable propagation the wave speed of each lap is about 95% of the C-J speed (red dotted line), whereas during unstable propagation, the wave speed is highest at the highest pressure point of the packet and is about 90% of the C-J speed. The average wave speed, however, is considerably lower due to the waxing and waning on either sides of the maximum pressure value. Of heightened importance are the

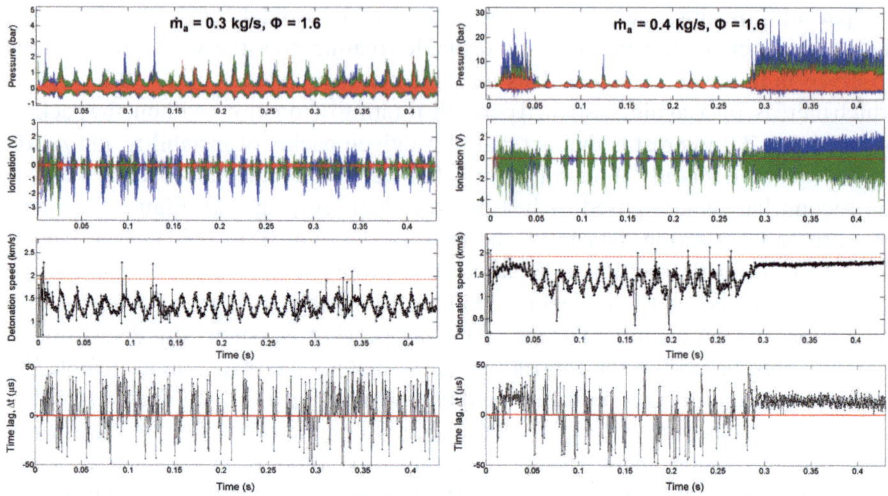

Fig. 8 Pressure traces (first row), Ionization traces (second row), wave speed (third row) and time lag between pressure and ionization wave (fourth row)

time lag characteristics across the different test cases. It is seen that when the propagation is stable, the time lag is predominantly positive across the approximately thousand laps, i.e. peak pressure precedes ionization. Initially, after ignition, combustion wave precedes the pressure wave (region indicated by blue arrow). However, after this initial period, the shock wave supersedes ionization events, thereby exhibiting the commonly known behavior of stable detonation wave propagation. When the propagation is unstable, such an observation is not seen. In fact, during unstable propagation, for a given packet of instability, the number of laps with ionization preceding the pressure wave is almost equal to the number of laps with the opposing trend. A maximum of ± 50 μs is set to remove fallacious values that are an artifact of the algorithm.

While this result is seemingly contradictory at first glance, it can be readily explained to be due to the flame acceleration and transition to detonation phenomenon, most commonly observed in ducts with deflagration-to-detonation inducing obstacles [43]. In their comprehensive review of the said phenomenon, Ciccarelli and Dorofeev state that "flame propagation in an enclosure generates acoustic waves that, after reflections from walls and obstacles, can interact with the flame front and develop flame perturbations through a variety of instability mechanisms" and say that if such a flame propagation is fast enough, it "can result in severe flame distortion which can induce flame acceleration and, in extreme cases, cause transition to detonation" [43]. One could thus postulate that the concave surface of the RDC outer wall acts like a reflecting obstacle thereby sustaining the propagation process. It is a well-known property of detonations in curved channels to have pronounced collision and subsequently stronger ignitions at the outer concave wall [44, 45]. Thus, finally we can attribute the low magnitude rotating pressure waves to be a type of unstable detonation wave propagation, where ignition seems to be continual and caused due to a probable flame acceleration mechanism sustained by the concave wall. Of course, such a claim is a preliminary theory and needs to be validated through focused, future studies. At present, however, it seems reasonable to attribute two different propagation mechanisms—conventional detonation propagation / flame acceleration causing DDT—to stable / unstable AM propagation. An analogous behavior to this would be the change from steady propagation to galloping and eventually stuttering detonations, in planar detonations in tubes, which exhibits a similar type of low frequency oscillation in pressure and wave speed [46].

4 Conclusion

The current chapter dealt with the identification and classification of three types of low frequency instabilities in rotating detonation combustors. While all three types are characterized by a sinusoidal amplitude modulation of the rotating detonation wave's peak pressure, their driving mechanisms appear to be divergent. The first two types of LFI are linked to the supply plenum dynamics. For the first case, Helmholtz resonance appears to the driving factor that causes acoustic oscillation in the air inlet which in turn produces spatially homogeneous low frequency instability in the combustor. This type of LFI is similar to the chugging instability widely reported in rocket engines, which is also predicated to depend on supply line and plenum acoustics. The second type of LFI is spatially non-homogenous and is characterized by a slow-moving rotary event in the combustor and the air inlet. This type of LFI is also alluded to in rocket engine instabilities' literature and is termed a precessing tangential mode. We show evidence that this rotary sinusoidal oscillation in detonation peak pressure moves in a direction opposite to the direction of the rotating detonation wave, and subsequently hypothesized it to be caused due to a complex constructive and destructive interference of the shock waves leaked into the supply plenum. This needs to be verified in future studies. Finally, the last LFI type is linked to the combustion wave dynamics itself and is spatially homogenous. It is shown that such LFI is distinguished by a complex interplay between the pressure wave and the combustion wave that together make a revolving front. During stable operation in a hollow RDC, the shock wave precedes the combustion front, whereas during LFI-defined unstable operation in the same RDC, the combustion wave precedes the pressure wave implying the presence of the phenomenon of flame-acceleration in ducts causing a detonation wave, rather than a detonation event itself. This process is similar to the alternative coupling and decoupling of the shock wave and combustion front, which is observed widely in unstable detonation propagation at near-limits in tubes. As noted in the introduction, the field of instabilities analysis in RDCs is nascent in comparison to the decades of research on the same in rocket engines, where there are considerable unknowns even at present. Moving forward, it is imperative to apply a similar impetus in understanding the off-design functioning of RDCs to eventually enable its implementation in real-world applications.

References

1. Turns, S.: An Introduction to Combustion. McGraw-Hill International, Singapore (2006)
2. Wolański, P.: Detonative propulsion. Proc. Combust. Inst. **34**, 125–158 (2013)
3. Yi, T.-H., Lou, J., Turangan, C., Choi, J.-Y., Wolanski, P.: Propulsive performance of a continuously rotating detonation engine. J. Propuls. Power. **27**, 171–181 (2011)
4. Wintenberger, E., Sheperd, J.E.: Thermodynamic cycle analysis for propagating detonations. J. Propuls. Power. **22**, 694–697 (2006)
5. Jones, S., Paxson, D.: Potential benefits to commercial propulsion systems from pressure gain combustion. In: Joint Propulsion Conference, San Jose, CA (2013)
6. Frolov, Dubrovskii, A.V., Ivanov, V.S.: Three-dimensional numerical simulation of operation process in rotating detonation engine. Prog. Propuls. Phys. **4**, 467–488 (2013)
7. Sousa, J., Paniagua, G., Collado Morata, E.: Thermodynamic analysis of a gas turbine engine with a rotating detonation combustor. Appl. Energy. **195**, 247–256 (2017)
8. Strakey, P., Ferguson, D., Sisler, A., Nix, A.: Computationally quantifying loss mechanisms in a rotating detonation engine. In: 54th AIAA Aerospace Sciences Meeting, pp. 1–14 (2016)
9. Sonwane, C., Clafli, S.: Recent advances in power cycles using rotating detonation engines with subcritical and supercritical CO2. In: The 4th International Symposium—Supercritical CO2 Power Cycles, Pittsburgh, Pennsylvania (2014)
10. Harrje, D.T.: Liquid propellant rocket combustion instability, Washington D.C. (1972)
11. Anand, V., St. George, A., Farbos de Luzan, C., Gutmark, E.: Rotating detonation wave mechanics through ethylene-air mixtures in hollow combustors, and implications to high frequency combustion instabilities. Exp. Therm. Fluid Sci. 92 (2018)
12. Anand, V., St. George, A., Driscoll, R., Gutmark, E.: Longitudinal pulsed detonation instability in a rotating detonation combustor. Exp. Therm. Fluid Sci. **77**, 212–225 (2016)
13. Anand, V., St. George, A., Driscoll, R., Gutmark, E.: Characterization of instabilities in a rotating detonation combustor. Int. J. Hydrogen Energy. **40**, 16649–16659 (2015)
14. Anand, V., George, A.S., Driscoll, R., Randall, S., Gutmark, E.J.: Statistical treatment of wave instability in rotating detonation combustors. In: 53rd AIAA Aerospace Sciences Meeting (2015)
15. Liu, Y., Wang, Y., Li, Y., Li, Y., Wang, J.: Spectral analysis and self-adjusting mechanism for oscillation phenomenon in hydrogen-oxygen continuously rotating detonation engine. Chinese J. Aeronaut. **28**, 669–675 (2015)
16. Li, Y., Wang, Y., Wang, J., Li, Y.: Detonation instability of continuously rotating detonation engines for H 2 -Air mixture **1**, 1–7 (2014)
17. Frolov, S.M., Aksenov, V.S., Ivanov, V.S., Shamshin, I.O.: Large-scale hydrogen-air continuous detonation combustor. Int. J. Hydrogen Energy. **40**, 1616–1623 (2015)
18. Suchocki, J., Yu, S.-T., Hoke, J., Naples, A., Schauer, F., Russo, R.: Rotating detonation engine operation. In: 50th AIAA Aerospace Sciences Meeting including the New Horizons Forum and Aerospace Exposition. American Institute of Aeronautics and Astronautics, Reston, Virigina (2012)
19. Dyer, R., Naples, A., Kaemming, T., Hoke, J., Schauer, F.: Parametric testing of a unique rotating detonation engine design. In: 50th AIAA Aerospace Sciences Meeting including the New Horizons Forum and Aerospace Exposition, Nashville, Tennessee (2012)
20. Shank, J.C., King, P.I., Karnesky, J., Schauer, F.R., Hoke, J.L.: Development and testing of a modular rotating detonation engine. In: 50th AIAA Aerospace Sciences Meeting, Nashville, TN (2012)
21. Kindracki, J.: Experimental research on rotating detonation in liquid fuel-gaseous air mixtures. Aerosp. Sci. Technol. **43**, 445–453 (2015)
22. Anand, V., George, A., Driscoll, R., Gutmark, E.: Analysis of air inlet and fuel plenum behavior in a rotating detonation combustor. Exp. Therm. Fluid Sci. **70**, 408–416 (2016)
23. Yang, C., Wu, X., Ma, H., Peng, L., Gao, J.: Experimental research on initiation characteristics of a rotating detonation engine. Exp. Therm. Fluid Sci. **71**, 154–163 (2016)

24. Wang, C., Liu, W., Liu, S., Jiang, L., Lin, Z.: Experimental verification of air-breathing continuous rotating detonation fueled by hydrogen. Int. J. Hydrogen Energy. **40**, 9530–9538 (2015)
25. Lin, W., Zhou, J., Liu, S., Lin, Z., Zhuang, F.: Experimental study on propagation mode of H2/Air continuously rotating detonation wave. Int. J. Hydrogen Energy. **40**, 1980–1993 (2015)
26. Kindracki, J., Kobiera, A., Wolański, P., Gut, Z., Folusiak, M., Swiderski, K.: Experimental and numerical study of the rotating detonation engine in hydrogen-air mixtures. Prog. Propuls. Phys. **2**, 555–582 (2012)
27. Peng, L., Wang, D., Wu, X., Ma, H., Yang, C.: Ignition experiment with automotive spark on rotating detonation engine. Int. J. Hydrogen Energy. **40**, 8465–8474 (2015)
28. Clayton, R., Rogero, R., Sotter, J.: An experimental description of destructive liquid rocket resonant combustion. AIAA J. **6**, 1252–1259 (1968)
29. Narayanaswamy, V., Raja, L.L., Clemens, N.T.: Control of unsteadiness of a shock wave/turbulent boundary layer interaction by using a pulsed-plasma-jet actuator. Phys. Fluids. **24** (2012)
30. Boshoff-Mostert, L., Viljoen, H.J.: Analysis of combustion-driven acoustics. Chem. Eng. Sci. **53**, 1679–1687 (1998)
31. Anand, V., St. George, A.: Amplitude modulated instability in reactants plenum of a rotating detonation combustor. Int. J. Hydrogen Energy. **42**, 12629–12644 (2017)
32. St. George, A.C., Driscoll, R.B., Anand, V., Munday, D.E., Gutmark, E.J.: Development of a rotating detonation engine facility at the university of cincinnati. In: 53rd AIAA Aerospace Sciences Meeting. American Institute of Aeronautics and Astronautics, Reston, Virginia (2015)
33. Anand, V., St. George, A., Driscoll, R., Gutmark, E.: Investigation of rotating detonation combustor operation with H 2 -Air mixtures. Int. J. Hydrogen Energy. **41**, 1281–1292 (2016)
34. St. George, A., Randall, S., Anand, V., Driscoll, R., Gutmark, E.: Characterization of initiator dynamics in a rotating detonation combustor. Exp. Therm. Fluid Sci. **72**, 171–181 (2016)
35. Stevens, C.A., Fotia, M.L., Hoke, J.L., Schauer, F.R.: Comparison of transient response of pressure measurement techniques with application to detonation waves. AIAA Aerosp. Sci. Meet. **5**, 1102–2015 (2015)
36. Hulka, J.R., Jones, G.W.: Performance and stability analyses of rocket thrust. In: 46th AIAA/ASME/SAE/ASEE Joint Propulsion Conference, Nashville, TN (2010)
37. Alster, M.: Improved calculation of resonant frequencies of Helmholtz resonators. J. Sound Vib. **24**, 63–85 (1972)
38. Mariappan, S., Sujith, R.I.: Thermoacoustic instability in a solid rocket motor: non-normality and nonlinear instabilities. J. Fluid Mech. **653**, 1–33 (2010)
39. Fotia, M.L., Hoke, J.L., Scientific, I.: Propellant plenum dynamics in a two-dimensional rotating detonation experiment. In: 52nd Aerospace Sciences Meeting, pp. 1–10 (2014)
40. Schwer, D., Kailasanath, K.: Feedback into mixture plenums in rotating detonation engines. In: 50th AIAA Aerospace Sciences Meeting including the New Horizons Forum and Aerospace Exposition, Nashville, TN (2012)
41. Gruber, S., Skews, B.: Weak shock wave reflection from concave surfaces. Exp. Fluids. **54**, 1571 (2013)
42. Bedick, C., Sisler, A., Ferguson, D., Strakey, P., Nix, A., Billips, D.: Development of a lab–scale experimental testing platform for rotating detonation engine inlets. In: 55th AIAA Aerospace Sciences Meeting, Grapevine, TX (2017)
43. Ciccarelli, G., Dorofeev, S.: Flame acceleration and transition to detonation in ducts. Prog. Energy Combust. Sci. **34**, 499–550 (2008)
44. Nakayama, H., Moriya, T., Kasahara, J., Matsuo, A., Sasamoto, Y., Funaki, I.: Stable detonation wave propagation in rectangular-cross-section curved channels. Combust. Flame. **159**, 859–869 (2012)
45. Pan, Z., Fan, B., Zhang, X., Gui, M., Dong, G.: Wavelet pattern and self-sustained mechanism of gaseous detonation rotating in a coaxial cylinder. Combust. Flame. **158**, 2220–2228 (2011)
46. Lee, J.H.S.: The Detonation Phenomenon. Cambridge University Press (2008)

Influence of Operating Conditions and Residual Burned Gas Properties on Cyclic Operation of Constant-Volume Combustion

Quentin Michalski, Bastien Boust and Marc Bellenoue

Abstract The pressure-gain combustion concept is a solution envisioned to increase the thermodynamic efficiency of gas turbines. This article addresses the behaviour of piston-less constant-volume combustion in relevant conditions of engine application. For this purpose, a lab-scale combustion vessel (0.3 L) is run in cyclic operation (10 Hz) with an improved control over the boundary conditions. This facility features the spark-ignited, turbulent combustion of n-decane directly injected in preheated air (423 K, 0.4 MPa), with an overall equivalence ratio of 0.9. Solenoid valves are used to perform the air intake and burnt gas exhaust. A 0D analysis is developed and used to compute the gas thermodynamic evolution based on the experimental pressure traces. The effect of the main operating parameters on the combustion process is discussed: ignition delay, exhaust pressure and wall temperature. The vessel is operated without scavenging, hence the exhaust pressure drives the amount and the temperature of residual burnt gas (16–39% according to the 0D analysis). Highly diluted cycles (exhaust pressure 0.2 MPa) exhibit a higher combustion efficiency, but have a longer combustion duration (3 times more) than those of low dilution (exhaust pressure 0.07 MPa). For a higher wall temperature representative of engine combustor (1000 K), the heat losses are directly reduced, which affects the residual burnt gas properties. This also influences the residual gas temperature (870–1030 K) as well as dilution (10–26%).

Keywords Pressure-gain combustion · Thermodynamic analysis
Cycle optimization

Q. Michalski (✉) · B. Boust · M. Bellenoue
Institut Pprime, (CNRS, ISAE-ENSMA, Universite de Poitiers),
1 Avenue Clement Ader, Teleport 2, BP 40109, 86961 Futuroscope Chasseneuil, France
e-mail: quentin.michalski@ensma.fr

M. Bellenoue
e-mail: marc.bellenoue@ensma.fr

R. King (ed.), *Active Flow and Combustion Control 2018*,
Notes on Numerical Fluid Mechanics and Multidisciplinary Design 141,
https://doi.org/10.1007/978-3-319-98177-2_14

1 Introduction

Future turbomachine generation requires multiple breakthrough to achieve the targets decided for 2050 [1]. Pressure-gain combustion is a direct way to improve the thermal efficiency of aircraft combustor. Different technical solutions are currently developed such as pulsed or rotating detonation, controlled auto-ignition and confined deflagration. The latter, which is referred to as "constant-volume combustion" (CVC) was among the first solutions implemented on ground-based turbomachines through the work of Holzwarth [2]. At the time, large 200 L chambers were fed by an external compressor at 1 to 1.5 bar of initial pressure for a constant-volume combustion working up to 1 Hz with an overall efficiency reported as high as 20% (which is significantly high given the reported operating initial pressure is 1 atm). Recently a simplified prototype of deflagrative chamber with rotating valves, aimed at aircraft application, was operated at higher frequency (25 to 60 Hz) and smaller volume (0.65 L) for improved power density [3] and its combustion stability was investigated [4]. On such device (as on other devices based on deflagration such as wave rotors [5]), a minimum operating frequency is of importance mainly because it reduces the leaks at the clearance gaps between the rotating parts. Regarding the combustor thermodynamic efficiency, several studies discuss the ideal performances of pressure-gain thermodynamic cycles (for detonation as well as for deflagration [5–7]). However few studies performed on real-operated devices discuss non-ideal performances based on energy losses analysis [8]. The pressure rise is intrinsically related to the initial density and dilution of the fresh mixture. For piston-less combustors the exhaust happens at peak pressure and thus follows an adiabatic expansion at constant volume and decreasing mass. The expansion must start directly at the end of combustion. Consequently, the residual burnt gas (RBG) temperature can be considerably higher than that found on four-stroke internal combustion engines (ICE) and close to that found on two-stroke ICE. Since such devices should operate upstream of a turbine, the RBG pressure should also be higher, thus directly affecting the RBG mass (amount and temperature) as well as the expansion ratio. To pursue the investigation of such conditions, a new facility referred to as CV2 was designed. It allows for a simplified operation, a study of performances and a detailed analysis of key processes found in such combustors. Two solenoid valves handle the intake and exhaust, and ensure negligible leaks during combustion. Such choice of technology involves a limited operating frequency (approximately 10 Hz) but allows for a flexible control over cycles time charts. Two dedicated tanks allow for a control over the inlet and exhaust boundary conditions, e.g. exhaust pressure, intake temperature and pressure. Moreover, a 0D multi-zone model is developed to perform the thermodynamic analysis of the experimental combustion cycles. A previous study based on direct injection of gaseous propane evidenced ignition stability issues that were shown to be statistically related to the velocity at ignition [9]. For more representative conditions of practical aircraft combustor application, direct injection (DI) of liquid n-decane is used. Air is preheated up to 423 K and pressure in the chamber is set to 0.4 MPa prior to combustion. In this article, the effect of two crucial boundary conditions is investigated, namely the exhaust pressure and the wall temperature. Their influence on the thermodynamic efficiency is discussed with respect to the different energy loss phenomena evidenced in the cycle. Selected operating points are optimized and the energy losses related to the different phases of the cycle are compared.

2 Experimental Setup

The CV2 rig is designed to produce cyclic CVC conditions representative of an air-breathing application. The rig is composed of a dedicated intake and exhaust system connected to a single combustion vessel (see Fig. 1). The conditions of intake and exhaust are set in dedicated tanks where the gas composition, pressure and temperature are measured and regulated. Intake and exhaust control systems consist in high-speed valves (COAX MK10) equipped with position sensors (5 kHz). Their response time and transient behaviour were carefully characterized and tabulated (see cross-section in Fig. 4). The thermal load on the exhaust valve, due to burnt gases flowing through, increases its response time. To limit this thermal drift and ensure the repeatability between consecutive sequences, the exhaust valve temperature is regulated at 40 °C (Kistler 2621 F conditioning unit). During an experimental sequence, from the first cycle to the twelfth one, it corresponds to a 3–4 ms maximum delay at opening. The valve is spring-loaded, thus the closing is not affected by the thermal load. Consequently, the exhaust duration is reduced by the same amount of time following the opening delay. This drift is recorded by the position sensors and taken into account in the following analysis. The tanks are connected to the valves through

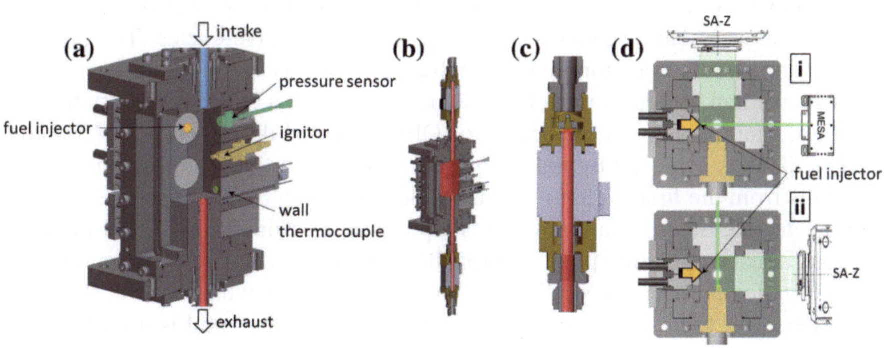

Fig. 1 CV2 in the configuration of interest: **a** combustion chamber and key components; **b** constant-volume parts of the chamber and connections highlighted in red. The wall thermocouple is symmetrically positioned relatively to the cut-plane to the second pressure sensor (in green). **c** High-speed valve where the constant-volume part is highlighted in red. **d** The two configurations (i) and (ii) of the laser sheet and camera: (i) the laser sheet is in the injector symmetry plane, (ii) the laser sheet is in the ignitor plane

flexible hoses of internal diameter 16 mm, length 1 m. The chamber is rectangular with internal dimensions of 50 × 50 × 100 mm. Its aspect ratio λ (length/width) is equal to 2, therefore internal wave dynamics is not as dominant as in wave rotor ($\lambda =$ 13 in [8]). In the present configuration, the closed volume ranges from the intake to the exhaust valve seat, including the connection tubes of internal diameter 12 mm. Overall those volumes account for an additional 0.05 L (see Fig. 1b, c). The complete volume, measured precisely with water, is 0.32 L (0.5% uncertainty) including dead cavities. Intake air is supplied by a 20.76 L (0.3% uncertainty) intake tank at a pressure of 1.0 MPa (0.2% uncertainty). Its volume, including the connecting hose and connecting parts, was precisely measured by weighing the mass variation of a 50 L air bottle while pressurizing the tank. The fuel is injected directly into the chamber through one liquid injector (Bosch 0 261 500 029: 7 holes, 15 ° bias angle) fitted to the wall. The injector is fed with liquid n-decane pressurized at 5.0 MPa. The injector mass flow rate was measured separately, at atmospheric pressure, by weighing the fuel quantity injected for different injection durations, which yields a linearized effective mass flow rate of 6.74 mg per ms of injection (1.1% uncertainty). The initial pressure in the chamber as well as the fuel quantity injected are controlled by varying respectively the intake and injection durations. Ignition is triggered by a non-resistive sparkplug topped with a conventional ignition coil charged up to 2.5 ms (maximum charge), thus delivering an inductive discharge of approximately 40 ± 10 mJ lasting for 2.5 ms. The command, synchronization and acquisition are handled by a National Instrument controller (PXIe-8820 with a PXIe-6363) working up to 100 kHz for fast sensors and 10 kHz for slow sensors. The intake system is thermally insulated and its temperature is regulated by PID-controlled heating collars. The intake hose has its own integrated heating system regulated by its own PID. The chamber is non-insulated and heated with standard cartridge heaters fitted into the walls of the vessel; its temperature is PID-regulated based on the eroding junction thermocouple (Nanmac) used for the local measurement of wall surface temperature. Each experimental sequence is a finite series of twelve cycles. The pressure decrease in the fuel injection tank is negligible, whereas the pressure decrease in the air intake tank is approximately 0.7% per cycle, thus a total 8% of variation for 12 cycles. The overall equivalence ratio (OER) injected during each cycle is kept constant. This is achieved by reducing the fuel injection duration, by a maximum of 1–2 ms from the first cycle to the last cycle depending on the experimental conditions selected. Such variation requires a preliminary calibration of fuel injection duration for each operating point. A broadband oxygen sensor (lambda sensor LSU 4.9) is installed at the exhaust, directly downstream of the exhaust valve. It gives a measurement of the oxygen concentration in the burnt gases and is used to verify that the oxygen concentration found in the exhaust gas matches the one associated with the injected OER. An absolute piezoresistive sensor (Kistler 4049B, 60 kHz bandwidth) is positioned near the oxygen sensor for transient pressure recording. This measurement is used to compensate the lambda measurement as the exhaust pressure varies [10]. A piezoelectric sensor (Kistler 6067C, 90 kHz bandwidth) is used for pressure recording in the combustion vessel. A silicon layer is deposited on its surface, approximately 1 mm thick, to prevent thermal drift. For combustion pressure measurement, an absolute piezoresistive sensor (Kistler 4049B) is used in the chamber as well to compensate for the piezoelectric thermal drift. Those two sensors are temperature-regulated (70 °C) by a dedicated Kistler 2621 F conditioning unit. The OER is computed by measuring the pressure variation in the air tank for each cycle and by the a priori calibration of the injector mass-flow rate. Using the perfect gas law and assuming isentropic expansion in the air tank, the OER of the fresh charge introduced in the chamber during the cycle reads

$$OER = D_{st} \frac{\gamma_a m_f}{M_a \Delta p_a V_a} \tag{1}$$

where M, Δp, m and V are respectively the molar mass, pressure variation, mass variation and tank volume, a and f stand respectively for air and fuel tanks and D_{st} is the mass air-fuel ratio of a n-decane air stoichiometric mixture. The injection of both air and fuel starts after the closing of the exhaust valve, that way there is no scavenging phase. The effective OER in the chamber thus corresponds to the injected OER. The uncertainty in the OER for one cycle is mainly related to the air tank pressure measurement, Δp (16 mbar of uncertainty). However, given the high reproducibility of the valve opening cross-section (measured optically), one can suppose the linearization of the pressure loss starting from the third cycle. Doing so, the uncertainty in Δp for a single cycle drops to 5.4%, which yields an uncertainty of 6.8% in the OER calculated for a single cycle, e.g. at OER = 0.9, it corresponds to an uncertainty of 0.06. 2D time-resolved particle image velocimetry (PIV) is performed to characterize the aerodynamic conditions. The air injected in Fig. 2b was oil-seeded (2.0 μm SMD), which can be seen in the first image (SOI+0.2 ms). As for Fig. 2a, air was not oil-seeded although residual particles can be seen in the air flow. Additionally, Mie tomography pictures give an insight into the fuel atomization during the intake phase. PIV is performed at 20 kHz (50 μs time between two velocity fields), which means the laser emission (Nd:YAG, 532 nm MESA PIV) works at 20 kHz and the camera (SA-Z Photron) works at 40 kHz. To account for the velocity variation encountered in one cycle, a variable PIV interframe delay (from 2 μs to 25 μs) is used (PTU X from Lavision). The filling process is highly repeatable and the interframe profile is tuned after a few iterations on experiments. PIV data is processed with a multi-pass iterative calculation starting from 64 × 64 px down to 16 × 16 px with a 50% overlap (Davis 8.4.0). This yields a physical velocity resolution of 1.3 × 1.3 mm.

The PIV velocity fields highlight an internal aerodynamics structured around a highly compressible flow which penetrates the chamber a few milliseconds after the opening of the valve and generates intense shear flow on the walls [11]. Air and fuel are injected simultaneously for enhanced mixing and atomization (see Fig. 2) during a fully closed exhaust phase. The injection process is apparently delayed, due to the time induced by the 150 mm distance from the valve seat to the chamber. The same delay occurs after the closing of the valve, as flow is still observed on the pictures. Finally the velocity decay is exponential because of a free-turbulence decay during this later phase. Regarding the atomization process, the PIV yields intake velocities of

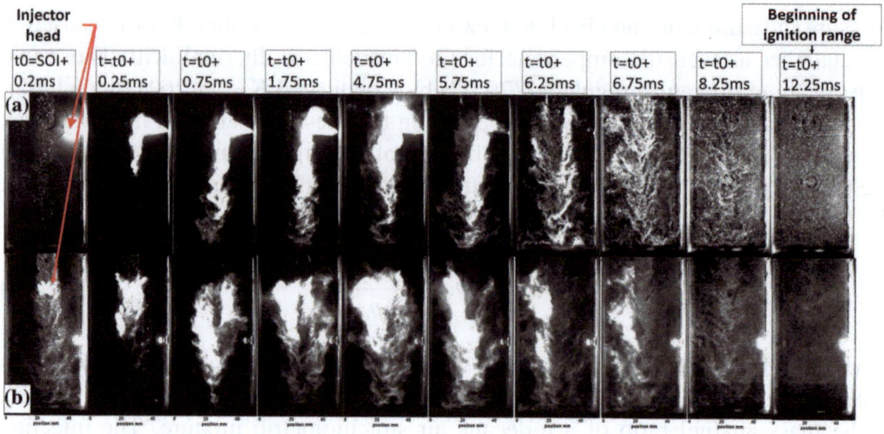

Fig. 2 Mie tomography images of the intake phase during a reactive combustion cycle, which shows the atomization of the spray on the main air mass flow from the centre upper part of the chamber. A. PIV in injector plane (see Fig. 1a), B. PIV in ignitor plane (see Fig. 1b)

approximately 250 m/s (confirmed by LES [11]) at air mass flow rates ranging from 60 to 75 g/s. Phase-Doppler interferometry (PDI) measurements were performed on a cold injector (21.5 °C) fueled with n-decane outside the chamber at ambient conditions, using an Artium PDI-200MD equipped with focal lenses of 500 mm for both emission and reception (1.6–180 μm detection range). Those measurements performed 30 mm downstream of the injection head yield an average droplet velocity of 18.5 m/s and a 30 μm Sauter Mean Diameter (SMD), for a mass flow rate of 6.74 g/s. The fuel spray and the air are injected simultaneously. Obviously, the spray and the central air plume have a similar dimension at their impingement point. This behaviour prevents the fuel spray to reach the wall and promotes a uniform distribution of the fuel droplets in the chamber (as seen on the last MIE tomography pictures of Fig. 2). Regarding this spray-air plume interaction, based on a fuel density of 643 kg/m3 (403 K and 1 bar), an air viscosity of 24.1 μPa.s (423 K and 1 bar), the fuel particles are estimated to penetrate up to 0.1 mm into the air plume which corresponds to 1 % of its total width [12]. Along their trajectory, the droplets evaporate so that the penetration length should then be lower. Such behavior is thus consistent with the Mie tomography pictures recorded for PIV measurements during the simultaneous fuel and air injection (Fig. 2). In the following, the procedure for the 0D calculation fit to the acquisition is detailed.

3 0D Numerical Setup

A specific 0D multi-zone code was developed to compute the time-resolved evolution of non-directly measurable properties such as mixture density, exhaust gas temperature and composition. The code is developed under Matlab R2015b. The chemical description of the fluid is based on the Cantera (2.3.0) Matlab library [13] using a reduced number of species to speed up the computation. The species included in the description are $H_2, O_2, OH, H_2O, CO, CO_2, NO, N_2, nC_{10}H_{22}$ which thermodynamic properties are taken from standard JANAF table. The model includes a separate representation of each vessel with regard to heat or mass transfer, including the intake tank and line (AD), the fuel injection tank (INJ T), the combustion chamber (CC), the exhaust tank (EJ) and the exhaust line (EJL) (Fig. 3). The isochoric combustion phase is modeled through a three-zone description. Combustion happens in 3 phases: the flame propagates over a slice of fresh gases, this reacting slice (referred to as zone 13 in Fig. 3) is set to chemical isobaric equilibrium at fixed enthalpy, which increases its volume. The overall volume of gas in the chamber is temporarily higher than the geometrical chamber volume. The 3 volumes of gas in the chamber are compressed (without mixing) at fixed entropy until their cumulated volume matches the chamber volume. At each integration step, the burning slice (zone 13 see Fig. 3) is mixed with the burnt gases (zone 12 see Fig. 3). The combustion duration is set as measured from experimental data for each cycle. Unburnt gases are accounted for by stopping the combustion over a volume including the gaps of the chamber assembly, as well as the quenching distance estimated at 100 μm according to high-pressure measurements [14]. Overall unburnt gas accounts for 3.0% of the total volume with 2% due to the cavity quenching and the remaining 1% due to the flame-wall quenching. At the end of the combustion phase, fresh gases and burnt gases are mixed, thus the chemistry is frozen during the exhaust phase until the combustion of the following cycle. For each cycle, the intake and exhaust durations are set as measured on their respective position sensors. The intake flow is choked during most of the

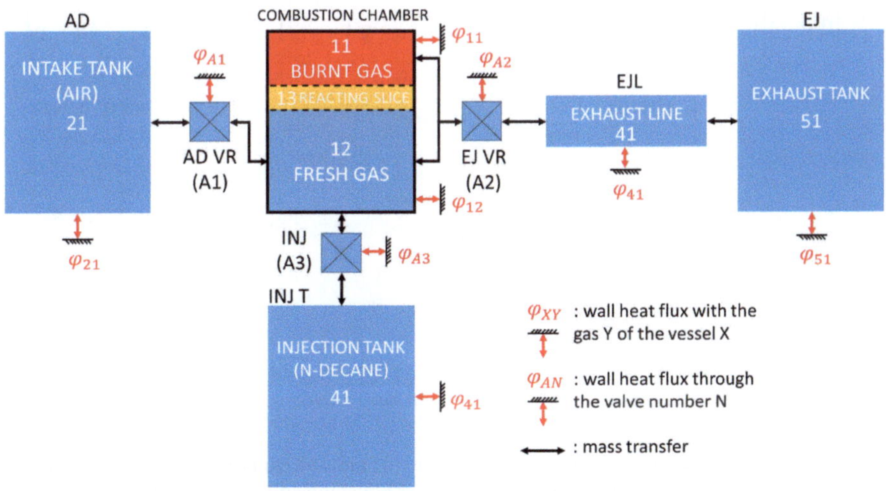

Fig. 3 Graphical description of the vessels described in the 0D analysis with detailed number of heat (ϕ with index relative to the gas phase number, red arrows) and mass transfer (black arrows) connections set between the vessels

intake phase, a constant discharge coefficient is thus set until the mass drop measured experimentally in the intake tank matches the 0D simulated one that sets the mass of air injected per cycle into the chamber. During the first cycle, the sensible enthalpy loss is set for the air that flows through the valve (which internal temperature is lower than the pre-heated air) until the initial pressure in the chamber is matched. Wall heat transfer coefficient h (in $W/m^2/K$) is modeled using Hohenberg's correlation [15]

$$h = 130 V_{cc}^{-0.06} p_{1X}^{0.8} T_{1X}^{-0.4} (v + 1.4)^{0.8} \tag{2}$$

where V_{cc} is the volume of the chamber, p_{1X} and T_{1X} are the pressure and temperature of the gas phase for $X = 1$ or 2 (either fresh or burnt gases). The representative velocity v is computed based on the chamber cross-section and on the instantaneous mass-flow rate. The experimental increase in wall temperature is less than 10 K during combustion (see Fig. 4); there is enough time for heat conduction between each cycle, so that the cumulated wall temperature rise at the end of the 12th cycle is of 28 K. The wall temperature for the 0D calculation is thus considered constant over a complete sequence. The overall pressure shape is matched for the combustion phase and the exhaust phase on most cycles by adjusting the velocity (during the

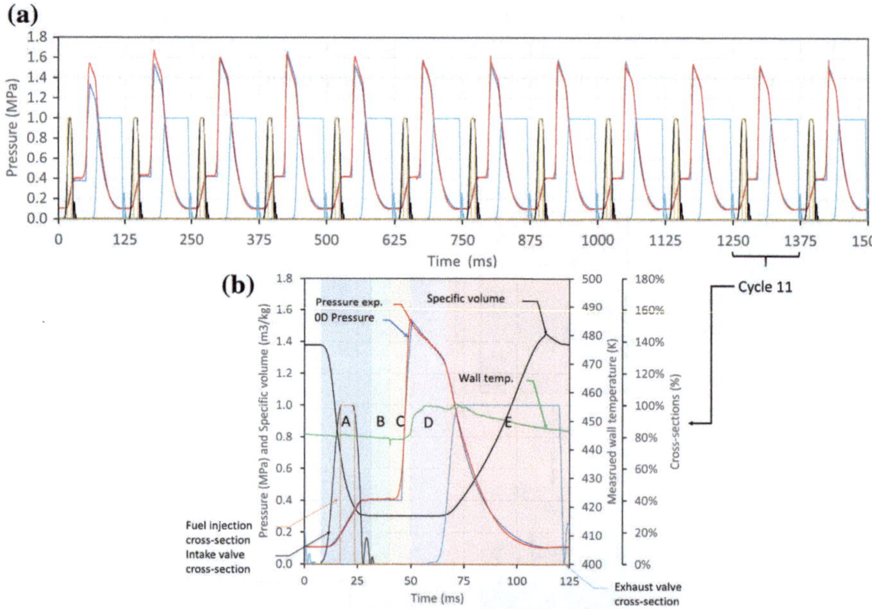

Fig. 4 **a** Comparison between numerical 0D pressure (in blue) and experimental pressure signal (in red) for a single sequence (12 cycles) with intake (black), fuel injection (orange) and exhaust (blue) cross-sections. **b** Best fitted cycle selected for detailed analysis, the specific volume is figured as well. The different phases of the experimental cycle are figured in color (A the intake, B the relaxing phase, C the combustion, D the isochoric cooling and E the exhaust). The wall temperature is figured in green

combustion and the isochoric cooling phase) and the exhaust discharge coefficient. Only experimental conditions with consecutive successful ignitions are analyzed. In the present cases, the cycle-to-cycle variation is low enough so that one cycle can be considered as representative of the whole set.

4 Detailed Analysis on One Operating Point

More details are given on the comparison between the experimental pressure signal and its respective computed 0D value.

The pressure signal is well matched during the intake phase until the opening of the exhaust (Fig. 4). The exhaust is more complex to simulate for several reasons. First the flow is only choked for part of its duration and the stagnation pressure and temperature vary during the exhaust phase. Also, the temperature of the exhaust gas, higher than that of the intake, increases the volume flow rate. Even with the large exhaust pipes (16 mm inner diameter), the pressure rises slightly in their residual volume. This effect is accounted for by using an intermediate volume EJL between the exhaust tank and the chamber. The unburnt gas influence is quantified as follows. From the conditions of Fig. 4b cycle, the unburnt volume is varied from 0 to 6.0%. The mass fraction of unburnt gas depends on the maximum pressure reached. At 3.0% unburnt volume, up to 8.7% of the end-compression gases remain unburnt which accounts for a 0.10 MPa straight pressure loss (reference value of 0 MPa in Table 1). If this unburnt volume is brought to 0%, the resulting maximum pressure is thus raised by 0.1 MPa (hence with a positive sign in Table 1). Such unburnt volume can indeed be avoided by optimizing the chamber design. The assembly gaps (required in our modular conception) can be avoided in the case of real application designs, possibly made by additive manufacturing. As well, fuel-charge stratification can limit the effect of flame-wall interaction by isolating the combustion in the centre of the chamber, thus reducing the amount of unburnt gas due to quenching.

A constant value of 3.0% of unburnt gases in volume (among which 1.0% is due to quenching distance) corresponding to our estimation is considered in the following analysis. Once properly tuned, from one sequence to another with different experimental settings, e.g. varying exhaust pressure or ignition timing, only small adjustments to the heat transfer model parameters are made to match the pressure traces. The heat lost to the wall is then compared to the ideal heat of combustion $Q_c = \Delta m_f I_p$, with $I_p = 44.2$ MJ/kg being the lower heating value of n-decane. During

Table 1 Effect of unburnt gases on the maximum pressure, the reference volume values are in bold

Percentage of unburnt volume (%)	0	**3.0**	6.0
Unburnt mass (%)	0	**8.7**	16.5
Induced relative pressure variation (MPa)	+0.10	**0**	−0.09

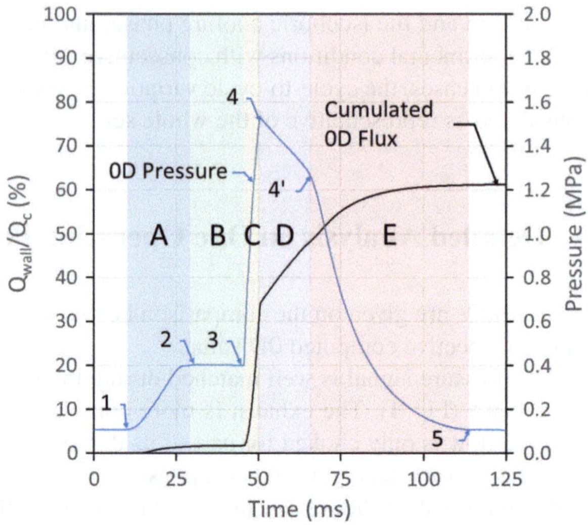

Fig. 5 Experimental and 0D simulated cumulated wall heat flux Qw in percentage of Qc, with its respective 0D pressure signal used to figure the different phases of the 0D cycle. The reference points (1, 2, 3, 4, 4' and 5) for energy losses calculation during each phase are figured

the intake phase, turbulent mixing occurs between the RBG and the fresh gases. This partially premixed charge of RBG, air and fuel, is continuously cooled by the walls at a rate driven by near-wall turbulence. A fraction of the total thermal energy (2.0% in Fig. 5) is thus lost during the intake phase (A-B). Approximately 30% of Q_c is transferred to the wall during the combustion phase (C), 18% after the combustion is complete and before the exhaust full opening (D), and 11% during the exhaust phase (E) until the pressure equilibrium is completed. Indeed, those values depend on the unburnt volume taken in the computation. Considering the wall heat flux is adjusted to match the pressure peak, a lower unburnt mass fraction would require a higher wall heat flux and vice-versa. Nevertheless, the same amount of energy-loss must be set to match the pressure.

The intake phase (A) starts at point 1 and ends at point 2. Following a relaxation phase (B) which duration depends on the chosen ignition timing, the isochoric combustion phase (C) then starts from point 3. On that cycle, the maximum combustion pressure reaches 1.53 MPa at point 4. The beginning of the exhaust phase is delayed to make sure the whole combustion process is isochoric. Therefore, the combustion phase is followed by a 25 ms isochoric cooling phase that comprises a true isochoric phase as well as part of the unsteady opening of the valve (approximately 15 ms). The isochoric cooling phase (D), from 4 to 4', is a straight energy loss whereas the exhaust phase (E) is a non-adiabatic expansion that starts from 4' to pressure equalization with the exhaust tank. A post-processing is made from the condition 3, to compute the associated ideal isochoric and adiabatic combustion. Under constant heat capacity and perfect gas assumptions, the adiabatic pressure ratio π_a is directly related to the energy-loss and computed as:

$$\pi_a = \frac{p_4 - p_3}{p_{4a} - p_3} \tag{3}$$

Table 2 Cycle thermodynamic properties and detailed balance of energy losses. The dilution is given in mass percentage relative to the mass at ignition

	Unburnt volume (%)	Maximum pressure (MPa)	RBG Dilution (%)	Residual temperature before intake (K)	Temperature after intake (K)
Cycle properties	3.0	1.53	22.0	500	434
	Intake losses	Unburnt gases	Combustion losses	Isochoric cooling heat losses	Exhaust heat loss
Energy loss Q/Q_c (%)	1.8	8.7	32.4	12.6	14.5

where p_{4a} is the adiabatic and isochoric combustion pressure. In Fig. 5, this pressure ratio is 67% which is consistent with the 41 % energy-loss simulated during the combustion phase from point 3 to 4 (see Table 2): 8.7% corresponding to unburnt gases and 32.4% to heat losses. In the following, this energy-loss balance is used to evaluate the effect of initial conditions (at point 3) on combustion through RBG dilution, both in mass and temperature. At fixed boundary conditions, the influence of the ignition timing is discussed by combining experiments and 0D simulation analysis.

5 Influence of the Ignition Timing on the Cycle Performances

To improve the power density, high operating frequency is required. Consequently, combustion cycles need to be as short as possible. For combustion to be as fast as possible after ignition, one requires both high turbulence as well as favourable local mixture conditions in equivalence ratio and dilution to promote ignition success and flame velocity. The local variation of those properties has a direct impact on the ignition probability and on the ignition kernel growth towards the rest of the chamber. A previous study in direct gaseous injection with non-preheated air evidenced the sensitivity of ignition success to the local velocity properties; for a given dilution and equivalence ratio, statistical limits for the velocity were evidenced [9]. Only stable conditions are considered in the following. For that purpose, the overall equivalence ratio selected is 0.90 ± 0.05, and the exhaust backpressure is set to 1 bar to limit the RBG dilution to approximately 20%. The ignition timing is varied from the end of the intake phase, to 15 ms later (12% of total cycle duration).

During that time, the mean velocity in the chamber drops from approximately 50 m/s, at the beginning of the pressure plateau, to 15 m/s in 3 ms, at the end of the first rebound, and then decays from 15 m/s to less than 5 m/s in the following

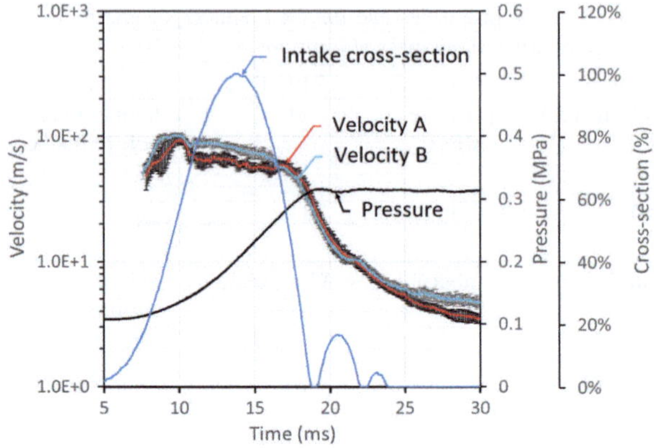

Fig. 6 Ensemble-averaged mean velocity in the chamber (with the cycle-to-cycle standard deviation as bars) from two separate sequences, for a PIV plane on the injector head (red) and in front of the injector head (blue). A and B refers to the cases presented on Fig. 2. The intake cross-section and pressure signal are figured as well. The initial pressure is slightly lower (3.2 MPa) than for the rest of the studied sequences (4.0 MPa) which affects the velocity magnitude but not the timings of events

Fig. 7 Instantaneous experimental pressure signals for different cycles with varying ignition timing relative to the beginning of the pressure plateau (closing of the intake valve)

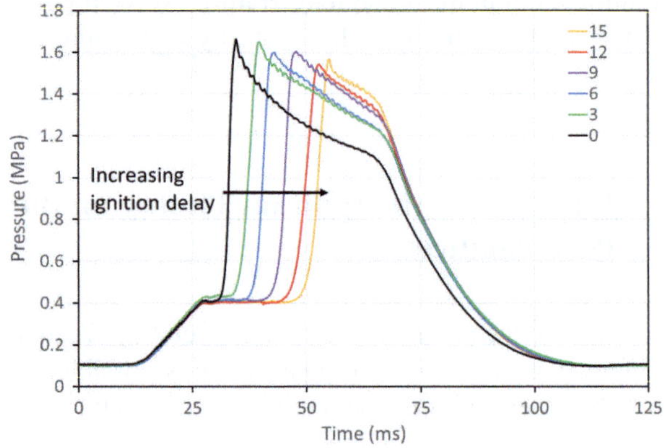

10 ms (Fig. 6). At OER = 0.9, the ignition timing is varied from the beginning of the pressure plateau to 15 ms by step of 3 ms (Fig. 7).

Characteristic combustion durations based on the pressure signal are usually computed to describe the different phases of the combustion process. Those durations are defined based on a percentage of the non-adiabatic combustion pressure gain, π, between the pressure at ignition and the maximum combustion pressure reached:

$$t_X = t(\pi = \frac{p_x - p_3}{p_4 - p_3} = X\,\%) \tag{4}$$

$$t_{X-Y} = t_Y - t_X \tag{5}$$

Table 3 Combustion characteristic average times with their respective standard deviation given for varying ignition delays as well as RBG dilution properties

Ignition delay	15 ms	12 ms	9 ms	6 ms	3 ms	0 ms
t_{0-10} (ms)	7.4 ± 0.6	6.8 ± 0.4	6.5 ± 0.4	5.5 ± 0.4	4.7 ± 0.5	3.3 ± 0.6
t_{10-90} (ms)	4.2 ± 0.4	3.8 ± 0.6	3.4 ± 0.2	2.9 ± 0.2	2.7 ± 0.3	1.8 ± 0.5
t_{90-100} (ms)	1.5 ± 0.4	1.3 ± 0.3	1.2 ± 0.4	1.1 ± 0.4	0.7 ± 0.4	0.7 ± 0.2
RBG Temperature (K)	527	514	512	497	486	460
RBG Dilution (%)	21.4	21.7	22.0	22.4	22.2	23.0
p_4 (MPa)	1.54	1.52	1.59	1.59	1.61	1.66
Expansion ratio (-)	12.8	12.4	11.9	11.4	11.2	10.5

According to these formulas, the duration t_{0-10} can be considered as representative of the time taken for ignition kernel growth to free flame propagation, t_{10-90} corresponds to free flame propagation and t_{90-100} corresponds to the end of propagation at which the flame-wall interaction is maximum (Table 3).

For every ignition timing, the exhaust timing is kept constant. The overall combustion duration is strictly decreasing (reduced by almost a factor of 2) towards earlier ignition timing. At a given OER and almost constant RBG dilution rate, the flame velocity depends on the turbulent velocity fluctuation, in first-order approximation [3, 16]. At the closing of the intake valve, turbulent velocity rapidly decays following the mean velocity evolution (Fig. 6). Earlier ignition yields higher flame velocity, which thus results in shorter combustion duration. The following isochoric cooling phase duration then increases for earlier ignition, which in turn increases the heat losses. The consequence is a lower initial pressure for the exhaust expansion, which can be seen directly on the pressure signal (Fig. 7). At the earliest ignition timing (0 ms), 33% of the fuel energy is lost to the wall during the isochoric cooling phase. Those losses drop to 8.6% for the latest ignition (15 ms). As the peak combustion pressure is almost the same (1.54 MPa at 15 ms to 1.66 MPa at 0 ms), the more cooling there is during that phase, the lesser the initial exhaust pressure, the less efficient the cycle will be by lowering the potential for work recovery. However, higher initial pressure prior to expansion means higher velocity and temperature at the opening of the exhaust valve, which yields slightly higher heat losses during the exhaust phase. Considering any actuator used to control the isochoric phase will have their own time response, a trade-off must be found between the isochoric cooling phase losses and a fully isochoric combustion for optimal work recovery. Following Fourier's law, in first-order approximation, the heat losses are directly related to the wall temperature. The rig is not designed to reproduce stationary thermal equilibrium conditions. Therefore, the wall temperature is much lower than what could be expected in industrial applications, hence higher heat losses. In the following, we discuss the influence of the boundary conditions on the combustion properties through an experimental variation of the exhaust pressure. The effect of the wall temperature is also discussed through the 0D simulation.

6 Influence of the Boundary Conditions on Experimental Cycle Performances

In turbomachines, the combustor backpressure is a design parameter that can be controlled by the cross-section of the turbine first stage diffuser. Turbine efficiency is related to its pressure expansion ratio, and choked conditions are usually set at the first stage diffuser for best efficiency. In piston-less CVC processes, the exhaust back-pressure directly drives the expansion ratio from the pressure at the beginning of the exhaust (point 4'), to the exhaust backpressure (point 5). The expansion ratio, in turn, affects the RBG density through the resulting pressure and temperature at the end of exhaust phase. The operating points experimentally investigated start at an initial pressure of 4.0 MPa for a constant OER = 0.9. With increasing backpressure, an increased mass of RBG is trapped into the chamber and a lower air mass is introduced into the chamber. This is achieved through a reduction in intake valve opening duration (1 ms of opening corresponds to an increase of 250 mbar). The energy density is thus reduced for higher backpressure. The ignition timing (12 ms) as well as the exhaust opening duration are kept the same (see Fig. 8). Under those conditions, the higher the dilution, the slower the combustion, so that combustion duration is affected by a factor of 2.8 between an exhaust pressure of 0.07 MPa and 0.21 MPa (Table 4). Given the exhaust valve is opened at the same timing for each operating point, the isochoric cooling phase is long enough for each operating point

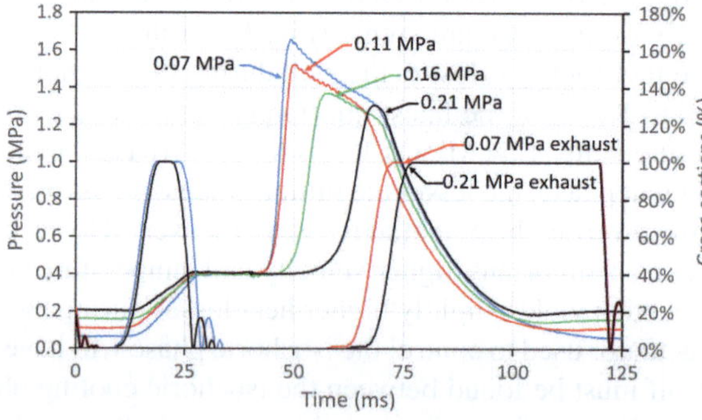

Fig. 8 Experimental pressure signals of single cycles with the same ignition timing (12 ms) for the 4 different backpressure investigated, 0.07 MPa (blue), 0.11 MPa (red), 0.16 MPa (green) and 0.21 MPa (black). The two extreme traces of intake (imposed) and exhaust (thermal drift) cross-sections are represented following the colour code as the one for their associated backpressure

to have the same initial exhaust expansion pressure (1.3 MPa). The expansion ratio decreases with the exhaust pressure from 18 to 6.3, hence higher temperature for higher exhaust backpressure. The trapped residual mass is influenced by the residual density through the expansion ratio and the initial exhaust pressure. Under those conditions, the exhaust gas quantity increases by 16.6% in mass every 0.1 MPa of backpressure. The fundamental flame velocity is known to decrease with increasing dilution (see [17]) for experimental measurements on diluted iso-octane/air flame) and to increase with the temperature. However, considering the important cooling of the exhaust gases, both during combustion (Fig. 9) and exhaust (Table 4), the increase in initial temperature (after mixing with the fresh mixture) is not high enough to compensate for the increase in dilution, hence a lower fundamental flame velocity. The energy density for diluted cycles is lower. They exhibit lower maximum pressure, and slightly lower combustion temperatures. The final density in the fresh gas when combustion stops (see point 4 on Fig. 5) is thus lower, which yields a reduced unburnt mass fraction and a higher combustion efficiency. In the present experiments, the heat losses are high because of the low wall temperature (423 K). In turbojet engines, the thermal management of the combustor is based on compressed air bypassed from the compressor stages. That means higher working temperatures are envisioned. The same cycle is simulated keeping the heat loss coefficient, combustion duration (as the real duration is difficult to estimate) and intake gas temperature constant, but at a wall temperature of 1000 K (which is compatible with most Inconel alloys). This lowers the combustion losses by 10% and isochoric cooling phase losses by a few percent (at which the temperature difference is maximum). The temperature of the cycle is overall higher, heat is gained by the charge during the intake, and the heat losses during the exhaust are lower, hence a higher initial temperature (Table 5). The flame velocity should thus be higher, which should reduce the heat losses during combustion (as seen with the variation of ignition timing). The increase of wall temperature increases the gas temperature in each phase of the cycle. In the present conditions, an increase of the exhaust pressure yields spontaneous re-ignition during fuel injection

Table 4 Combustion characteristic conditions (exhaust expansion ratio, RBG dilution of the mixture, the initial residual temperature, initial temperature of the cycle fresh mixture) and average times with their respective standard deviation given for varying backpressure

Exhaust backpressure	0.21 (MPa)	0.16 (MPa)	0.11 (MPa)	0.07 (MPa)
Wall temperature (K)	423			
Dilution (%)	38.7	31.3	22	15.8
Expansion ratio (-)	6.3	7.9	12.3	18
Residual temperature (point 5) (K)	577	547	500	469
Initial temperature (point 2) (K)	439	436	434	436
t_{0-10} (ms)	14.1 ± 1.1	10.2 ± 1.1	6.7 ± 0.6	5.1 ± 0.3
t_{10-90} (ms)	7.3 ± 0.6	5.4 ± 0.6	4.0 ± 0.6	2.7 ± 0.3
t_{90-100} (ms)	2.6 ± 0.7	2.2 ± 0.7	1.2 ± 0.6	0.8 ± 0.6

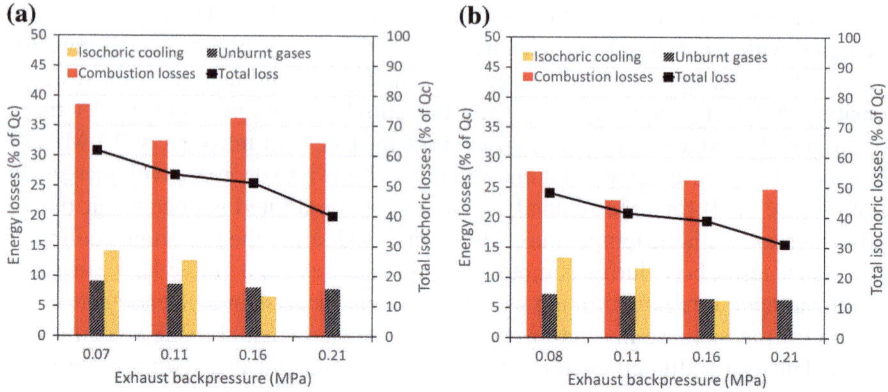

Fig. 9 **a** Energy balance during the combustion phase with heat losses (coloured background) and unburnt gas losses, for 4 different backpressures corresponding to the experimental boundary conditions (T_w = 423 K). **b** Same energy balance for a 0D simulation at T_w = 1000 K

[18]. This behaviour is closely related to the RBG temperature, hence spark-ignited cycles operating at wall thermal equilibrium conditions would certainly evidence such phenomena. Moreover, it would start at a lower exhaust backpressure, since the vessel is operated without scavenging. In other pressure-gain combustion applications, this behaviour is also under consideration, in order to avoid pre-ignition in high-frequency PDE (relative to spark-ignition) [19] or to control it in pulsed combustor [20]. The optimization of such cycles requires the reduction of the isochoric cooling phase duration to a minimum. This requires an optimization of the opening of the exhaust valve. An early opening of the exhaust valve yields a truncated pressure at the end of combustion which is then isobaric (although occurring at final pressure). For an ignition timing of 3.0 ms (Table 3), an optimization is made on the exhaust timing (Fig. 10a). The cycle duration is adjusted (reduced by 25 ms) to avoid further heat losses after the expansion, while still avoiding scavenging. For the two compared cycles (Fig. 10a), the OER (0.87) is the same with a slightly higher initial pressure for the unoptimized cycle (0.43 MPa) compared to the optimized cycle (0.40 MPa). In both cases, the dilution is identical (0.3% difference). With the optimization, the isochoric cooling phase is shorter, resulting in a 3.4% isochoric cooling loss compared to 21.0% for the unoptimized cycle. As the pressure ratio is higher, the mass flow rate increases, hence higher exhaust velocities that result in higher heat fluxes. The same optimized cycle is processed with an increased 1000 K wall temperature (same as Fig. 9b). Doing so, the combustion heat losses are 9% lower, the unburnt gas mass is 2% lower; but as for unoptimized conditions, the heat losses during the isochoric phase remain the same. Working in partially premixed conditions, the burnt gases at the end of combustion are directly exposed to the chamber walls, hence a high temperature gradient (still approximately a 1000 K difference) once the flame

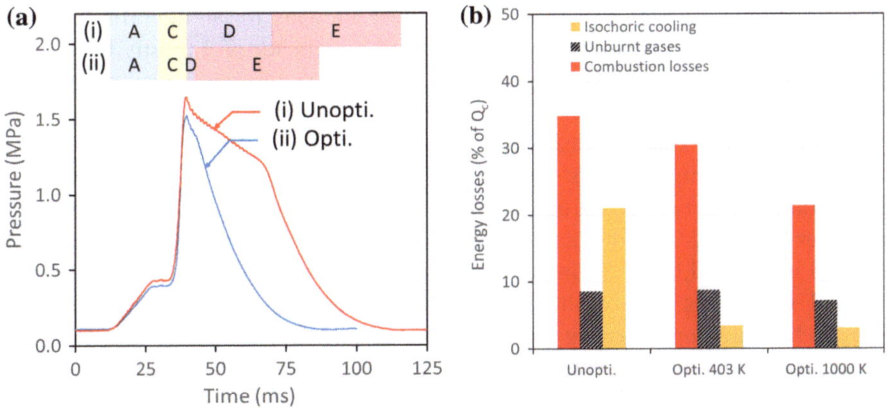

Fig. 10 **a** Comparison of experimental pressure signals for a cycle with optimized cycle (blue), with almost no isochoric phase and shorter cycle duration (100 ms instead of 125 ms) and unoptimized cycle (red). The different phases for the two cycles are figured as in Fig. 5. The ignition timing being early (at 31 ms), there is no relaxation phase“ **b**

Table 5 Combustion characteristic conditions (exhaust expansion ratio, RBG dilution of the mixture, the initial residual temperature, initial temperature of the cycle fresh mixture) for the extrapolation at T_w = 1000K

Exhaust backpressure	0.21 (MPa)	0.16 (MPa)	0.11 (MPa)	0.07 (MPa)
Wall temperature (K)	1000			
Dilution (%)	26.4	20.7	13.8	10.2
Expansion ratio (-)	7.2	9.3	15	21.1
Residual temperature (point 5) (K)	1028	975	911	871
Initial temperature (point 2) (K)	756	708	677	652

reaches the wall, which yields high heat flux. Such behaviour can be avoided by using a convenient air-fuel stratification strategy similar to spark-ignited ICE, where the mixture is fuel-rich at the centre of the cylinder and fuel-lean near the walls.

7 Conclusion

A constant-volume combustion chamber is operated with pre-heated air (423 K at an initial pressure of 0.4 MPa) and a directly injected liquid n-decane with a cyclic operation reaching up to 12 cycles. The combustion characteristics are shown to be strongly dependent on the residual burnt gas properties. Those properties are influenced by the two main boundary conditions of the cycle, namely the exhaust pressure and the wall temperature. In the current setup, the walls are heated up to 423 K which is sufficient to prevent the condensation of n-decane but is far too low to be representative of thermal equilibrium, thus allowing for an important cooling of the burnt gases. The expansion ratio is varied by a factor of 3 mostly by varying the exhaust backpressure. With higher backpressure, the dilution as well as the RBG temperature are increased. Nevertheless, due to an important cooling of those burnt gases during the exhaust phase, and due to the addition of the fresh charge, the resulting initial temperature is not affected. This results in a slower combustion, as the characteristic combustion time increases by a factor of 2.8 as the RBG dilution increases from 15.8% to 38.7%. In the case of a higher wall temperature, the cooling of RBG is reduced hence an increased temperature and a reduced mass fraction of residual burnt gases. As a result, the initial gas temperature increases with increasing exhaust backpressure. Among the losses which are accounted for on the present device, a certain amount is expected to be easily reduced further in an industrially designed chamber. Premixed combustion in internal combustion engines leads to higher heat losses as the burnt gases and the flame are directly in contact with the walls. Stratified combustion would thus help reduce both the heat losses during the combustion and the unburnt gas generated by quenching to the walls and inside mechanical gaps; hence leading to higher combustion efficiency.

Acknowledgements This work is part of the CAPA Chair research program on Alternative Combustion modes for Air-breathing Propulsion supported by SAFRAN Tech, MBDA France and ANR (French National Research Agency).

References

1. European Commissio: Flighpath 2050—Europe's Vision for Aviation (2011). https://doi.org/10.2777/50266
2. Holzwarth, H.: The gas turbine: theory, construction and records of the results obtained from two actual machines (1912)
3. Boust, B., Michalski, Q., Bellenoue, M.: Experimental investigation of ignition and combustion processes in a constant volume combustion chamber for airbreathing propulsion. In: AIAA (2016). https://doi.org/10.2514/6.2016-4699
4. Labarrere, L., et al.: Experimental and numerical study of cyclic variations in a constant volume combustion chamber. Combust. Flame **172**, 49–61 (2016). https://doi.org/10.1016/j.combustflame.2016.06.027
5. Akbari, P., Nalim, M.R., Donovan, E.S., Snyder, P.H.: Leakage assessment of pressure-exchange wave rotors. J. Propul. Power **24**(4) (2008). https://doi.org/10.2514/1.31725
6. Zockel, M.: The constant volume gas turbine—a thermodynamic assessment. In: ASME (1974)
7. Wintenberger, E., Shepherd, J.E.: Thermodynamic cycle analysis for propagating detonations. J. Propuls. Power **22**(3), 694–698 (2006). https://doi.org/10.2514/1.12775
8. Nalim, M.R., Snyder, P.H., Kowalkowski, M.: Experimental test, model validation, and viability assessment of a wave-rotor constant-volume combustor. J. Propuls. Power **33**(1) (2017). https://doi.org/10.2514/1.B36174
9. Michalski, Q., Boust, B., Bellenoue, M.: In Revision. Experimental investigation of ignition stability in a cyclic constant-volume combustion chamber featuring relevant conditions for air-breathing propulsion. Flow Turbul Combust
10. Asad, U., Tjong, J., Zheng, M.: Exhaust gas recirculation—zero dimensional modelling and characterization for transient diesel combustion control. Energy Convers. Manag. **86**, 309–324 (2014). https://doi.org/10.1016/j.enconman.2014.05.035

11. Michalski, Q. et al.: Joint numerical and experimental characterization of the turbulent reactive flow within a constant volume vessel. In: AIAA JPC (2018)
12. Rudinger, G.: Penetration of particles injected into a constant cross flow. AIAA J. **12**(8), 1138–1140 (1974). https://doi.org/10.2514/3.49429
13. Goodwin, D.G., Moffat, H.K., Speth, R.L.: Cantera: An object-oriented software toolkit for chemical kinetics, thermodynamics, and transport processes. Version 2.3.0. (2017). http://cantera.org
14. Labuda, S., Karrer, M., Sotton, J., Bellenoue, M.: Experimental study of single-wall flame quenching at high pressures. Combust. Sci. Tech. **183**(5), 409–426 (2010). https://doi.org/10.1080/00102202.2010.528815
15. Hohenberg, H.: Advanced approaches for heat transfer calculations. SAE Tehnical Paper, Issue 790825 (1979). https://doi.org/10.4271/790825
16. Lipatnikov, A.N., Chomiak, J.: Turbulent flame speed and thickness: phenomenology, evaluation, and application in multi-dimensional simulations. Prog. Energy Combust. Sci. **28**, 1–74 (2002)
17. Endouard, C., Halter, F., Chauveau, C., Foucher, F.: Effect of CO2, H2O, and exhaust gas recirculation dilution on laminar burning velocities and markstein lengths of iso-octane/air mixtures. Combust. Sci. Tech. (2016). https://doi.org/10.1021/ef101482d
18. Michalski, Q., Boust, B., Bellenoue, M.: Toward a cyclic self-ignited constant-volume combustion for airbreathing propulsion applications. In: AIAA—Joint Propulsion Conference (2018)
19. Mastuoka, K., Takagi, S., Kasahara, J.: Validation of pulse-detonation operation in low-ambient-pressure environnement. J. Propuls. Power **34**(1), 116–124 (2018). https://doi.org/10.2514/1.B36401
20. Yungster, S., Paxson, D.E., Perkins, H.D.: Effect of fuel injection and mixing characteristics on pulse-combustor performance at high-pressure. In: AIAA Propulsion and Energy Forum (2014). https://doi.org/10.2514/6.2014-3728

Part IV
Data Assimilation and Model Reduction

Validation of Under-Resolved Numerical Simulations of the PDC Exhaust Flow Based on High Speed Schlieren

M. Nadolski, M. Rezay Haghdoost, J. A. T. Gray, D. Edgington-Mitchell, K. Oberleithner and R. Klein

Abstract Owing to their high thermodynamic efficiency, pulsating combustion cycles have become an attractive option for future gas turbine designs. Yet, their potential gains should not be outweighed by losses due to unsteady pressure wave interactions between engine components. Consequently, the geometric engine design moves into focus. Ideally, one would quickly test several different principal layouts with respect to their qualitative behavior, select the most promising variants and then move on to detailed optimization. Computational fluid dynamics (CFD) appears as the methodology of choice for such preparatory testing. Yet, the inevitable geometric complexity of such engines makes fully resolved CFD an arduous and expensive task necessitating computations on top high-performance hardware, even with modern adaptive mesh refinement in place. In the present work we look at the initial flow field of a shock generated by a pulse detonation combustor (PDC) which leaves the combustion chamber and enters the plenum. We provide first indicators, however, that overall mechanical loads, represented by large-scale means of, e.g., mass, energy, and momentum fluxes can be well estimated on the basis of rather coarsely resolved CFD calculations. Comparing high-resolution simulations of the exit of a strong shock from a combustion tube with experimental schlieren photographs, we first

M. Nadolski (✉) · R. Klein
Freie Universität Berlin, Berlin, Germany
e-mail: nadolski@math.fu-berlin.de

R. Klein
e-mail: rupert.klein@math.fu-berlin.de

M. R. Haghdoost · J. A. T. Gray · K. Oberleithner
Technische Universität Berlin, Berlin, Germany
e-mail: mohammad.rezayhaghdoost@campus.tu-berlin.de

J. A. T. Gray
e-mail: joshua.gray@tu-berlin.de

K. Oberleithner
e-mail: oberleithner@tu-berlin.de

D. Edgington-Mitchell
Monash University, Melbourne, VIC, Australia
e-mail: daniel.m.mitchell@gmail.com

R. King (ed.), *Active Flow and Combustion Control 2018*,
Notes on Numerical Fluid Mechanics and Multidisciplinary Design 141,
https://doi.org/10.1007/978-3-319-98177-2_15

establish validity of fully resolved CFD. Next we compare several integral quantities representative of overall mechanical loads with a sequence of successively coarser grid simulations, thereby corroborating that the "quick and dirty" coarse-grained simulations indeed allow for good order of magnitude estimates.

Keywords Coarse resolution CFD code validation
Experimental Schlieren visualization · Expansion from detonation tube

1 Introduction

Pulsating combustion cycles of varying design are currently being discussed as a way to approximately achieve constant volume combustion (CVC) in gas turbines. Approximate CVC (aCVC) is desirable due to its considerably higher thermodynamic efficiency in comparison with the standard deflagrative combustion which progresses at approximately constant pressure. To accommodate pulsating combustion in a gas turbine, unsteady gasdynamic effects have to be carefully controlled to avoid excessive mechanical and thermal loads and, depending on the nature of the combustion cycle, to optimize its efficiency and robustness. Nonstationary compressible flow in enclosed domains depends crucially on the domain geometry, and we conclude that the design process for a new class of gas turbines involving pulsating aCVC requires flexible geometric design capabilities.

Support comes from modern Computational Fluid Dynamics (CFD) techniques and codes which allow the user to simulate flows in rather complex geometries utilizing, e.g., Cartesian grid cut cell methods [2, 3, 6–9]. Despite the general geometrical flexibility provided by these techniques, three-dimensional fully resolved simulations in complex geometries are highly demanding, and can hardly be used routinely for quick estimates of the principal consequences of major design alternatives. A central hypothesis of the present project (see the conclusions) is that the overall mechanical and thermal loads on enclosure confinements do not depend on the fine-scale structure of gasdynamic processes, but rather on their coarse-grained averaged behavior. If this hypothesis can be corroborated, then an efficient hierarchical approach to the geometrical design of pulsating combustion gas turbines may be adopted: One would first pursue a quick scan of various major geometrical design options utilizing coarse-resolution numerical approximations that can be run in tens of minutes on a PC, and only turn to high-resolution expensive super-computer-based simulations with excessive detail when the overall design has been fixed. To make such an approach feasible, however, we must first validate our base hypothesis and the present study represents a first step in this direction. This task is approached here through the following steps:

1. A laboratory experiment that involves strong multi-dimensional gasdynamic effects is reproduced by highly resolved numerical simulations.

2. A sequence of simulations with sucessively coarser resolution follows, the coarsest chosen such that all small-scale detail is smeared out by the finite volume averaging of the conservative numerical scheme.
3. Finally, several integral quantities, such as mean mass and momentum fluxes across some pre-defined large-scale surfaces, is monitored and compared between the members of the ensemble of coarser and coarser simulation runs.

These steps will provide estimates of how coarse a resolution, i.e., how efficient a computation, can be afforded to still obtain good quantitative or at least qualitative insights into the large-scale features of a highly unsteady flow.

In the present first test we consider experimental schlieren photographs of a strong shock wave exiting a cylindrical tube. The shock has been generated within a tube partially filled with a combustible mixture by firing a deflagration from its left end into an accelerator section. The ensuing high-speed turbulent flame generates a strong shock which then meets the inert column of air that covers the remainder of the tube. Figure 4 shows the geometrical setup and Fig. 5, left panel, shows the situation immediately after the shock has exited the tube (see also Sect. 4 for experimental details). We see the leading shock still being essentially plane near the axis of symmetry, but bending outwards where it expands into the surrounding air.

We follow the evolution of this flow over the first few hundred micro seconds and compare idealized two-dimensional axisymmetric and three-dimensional cartesian numerical simulations for this case in Sect. 6. That section also shows the comparison of some integral quantities computed for a high-resolution reference simulation and a sequence of *coarser* runs to demonstrate that such overall estimates can well be computed based on very rough, crudely resolved simulations. Before we get to this, however, we first describe the mathematical formulation of the flow problem in Sect. 2, the numerical method and code used in Sect. 3, and the specific numerical setup of the simulation runs in Sect. 5. Section 7 draws conclusions and provides future perspectives.

2 Governing Equations

2.1 Summary of the Equations in Cartesian Coordinates

We consider the homogeneous Euler equations in conservative form, i.e.,

$$\frac{\partial}{\partial t}\rho + \nabla \cdot (\rho \boldsymbol{V}) = 0 \quad \text{(Conservation of Mass)} \tag{1}$$

$$\frac{\partial}{\partial t}(\rho \boldsymbol{V}) + \nabla \cdot [\rho \boldsymbol{V} \otimes \boldsymbol{V} + p\,\mathbb{I}] = 0 \quad \text{(Conservation of Momentum)} \tag{2}$$

$$\frac{\partial}{\partial t}(\rho E) + \nabla \cdot [(\rho E + p)\boldsymbol{V}] = 0 \quad \text{(Conservation of Total Energy)} \tag{3}$$

where ρ, p, E and $\boldsymbol{V} = (u, v, w)$ are density, pressure, total energy, and the velocity vector respectively, $\otimes$ denotes the tensor product, and $\mathbb{I}$ the identity matrix. To close the system we choose the equations of state for perfect gases

$$p = \rho(\gamma - 1)\left(E - \frac{\boldsymbol{V}^2}{2}\right). \tag{4}$$

Here γ is the ratio of specific heat capacities, $\gamma = c_p/c_v$, and is set to a value of $\gamma = 1.4$ throughout this study. Equations (1)–(3) can also be written in a form useful when applying dimensional splitting as in the present study,

$$\boldsymbol{U}_t + \boldsymbol{F}(\boldsymbol{U})_x + \boldsymbol{G}(\boldsymbol{U})_y + \boldsymbol{H}(\boldsymbol{U})_z = 0 \tag{5}$$

with

$$(\boldsymbol{U}, \boldsymbol{F}, \boldsymbol{G}, \boldsymbol{H}) = \left(\begin{pmatrix} \rho \\ \rho u \\ \rho v \\ \rho w \\ \rho E \end{pmatrix}, \begin{pmatrix} \rho u \\ \rho u^2 + p \\ \rho u v \\ \rho u w \\ \rho H u \end{pmatrix}, \begin{pmatrix} \rho v \\ \rho u v \\ \rho v^2 + p \\ \rho v w \\ \rho H v \end{pmatrix}, \begin{pmatrix} \rho w \\ \rho u w \\ \rho v w \\ \rho w^2 + p \\ \rho H w \end{pmatrix}\right) \tag{6}$$

and the total specific enthalpy $H = E + p/\rho$.

2.2 Axi-Symmetric Flows

For axisymmetric arrangements, (1)–(3) are conveniently recast in terms of cylindrical coordinates. Let the x-axis represent the axis of symmetry, then (5) transforms to, [12],

$$\boldsymbol{U}_t + \boldsymbol{F}(\boldsymbol{U})_r + \boldsymbol{G}(\boldsymbol{U})_x = \boldsymbol{S}(\boldsymbol{U}), \tag{7}$$

with

$$(\boldsymbol{U}, \boldsymbol{F}, \boldsymbol{G}, \boldsymbol{S}) = \left(\begin{pmatrix} \rho \\ \rho u \\ \rho v \\ \rho E \end{pmatrix}, \begin{pmatrix} \rho u \\ \rho u^2 + p \\ \rho u v \\ \rho H u \end{pmatrix}, \begin{pmatrix} \rho v \\ \rho u v \\ \rho v^2 + p \\ \rho H v \end{pmatrix}, -\frac{1}{r}\begin{pmatrix} \rho u \\ \rho u^2 \\ \rho u v \\ \rho H u \end{pmatrix}\right). \tag{8}$$

Here u is now the radial and v the velocity along the x axis.

3 Numerical Methods

Pulsating combustion cycles for near stoichiometric hydrocarbon-air or hydrogen-air mixtures come with pressure wave amplitude ratios of p/p_{ref} ranging from 4 to 20, depending on the combustion mode. Thus, strong shocks ensue and we have no alternative but to use conservative numerical schemes to solve the Euler equations. In this study we use the MUSCL-Hancock, [13], conservative finite volume method. This scheme is second order accurate and employs slope-limiting near discontinuities to prevent the formation of spurious oscillations. The scheme is conveniently employed for multiple space dimensions using Strang splitting [11]. For the representation of complex geometries, we use a cut-cell technique [8]. To evolve the source term in (7) we use again Strang splitting, advancing the solution in the related split step by an adaptive Runge-Kutta-45 scheme.

The code, which is a joint development with the Centre for Scientific Computing, Cavendish Laboratory, Cambridge University, is capable of using block structured adaptive mesh refinement (AMR) similar to [1]. This capability is not utilized in the present study, however, as we intend to first understand the basic behavior of this numerical methodology under grid coarsening.

4 Experiment: Strong Shock Exiting a Straight Tube

In order to generate a shock wave propagating through a tube, a pulse detonation combustor (PDC) is used in this study. The schematic of the PDC and the instrumentation of the experimental setup is shown in Fig. 1. The PDC consists of two sections: the section in which the flame acceleration and subsequent deflagration-to-detonation transition (DDT) occurs and the exhaust tube. Hydrogen and air are injected into the upstream end of the tube. Combustion is initiated with a spark plug, positioned at the upstream end of the DDT section. The specific design of the hydrogen and air injection schemes is described in [5].

Once the mixture is ignited, orifices positioned in the DDT section accelerate the flame propagation. The speed of the leading shock wave is determined by five piezoelectric pressure sensors (PCB112A05), flush-mounted in the exhaust tube, as shown in Fig. 1. In order to assess whether the transition from deflagration to detonation takes place before entering the exhaust tube, three additional closely spaced pressure probes are placed in the downstream end of the DDT section.

The reaction front is tracked inside the tube by three flush-mounted ionization probes on the opposite side from the pressure sensors. The ionization probes, fabricated in-house, consist of two electrodes separated by a ceramic coating. When a potential difference is applied to the electrodes, the ionized species in the combustion region allow for an electric current to flow. The resulting voltage drop marks the arrival of the combustion front.

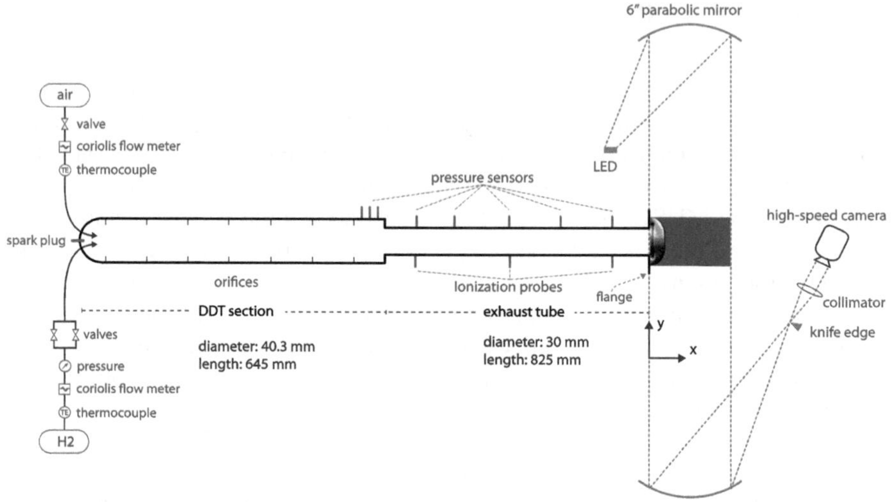

Fig. 1 Sketch of the experimental setup showing the pulse detonation combustor, pressure sensors and ionization probes for reaction front and pressure wave detection, and the high-speed schlieren setup at the tube exit

Data is acquired simultaneously on 11 channels using a National Instruments MXI-Express DAQ system at a sampling rate of 1 MHz. The mass flow rate of air and hydrogen is measured with Endress + Hauser Coriolis mass flow meters. Two type K thermocouples are used to measure the temperature of air and hydrogen. The pressure of the hydrogen fuel line upstream of the injection valves is measured with a Festo pressure transmitter (SPTW-P10R).

The high-speed schlieren images are acquired with a Photron SA-Z high-speed camera at frequencies up to 80 kHz. A standard z-type schlieren configuration is used with two 6-inch parabolic f/8 mirrors for collimating and refocusing of light. As a light source, a pulsed LED is used as suggested by Willert et al. [14]. The overdriven-operated LED provides a light pulse with very high intensity at very short time span (1 μs).

The operating condition of the PDC can be adjusted to maintain high conformity with the numerical simulations. Since the numerical models do not account for chemical reactions, it is desirable to minimize the impact of combustion on the flow evolution once the shock wave exits the tube. Therefore, the PDC is filled only partially with combustible mixture to allow for the shock wave to decouple from the reaction front and maintain a time gap between the shock wave and combustion products leaving the tube.

The shock wave velocity along the PDC is shown in Fig. 2. The wave velocity is captured at seven axial positions via eight pressure probes. The shock wave velocity of 1000 m/s within the DDT section is about half the Chapman-Jouguet (CJ) velocity. Because the tube is only partially filled, the deflagration wave does not transition to a detonation wave. The shock wave is transmitted to the air, while the reaction

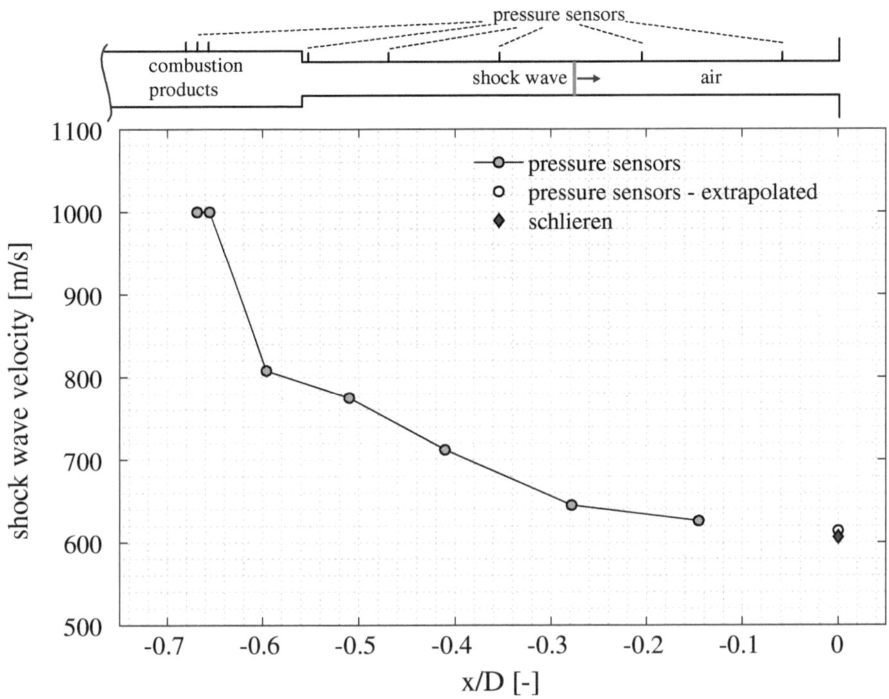

Fig. 2 Leading shock wave velocity determined by the pressure sensors in the tube and three consecutive schlieren images at the tube outlet

front is extinguished as the deflagration wave reaches the intersection of the reactive mixture with air. The decelerating shock wave propagates through the exhaust tube while compressing the air. The shock wave diffracts through the atmosphere once it reaches the tube exit. The flow evolution at the tube exit is captured by means of schlieren images. These images are used for validation of the numerical simulations.

The primary parameter that sets the initial condition for the numerical simulation is the shock wave Mach number at the exit of the tube. Hence, the shock wave velocity has been determined via two independent methods using schlieren images and pressure probes. Therefore, consecutive schlieren images captured at 80 kHz are used as shown in Fig. 3. The straight vertical shape of the leading shock in all three frames indicates that the information about the area expansion at the tube outlet has not yet reached the jet centerline. This implies that the shock velocity is still the same as directly after the tube exit without any deceleration due to shock wave diffraction [10]. Consequently the shock velocity and related Mach number can be unambiguously determined from the shock displacement shown in Fig. 3, using the time-of-flight. The determined shock wave velocity of 606 m/s corresponds to an exit Mach number of $M = 1.77$. The exit velocity of the shock wave at the tube exit can also be estimated from the pressure sensor signals. Therefore, the arrival time of the

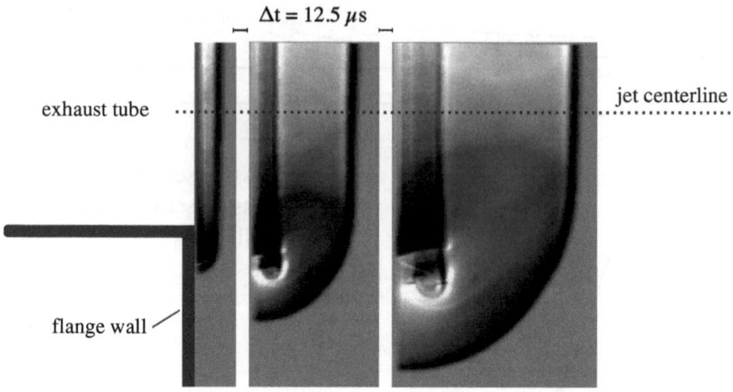

Fig. 3 Three consecutive schlieren images for determining the shock velocity at the exit of the tube

shock wave is extrapolated based on piecewise cubic hermite polynomials (PCHIP) [4] to the tube exit from the arrival times measured at six different streamwise locations. As shown in Fig. 2, the shock velocity determined from schlieren images and the one extrapolated from the pressure signals are in very good agreement with a discrepancy of 1.3%. For this reason, the experimental schlieren images can be used as a benchmark for validation of computational simulations.

5 The Numerical Setup

The geometry of the problem is given as a linear, cylindrical tube leading into a theoretically infinite half space filled with air at room temperature. Figure 4 displays the computational domain of our three-dimensional computation as well as the simplified configuration for the axisymmetric case.

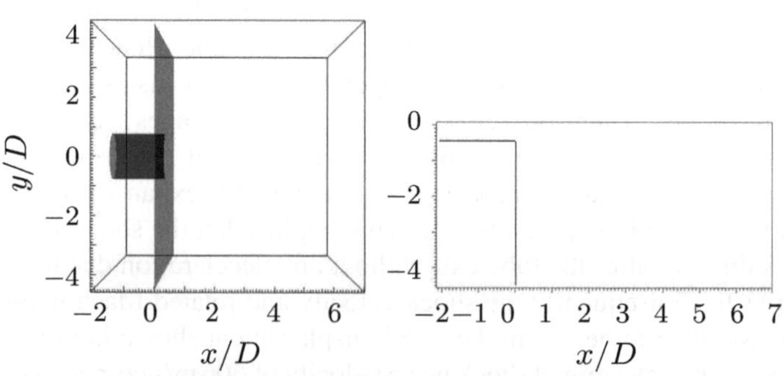

Fig. 4 Left: the computational domain for the three dimensional case, Right: the computational domain for the two dimensional case in cylindrical coordinates

For a simple approximation to the experiment described in Sect. 4 we estimated the velocity of the leading shock from measured data as described in the previous section. The estimate used here is $u_S \approx 585\frac{m}{s}$, equivalent to a Mach number of $M_S \approx 1.72$, and within a 3% window around the estimate given above. Thus, the simulation begins with an ideal one-dimensional shock of this strength traveling downstream the tube. The upstream boundary conditions in the tube are set to match the post-shock conditions, indicated by a $*$ subscript below. These result from a given state $\boldsymbol{U}_{\text{Air}}$ ahead of the shock and from the shock Mach number M_S via the Rankine-Hugoniot conditions reading

$$\frac{\rho_*}{\rho_{\text{Air}}} = \frac{(\gamma+1)(M_{\text{Air}} - M_S)^2}{(\gamma-1)(M_{\text{Air}} - M_S)^2 + 2} \tag{9}$$

$$\frac{p_*}{p_{\text{Air}}} = \frac{2\gamma(M_{\text{Air}} - M_S)^2 - (\gamma-1)}{(\gamma+1)} \tag{10}$$

$$u_* = u_S\left(1 - \frac{\rho_{\text{Air}}}{\rho_*}\right) + u_{\text{Air}}\frac{\rho_{\text{Air}}}{\rho_*} \tag{11}$$

Assuming $\gamma = 1.4$, $p_{\text{Air}} = 1.01 \cdot 10^5$ Pa, $\rho_{\text{Air}} = 1.22\,\text{kg/m}^3$ and $u_{\text{Air}} = 0\,\text{m/s}$ we obtain the post-shock state $p_* = 3.33 \cdot 10^5$ Pa, $\rho_* = 2.73\,\text{kg/m}^3$ and $u_* = 323\,\text{m/s}$.

The computational domain is necessarily limited. To match the experimental set-up we have to utilize approximate non-reflecting boundary conditions on all domain boundaries intersecting with the open half space into which the shock wave is expanding. We have adopted a simple version of such a numerical condition. A two cell wide layer of ghost cells surrounds the computational domain. At the beginning of each time step, these cells are assigned the same states found in their next neighbor cells within the computational domain in the sense of a zeroth order extrapolation. This procedure is simple and rather effective, although some weak wave reflection is seen when the shock hits the outer boundaries. For the targeted coarse-grained estimates of integral loads, such a straightforward boundary condition formulation seems appropriate.

To validate our code we set up multiple sets of numerical computations, one for each of the coordinate systems described above and several three dimensional computations in varying resolution. To compare the computations to each other we set the finest grid resolution to 512 cells in each direction, corresponding to cell sizes of $6 \cdot 10^{-4}$m, and simulate the first two milliseconds. For the three dimensional simulation in Cartesian coordinates this implies $512^3 = 134.217.728$ cells and $256 \cdot 512 = 131.072$ cells for the simulation in cylindrical coordinates. Experimental observations are then compared with these reference runs, while we compare several integral quantities indicative of mechanical loads between these highest resolution and other simulations based on coarser grids below.

6 Results

6.1 Comparison of High-Resolution Runs with the Experiment

Schlieren pictures from the experiment are available in intervals of 50 ms. For the subsequent discussion we set the clock to time $t = 0$ at the instance shown in Fig. 5, when the shock has just exited the tube. Simulation data are available for each consecutive 50 ms after this point in time.

Figure 6 compares experimental data with path integrated schlieren images post-processed from the simulation data of the three-dimensional simulation. To be precise, the figure displays grey scales of the quantity

$$S(x, y) = \int_{z_0}^{z_1} \frac{\partial \rho}{\partial y}(x, y, z)\,\mathrm{d}z. \tag{12}$$

The comparison in Fig. 6 shows that the early evolution is successfully captured in terms of key flow features such as the vortex position, the Mach disk, the reflected shocks, and the slip lines.

On the downside we also observe a mismatch in the position and width of the Mach disk, which increases with time. This discrepancy is likely due to the simplified computational set-up which assumes a shock travelling at constant velocity to exit from the tube. In the experiment, in contrast, the observed shock was generated by firing a high-speed deflagration in the tube and letting the generated strong shock pass over a buffer region of inert gas covering roughly the latter third of the tube length. A strong indication that the upstream conditions between experiment and simulation do not match well after some time is the exit of multiple shocks from the tube seen in the third panel on the left in Fig. 6. This feature is absent from the numerical simulation, because of the exact match of the post-shock conditions of the initial shock at the upstream end of the short tube section included in the simulation. A more realistic set-up that utilizes a one-dimensional version of the compressible

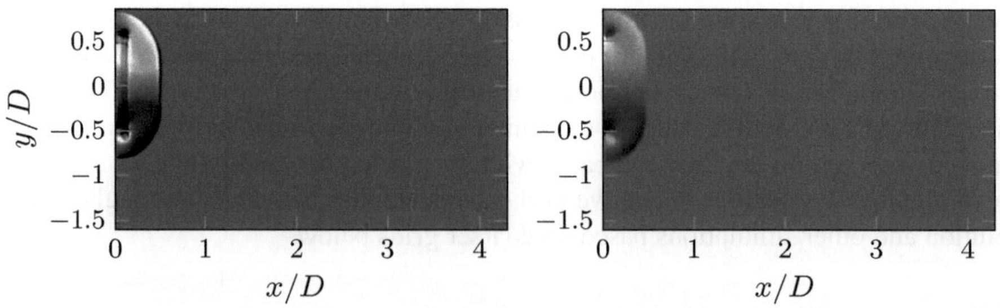

Fig. 5 Experimental schlieren versus path-integrated schlieren from numerical simulations at the reference time $t = 0$

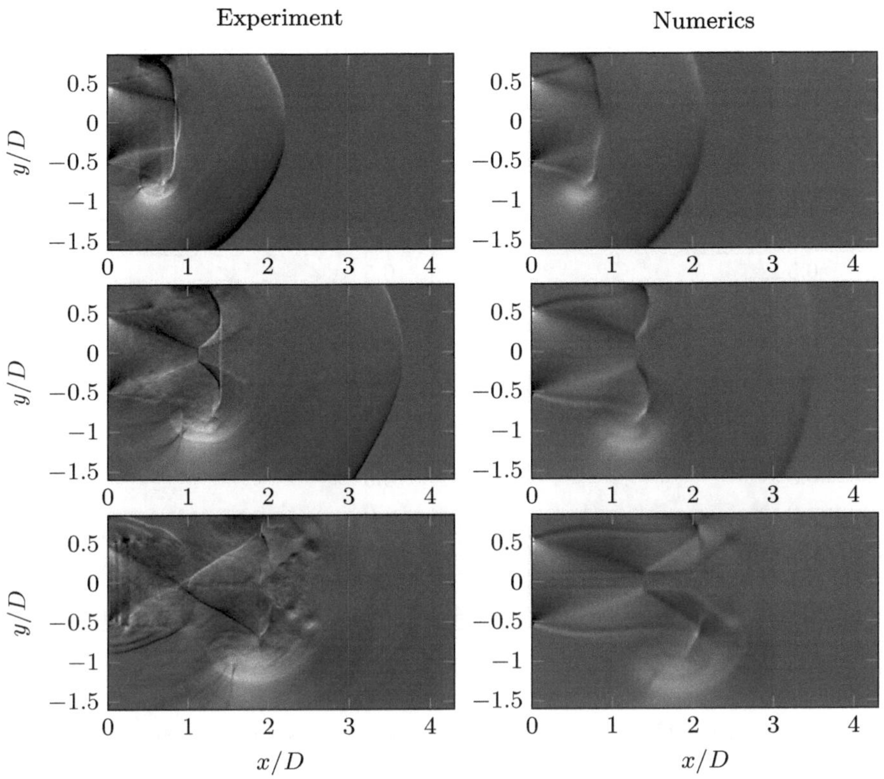

Fig. 6 Comparison with the experiment: path integrated schlieren pictures, at relative time points 100, 200 and 300 μs

(reactive) flow solver to also represent this non-trivial gasdynamic process in the tube prior to the shock wave exit is work in progress.

Nevertheless, the simulation up to time point 100 μ shows results which are close to the experiment. At later times all major flow features are qualitatively present. The quantitative disagreement may be hampered by overly simplified inflow boundary conditions in the numerical simulations.

6.2 *Comparison of Different Coordinate Systems*

We are interested in the differences between the computations with respect to the two different coordinate systems introduced in Sect. 2. From now on we consider planar density gradients $P(x, y) = |\nabla\rho(x, y, 0)|$, since these reveal more details of the flow. Figure 7 shows that the computation in cylindrical coordinates is in good

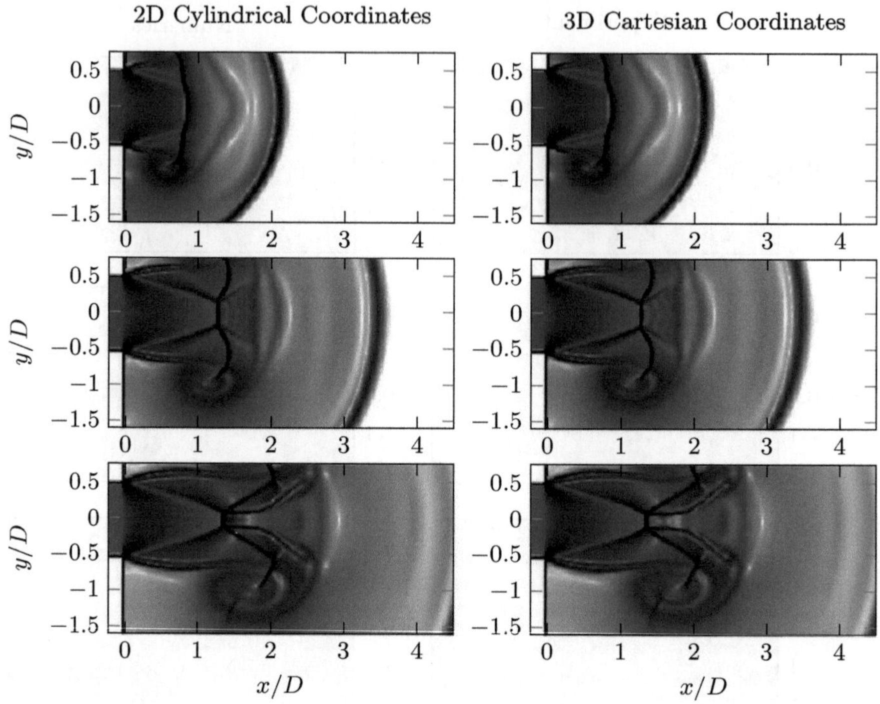

Fig. 7 Comparison of density gradients at relative time points 100, 200 and 300 μs

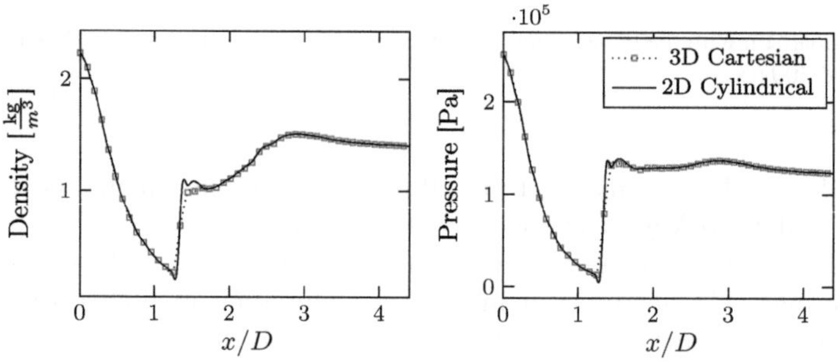

Fig. 8 Plots for density (left) and pressure (right) in $y = z = 0$ over the x position for the relative time point 300 μs

agreement with the three dimensional computation. Only beyond the relative time point of 300 μs a sizeable difference is seen in the region around the Mach disk.

Figure 8 displays the pressure and density distribution along the central axis in the flow direction. Good quantitative match between the simulations is obtained. As in the visualisation for the density gradients we see differences in the Mach disk area.

6.3 Effects of Resolution

As stated earlier, we are interested in utilizing quick coarse-resolution simulations for efficient testing of different engine geometries with respect to their influence on overall mechanical loads. Thus, in this section we test whether some globally integrated quantities can well be estimated from such coarse simulations.

Figure 9 illustrates the solution structure for varying resolutions. With a tube diameter of 0.03 m and a resolution of approximately $\Delta x \approx 0.0046$ m the coarsest simulation uses around 6 cells in each direction to cover the tube inlet. Despite being shifted basic structural properties such as shocks, Mach disk and vortex positions are still represented down to the resolution of 128 cells per dimension.

Figure 10 displays the time series of averaged radial mass flow

$$\dot{m}_{\mathrm{rad}} = \int_S \rho \boldsymbol{V} \cdot \boldsymbol{n}_S \, \mathrm{d}S \tag{13}$$

for these resolutions, where S is the prolongation of the tube wall beyond the tube exit. In analogy with the radial mass flow we also show the axial mass flow integrated over a plane vertical to the tube axis and placed at $x/D = 1.7$ downstream the tube exit.

Figures 11 and 12 show the amount of advected radial and axial momentum through these surfaces,

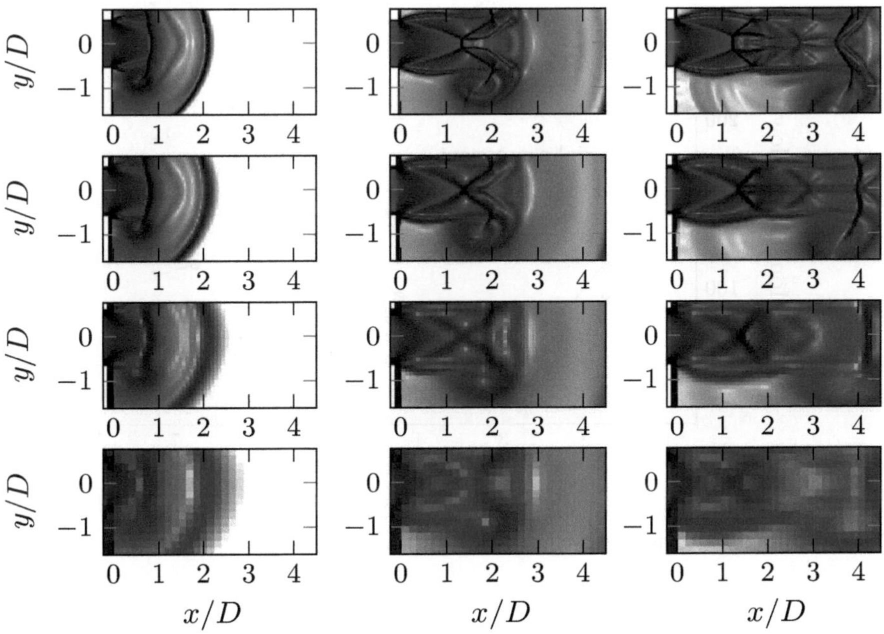

Fig. 9 Comparison of density gradients for a simulation with 512, 256, 128 and 64 cells per dimension at relative time points 100, 300 and 600 μs

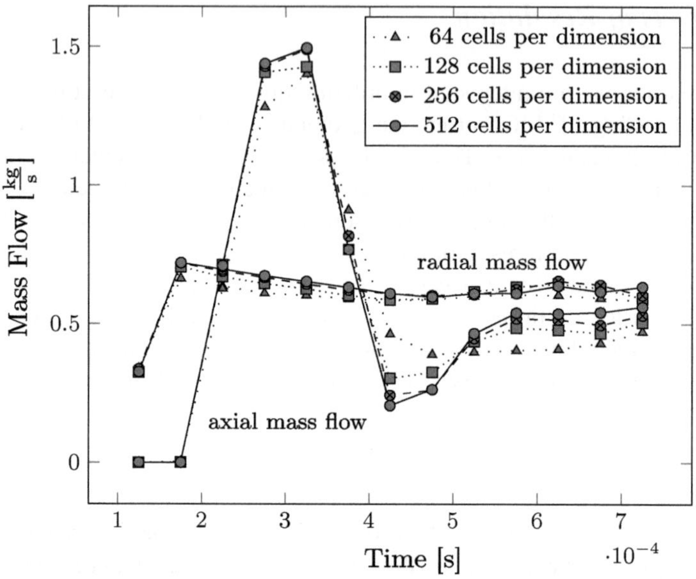

Fig. 10 Radial and axial massflow over time

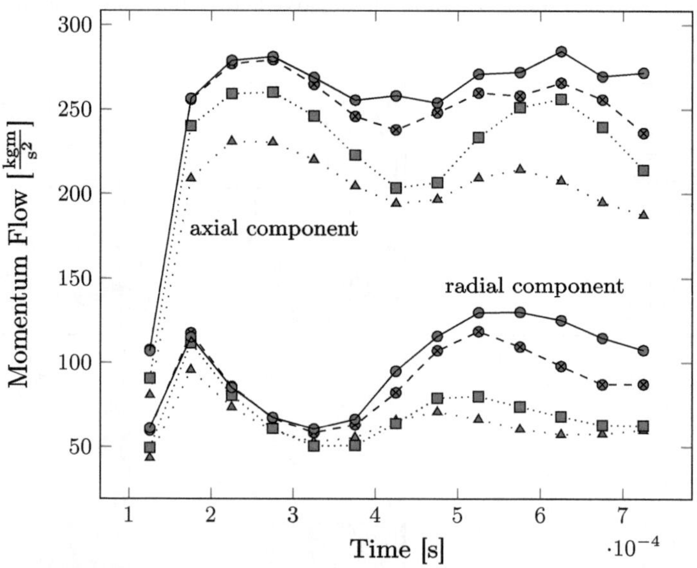

Fig. 11 Radial momentum flow over time

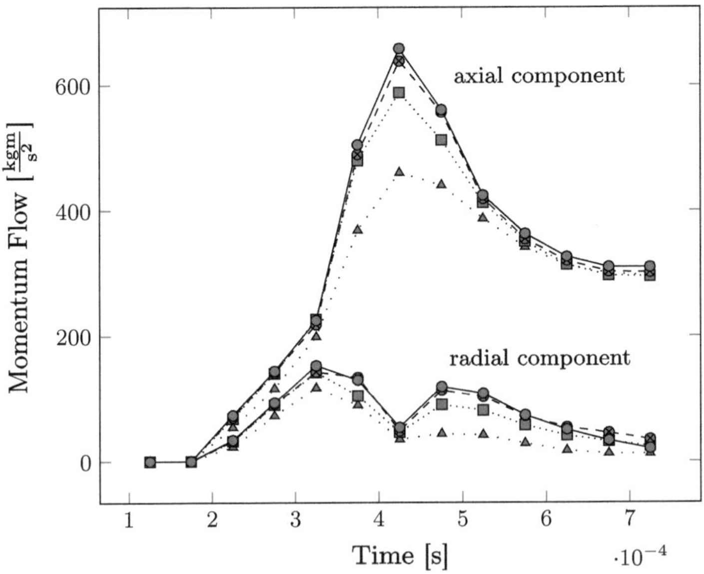

Fig. 12 Axial momentum flow over time

$$\dot{\boldsymbol{P}}_{\text{A}} = \int_A \rho \boldsymbol{V}(\boldsymbol{V} \cdot \boldsymbol{n}_A)\, \text{d}A, \tag{14}$$

where A is the prolonged tube or the vertical plane respectively. Note that while mass flow is a scalar, the momentum flow is a vector which we separated into its axial and radial components and in this study the axial component is equal to the projection onto the axis along the tube, the x-axis. For the radial component of the momentum flow we visualise the magnitude of its y and z components, i.e. let $\dot{\boldsymbol{P}}_{\text{A}}$ be written as $\dot{\boldsymbol{P}}_{\text{A}} = (p_{\text{A}}, q_{\text{A}}, r_{\text{A}})$, then the axial component of the momentum flow is

$$\text{axial } \dot{\boldsymbol{P}}_{\text{A}} = p_{\text{A}}$$

and its radial component is

$$\text{radial } \dot{\boldsymbol{P}}_{\text{A}} = \sqrt{q_{\text{A}}^2 + r_{\text{A}}^2}.$$

Figures 10 and 12 show good qualitative agreement across all resolutions but the coarsest resolution with 64 cells. Whereas we observe larger discrepancies in Fig. 11, where coarser simulations tend to underestimate the radially advected momentum. Comparing the outgoing stream throughout the computations, we also observe that it is substantially more smeared out for coarse grids. By choosing a relatively narrow cylindrical surface to compute the radial flows, we introduce additional numerical errors while interpolating values on that surface. This might result in not "seeing" the radial momentum for coarse computations for this particular choice and could be mitigated by choosing a larger radius for the radial test surface.

7 Conclusion

We used a simple and reproducible test case and experiments with numerical simulations. Despite the assumed simplified initial and boundary conditions the simulations resolve the dynamics of a shock exiting a combustion tube. An axisymmetric computational model produces essentially the same results as a full 3D code. We conclude that three-dimensional effects of turbulence are not relevant throughout this early stage. Coarse resolution runs reproduce large-scale averaged quantities potentially important for overall loads quite well in comparison with highly resolved computations. In fact, even key solution features are maintained qualitatively as well as quantitatively. This encourages the use of under-resolved simulations for efficient design studies of engine geometries. Subsequent studies will focus on more complex geometries, and improved modelling of boundary conditions. Adaptive mesh refinement will be employed to tune the local numerical resolution to the needs of geometry representation.

Acknowledgements The authors gratefully acknowledge support by the Deutsche Forschungsgemeinschaft (DFG) as part of collaborative research center SFB 1029 "Substantial efficiency increase in gas turbines through direct use of coupled unsteady combustion and flow dynamics" on project C01. M.N. and R.K. thank Phillip Blakeley and Nikos Nikiforakis (Cavendish Labs., Cambridge University) for very productive cooperation on code development.

References

1. Berger, M.J., Colella, P.: Local adaptive mesh refinement for shock hydrodynamics. J. Comput. Phys. **82**, 67–84 (1989)
2. Berger, M.J., Helzel, C.: A simplified h-box method for embedded boundary grids. SIAM J. Sci. Comput. **34**(2), A861–A888 (2012)
3. Colella, P., Graves, D.T., Keen, B.J., Modiano, D.: A Cartesian grid embedded boundary method for hyperbolic conservation laws. J. Comput. Phys **211**(1), 347–366 (2006)
4. Fritsch, F.N., Carlson, R.E.: Monotone piecewise cubic interpolation. SIAM J. Numer. Anal. **17**(2), 238–246 (1980)
5. Günther, Claudia, Meinke, Matthias, Schröder, Wolfgang: A flexible level-set approach for tracking multiple interacting interfaces in embedded boundary methods. Comput. Fluids **102**, 182–202 (2014)
6. Gray, J.A.T., Lemke, M., Reiss, J., Paschereit, C.O., Sesterhenn, J., Moeck, J.P.: A compact shock-focusing geometry for detonation initiation: experiments and adjoint-based variational data assimilation. Combust. Flame **183**, 144–156 (2017)
7. Hu, X.Y., Khoo, B.C., Adams, N.A., Huang, F.L.: A conservative interface method for compressible flows. J. Comput. Phys **219**, 553–578 (2006)

8. Klein, R., Bates, K.R., Nikiforakis, N.: Well-balanced compressible cut-cell simulation of atmospheric flow. Philos. Trans. R. Soc. Lond. A Math. Phys. Eng. Sci. **367**(1907), 4559–4575 (2009)
9. Meinke, M., Schneiders, L., Günther, C., Schröder, W.: A cut-cell method for sharp moving boundaries in Cartesian grids. Comput. Fluids **85**, 135–142 (2013)
10. Skews, B.W.: The shape of a diffracting shock wave. J. Fluid Mech. **29**(2), 297–304 (1967)
11. Strang, G.: On the construction and comparison of difference schemes. SIAM J. Num. Anal. **5**, 506–517 (1968)
12. Toro, E.F.: Riemann Solvers and Numerical Methods for Fluid Dynamics: A Practical Introduction. Springer, Heidelberg (2009)
13. van Leer, Bram: On the relation between the upwind-differencing schemes of Godunov, Engquist-Osher and Roe. SIAM J. Sci. Stat. Comput. **5**(1), 1–20 (1984)
14. Willert, C.E., Mitchell, D.M., Soria, J.: An assessment of high-power light-emitting diodes for high frame rate Schlieren imaging. Exp. Fluids **53**(2), 413–421 (2012)

On the Loewner Framework for Model Reduction of Burgers' Equation

Athanasios C. Antoulas, Ion Victor Gosea and Matthias Heinkenschloss

Abstract This paper addresses issues that originate in the extension of the Loewner framework to compute reduced order models (ROMs) of so-called quadratic-bilinear systems. The latter arise in semi-discretizations of fluid flow problems, such as Burgers' equation or the Navier-Stokes equations. In the linear case, the Loewner framework is data-driven and constructs a ROM from measurements of the transfer function; it does not explicitly require access to the system matrices, which is attractive in many settings. Research on extending the Loewner framework to quadratic-bilinear systems is ongoing. This paper presents one extension and provides details of its implementation that allow application to large-scale problems. This extension is applied to Burgers' equation. Numerical results show the potential of the Loewner framework, but also expose additional issues that need to be addressed to make it fully applicable. Possible approaches to deal with some of these issues are outlined.

Keywords Model reduction · Loewner framework · Rational interpolation
Burgers' equation

A. C. Antoulas
Department of Electrical an Computer Engineering, Rice University,
6100 Main Street MS-380, Houston, TX 77005-1892, USA
e-mail: aca@rice.edu

I. V. Gosea · A. C. Antoulas
Max Planck Institute for Dynamics of Complex Technical Systems Magdeburg,
Sandtorstrasse 1, 39106 Magdeburg, Germany
e-mail: gosea@mpi-magdeburg.mpg.de

M. Heinkenschloss (✉)
Department of Computational and Applied Mathematics, Rice University,
6100 Main Street MS-134, Houston, TX 77005-1892, USA
e-mail: heinken@rice.edu

R. King (ed.), *Active Flow and Combustion Control 2018*,
Notes on Numerical Fluid Mechanics and Multidisciplinary Design 141,
https://doi.org/10.1007/978-3-319-98177-2_16

1 Introduction

We study the extension of the Loewner framework to compute reduced order models (ROMs) of so-called quadratic-bilinear systems that arise in semi-discretizations of fluid flow problems, such as Burgers' equation or the Navier-Stokes equations. The attractive feature of the Loewner framework is that it is data-driven and constructs a ROM from measurements of transfer functions related to the system. In the linear case it does not explicitly require access to the matrices (such as $\mathbf{E}$, $\mathbf{A}$, etc. defined below) of the system. While the Loewner framework for linear systems is fairly developed, see e.g., the tutorial paper by Antoulas et al. [3], its extension to quadratic-bilinear systems is ongoing. Recently, Antoulas et al. [2, 8] have extended the Loewner framework to bilinear and quadratic-bilinear systems, but have not yet addressed several issues that need to be dealt with when applying the Loewner framework to fluid flow problems. This paper begins to address some of these issues. In particular, we present the Loewner framework from [8] in a way that allows its efficient implementation and scaling to large-scale problems. In addition, we present more extensive numerical results of an initial version of the Loewner framework to Burgers' equation. These numerical results show the potential of the Loewner framework, but also highlight open issues that need to be addressed. We will outline possible paths on how to deal with these issues.

Given

$$\mathbf{E}, \mathbf{A} \in \mathbb{R}^{n \times n}, \ \mathbf{b}, \mathbf{c} \in \mathbb{R}^n, \ d \in \mathbb{R}, \ \widetilde{\mathbf{N}} : \mathbb{R}^n \times \mathbb{R} \to \mathbb{R}^n, \ \mathbf{G} : \mathbb{R}^n \times \mathbb{R}^n \to \mathbb{R}^n, \tag{1}$$

we consider single input and single output (SISO) systems of the type

$$\mathbf{E}\tfrac{d}{dt}\mathbf{x}(t) = \mathbf{A}\mathbf{x}(t) + \mathbf{G}(\mathbf{x}(t), \mathbf{x}(t)) + \widetilde{\mathbf{N}}(\mathbf{x}(t), \mathbf{u}(t)) + \mathbf{b}\mathbf{u}(t), \quad t \in (0, T), \tag{2a}$$

$$\mathbf{y}(t) = \mathbf{c}^T\mathbf{x}(t) + d\mathbf{u}(t), \quad t \in (0, T), \tag{2b}$$

$$\mathbf{x}(0) = \mathbf{0}. \tag{2c}$$

We will make additional assumptions on (2) in Sect. 2. The system (2) is referred to as the full order model (FOM). The semi-discretized viscous Burgers' equations and the Navier-Stokes equations fit into the setting (2). Burgers' equation will be discussed in Sect. 4. We refer to, e.g., Elman et al. [6, Chap. 8], Layton [11] for the semi-discretized Navier Stokes equations, where (2a) is a system of differential algebraic equations for the velocity and pressure.

Our goal is to construct a small ($r \ll n$) dimensional model

$$\widehat{\mathbf{E}}\tfrac{d}{dt}\widehat{\mathbf{x}}(t) = \widehat{\mathbf{A}}\widehat{\mathbf{x}}(t) + \widehat{\mathbf{G}}(\widehat{\mathbf{x}}(t), \widehat{\mathbf{x}}(t)) + \widehat{\mathbf{N}}(\widehat{\mathbf{x}}(t), \mathbf{u}(t)) + \widehat{\mathbf{b}}\mathbf{u}(t), \quad t \in (0, T), \tag{3a}$$

$$\widehat{\mathbf{y}}(t) = \widehat{\mathbf{c}}^T\widehat{\mathbf{x}}(t) + d\mathbf{u}(t), \quad t \in (0, T), \tag{3b}$$

$$\widehat{\mathbf{x}}(0) = \mathbf{0}, \tag{3c}$$

with $\widehat{\mathbf{E}}, \widehat{\mathbf{A}} \in \mathbb{R}^{r\times r}, \widehat{\mathbf{b}}, \widehat{\mathbf{c}} \in \mathbb{R}^r, \widehat{\mathbf{N}} : \mathbb{R}^r \times \mathbb{R} \to \mathbb{R}^r,\ \widehat{\mathbf{G}} : \mathbb{R}^r \times \mathbb{R}^r \to \mathbb{R}^r$ such that the input-to-output map $\mathbf{u} \mapsto \widehat{\mathbf{y}}$ of the ROM (3) is a good approximation of the input-to-output map $\mathbf{u} \mapsto \mathbf{y}$ of the FOM (2).

The majority of ROM approaches use a Petrov-Galerkin projection. That is they compute matrices $\mathbf{V}, \mathbf{W} \in \mathbb{R}^{n\times r}$ of rank $r \ll n$, approximate the state of the FOM (2) by $\mathbf{x}(y) \approx \mathbf{V}\widehat{\mathbf{x}}(t)$, insert this approximation into (2), and multiply the resulting Eq. (2a) from the left by $\mathbf{W}^T$. This results in a ROM (3) with

$$\begin{aligned}\widehat{\mathbf{E}} = \mathbf{W}^T\mathbf{E}\mathbf{V},\quad \widehat{\mathbf{A}} = \mathbf{W}^T\mathbf{A}\mathbf{V},\quad \widehat{\mathbf{b}} = \mathbf{W}^T\mathbf{b},\quad \widehat{\mathbf{c}} = \mathbf{V}^T\mathbf{c},\\ \widehat{\mathbf{G}}(\widehat{\mathbf{x}}, \widehat{\mathbf{z}}) = \mathbf{W}^T\mathbf{G}(\mathbf{V}\widehat{\mathbf{x}}, \mathbf{V}\widehat{\mathbf{z}}),\quad \widehat{\mathbf{N}}(\widehat{\mathbf{x}}, \mathbf{u}) = \mathbf{W}^T\widetilde{\mathbf{N}}(\mathbf{V}\widehat{\mathbf{x}}, \mathbf{u}).\end{aligned} \tag{4}$$

Examples of projection based ROM approaches include proper orthogonal decomposition (POD), reduced basis (RB) methods, and Krylov subspace methods. These and other ROM approaches are described, e.g., in the books by Antoulas [1], Hesthaven et al. [10], Quarteroni et al. [13], or the review paper by Rozza et al. [15]. Rowley and Dawson [14] review ROMs for flow problems. These methods have in common that they compute $\mathbf{V}, \mathbf{W}$ and then project the system. Thus these methods need the matrices $\mathbf{E}, \mathbf{A}$, etc., to compute the ROM matrices $\widehat{\mathbf{E}}, \widehat{\mathbf{A}}$, etc. In contrast, for linear systems, the Loewner framework computes the ROM system matrices $\widehat{\mathbf{E}}, \widehat{\mathbf{A}}$, etc., directly from measurements of the transfer function, which in some cases can be obtained from experiments. In the case where values of the transfer function can be obtained from measurements, the Loewner framework does not require knowledge of the FOM matrices $\mathbf{E}, \mathbf{A}$, etc. See the tutorial paper by Antoulas et al. [3] and Sect. 2. We study the extension of the Loewner framework to compute ROMs of (2).

To focus our presentation we consider SISO systems (2). However, the Loewner framework and the material in this paper can be extended to multiple input and multiple output (MIMO) systems using so-called tangential interpolation, but this extension is technical and beyond the scope of this paper. For the reduction of linear MIMO systems using the Loewner framework see [3].

Since we have only one input, the bilinear map $\widetilde{\mathbf{N}}$ in (1) can be written in terms of a matrix $\mathbf{N} \in \mathbb{R}^{n\times n}$,

$$\widetilde{\mathbf{N}}(\mathbf{x}, \mathbf{u}) = \mathbf{N}\mathbf{x}\mathbf{u} \quad \text{for all } \mathbf{x} \in \mathbb{R}^n, \mathbf{u} \in \mathbb{R}. \tag{5a}$$

Furthermore, given the bilinear map $\mathbf{G}$ in (1) there exists a matrix $\mathbf{Q} \in \mathbb{R}^{n\times n^2}$ such that the bilinear map can be written using $\mathbf{Q}$ and the Kronecker product $\otimes$ as follows

$$\mathbf{G}(\mathbf{x}, \mathbf{z}) = \mathbf{Q}\,\mathbf{x} \otimes \mathbf{z} \quad \text{for all } \mathbf{x}, \mathbf{z} \in \mathbb{R}^n. \tag{5b}$$

Thus, the system (2) can be written as

$$\mathbf{E}\tfrac{d}{dt}\mathbf{x}(t) = \mathbf{A}\mathbf{x}(t) + \mathbf{Q}\,\mathbf{x}(t) \otimes \mathbf{x}(t) + \mathbf{N}\mathbf{x}(t)\mathbf{u}(t) + \mathbf{b}\mathbf{u}(t), \tag{6a}$$

$$\mathbf{y}(t) = \mathbf{c}^T\mathbf{x}(t) + d\,\mathbf{u}(t). \tag{6b}$$

The representation (6) of quadratic bilinear systems is used in previous model reduction approaches. See, e.g., the papers by Benner and Breiten [4], Breiten and Damm [5], Gosea and Antoulas [8], and Gu [9]. However, for some applications, incl. the Burgers' equation or the Navier-Stokes equations, the construction of $\mathbf{Q}$ is impractical and the use of (6) tends to be numerically inefficient. As mentioned before, the Burgers' equation or the Navier-Stokes equations are written in the form (2). Therefore we use (6) only to introduce the ideas, but use (2) for computations.

We use the system representation (6) to review the Loewner framework in Sect. 2. In Sect. 3 we show how the Loewner framework for (6) can be efficiently applied directly to (2). This makes the Loewner framework applicable to a number of flow problems. In addition, the implementation details given in Sect. 3 can also be used for other ROM approaches that are based on the representation (6), such as [4, 5], [9]. Section 4 illustrates the performance of the Loewner approach applied to Burgers' equation. The Conclusion Sect. 5 summarizes our findings and outlines some areas of current and future research.

2 The Loewner Framework

We begin with a review of the Loewner framework for the linear case ($\mathbf{Q} = \mathbf{0}$ and $\mathbf{N} = \mathbf{0}$) because it provides important insight into basic ideas and is the foundation for its extension to the quadratic bilinear case, which will be discussed in the second part of this section. The implementation of the Loewner framework in the context of the system (2) will be discussed later in Sect. 3.

To simplify our presentation, we assume that $\mathbf{E}$ is nonsingular and that $d = 0$. These assumptions can be relaxed, especially in the linear case, see [3].

2.1 Linear Systems

We consider linear systems (6) with $\mathbf{Q} = \mathbf{0}$ and $\mathbf{N} = \mathbf{0}$. The input-to-output map of the linear system in frequency domain is given by the transfer function

$$\mathbf{H}(s) = \mathbf{c}^T (s\mathbf{E} - \mathbf{A})^{-1}\mathbf{b}. \tag{7}$$

The Loewner framework uses measurements of the transfer function (7) to construct $\widehat{\mathbf{E}}, \widehat{\mathbf{A}} \in \mathbb{R}^{r\times r}$ and $\widehat{\mathbf{b}}, \widehat{\mathbf{c}} \in \mathbb{R}^r$ such that the transfer function

$$\widehat{\mathbf{H}}(s) = \widehat{\mathbf{c}}^T (s\widehat{\mathbf{E}} - \widehat{\mathbf{A}})^{-1}\widehat{\mathbf{b}} \tag{8}$$

of the corresponding ROM linear system matches or, depending on the amount of data given, approximates the original transfer function (7) at the measurements. Specifically, we want to construct a ROM so that its transfer function (8) matches or approximates the transfer function (7) at the left driving frequencies[1] $\mu_1, \ldots, \mu_k \in \mathbb{C}$ and at the right driving frequencies $\lambda_1, \ldots, \lambda_k \in \mathbb{C}$. Given distinct frequencies $\mu_1, \ldots, \mu_k$ and $\lambda_1, \ldots, \lambda_k$, and corresponding transfer function (7) measurements $\mathbf{H}(\mu_1), \ldots, \mathbf{H}(\mu_k)$ and $\mathbf{H}(\lambda_1), \ldots, \mathbf{H}(\lambda_k)$, we consider the Loewner matrix[2]

$$\mathbb{L} = \begin{bmatrix} \frac{\mathbf{H}(\mu_1)-\mathbf{H}(\lambda_1)}{\mu_1-\lambda_1} & \cdots & \frac{\mathbf{H}(\mu_1)-\mathbf{H}(\lambda_k)}{\mu_1-\lambda_k} \\ \vdots & \ddots & \vdots \\ \frac{\mathbf{H}(\mu_k)-\mathbf{H}(\lambda_1)}{\mu_k-\lambda_1} & \cdots & \frac{\mathbf{H}(\mu_k)-\mathbf{H}(\lambda_k)}{\mu_k-\lambda_k} \end{bmatrix} \in \mathbb{C}^{k\times k} \tag{9a}$$

and the shifted Loewner matrix

$$\mathbb{L}_s = \begin{bmatrix} \frac{\mu_1\mathbf{H}(\mu_1)-\mathbf{H}(\lambda_1)\lambda_1}{\mu_1-\lambda_1} & \cdots & \frac{\mu_1\mathbf{H}(\mu_1)-\mathbf{H}(\lambda_k)\lambda_k}{\mu_1-\lambda_k} \\ \vdots & \ddots & \vdots \\ \frac{\mu_k\mathbf{H}(\mu_k)-\mathbf{H}(\lambda_1)\lambda_1}{\mu_k-\lambda_1} & \cdots & \frac{\mu_k\mathbf{H}(\mu_k)-\mathbf{H}(\lambda_k)\lambda_k}{\mu_k-\lambda_k} \end{bmatrix} \in \mathbb{C}^{k\times k}. \tag{9b}$$

The desired ROM is constructed from the Loewner and shifted Loewner matrices. There are two cases: (1) The 'right' amount of data is available, and (2) a redundant amount of data is provided. We focus on the second case, which is more relevant in practice.

In the second case the pencil $(\mathbb{L}_s, \mathbb{L})$ is singular and we use the singular value decomposition (SVD) to reduce the data. Consider the (short) SVDs

$$[\mathbb{L} \;\; \mathbb{L}_s] = \mathbf{Y}_1\Sigma_1\mathbf{X}_1^*, \quad \begin{bmatrix} \mathbb{L} \\ \mathbb{L}_s \end{bmatrix} = \mathbf{Y}_2\Sigma_2\mathbf{X}_2^*, \tag{10}$$

where $\Sigma_1 \in \mathbb{R}^{k\times 2k}$, $\Sigma_2 \in \mathbb{R}^{2k\times k}$, $\mathbf{Y}_1, \mathbf{X}_2 \in \mathbb{C}^{k\times k}$. A ROM is now computed as follows. The matrices $\mathbf{Y}, \mathbf{X} \in \mathbb{C}^{k\times r}$ are obtained by selecting the first r columns of the matrices $\mathbf{Y}_1$ and $\mathbf{X}_2$. The ROM is

$$\begin{aligned} \widehat{\mathbf{E}} &= -\mathbf{Y}^*\mathbb{L}\mathbf{X}, \quad \widehat{\mathbf{A}} = -\mathbf{Y}^*\mathbb{L}_s\mathbf{X}, \\ \widehat{\mathbf{b}} &= \mathbf{Y}^*\big(\mathbf{H}(\mu_1), \ldots, \mathbf{H}(\mu_k)\big)^T, \quad \widehat{\mathbf{c}}^T = \big(\mathbf{H}(\lambda_1), \ldots, \mathbf{H}(\lambda_k)\big)\mathbf{X}. \end{aligned} \tag{11}$$

The transfer function (8) of the ROM (11) approximates the transfer function of the FOM at the driving frequencies, $\widehat{\mathbf{H}}(\mu_j) \approx \mathbf{H}(\mu_j)$, $\widehat{\mathbf{H}}(\lambda_j) \approx \mathbf{H}(\lambda_j)$, $j = 1, \ldots, k$. Expressions for the error $|\widehat{\mathbf{H}}(s) - \mathbf{H}(s)|$ are given in [3, p. 359]. The Loewner framework provides a trade-off between accuracy and complexity of the ROM by means of the singular values of the Loewner matrices $\mathbb{L}$ and $\mathbb{L}_s$.

---

[1]The reason for the terminology 'left' and 'right' driving frequencies only becomes clear in the MIMO case and is adopted here for consistency with [3].

[2]We point out an important difference in notation between this paper and [3]. In [3], $\mathbf{W} = (\mathbf{w}_1, \ldots, \mathbf{w}_k) \in \mathbb{C}^{p\times k}$ and $\mathbf{V}^T = (\mathbf{v}_1, \ldots, \mathbf{v}_q) \in \mathbb{C}^{m\times q}$ are matrices (in the SISO case $m = 1$, $p = 1$ vectors) of transfer function (7) measurements, whereas in our paper $\mathbf{V}, \mathbf{W} \in \mathbb{R}^{n\times r}$ are the Petrov-Galerkin projection matrices in projection based ROMs (4).

The ROM (11) is in general complex. However, if data $\mu_1, \ldots, \mu_k, \lambda_1, \ldots, \lambda_k$, and consequently $\mathbf{H}(\mu_1), \ldots, \mathbf{H}(\mu_k)$ and $\mathbf{H}(\lambda_1), \ldots, \mathbf{H}(\lambda_k)$, contain also the conjugate complex data $(\{\mu_j\}_{j=1}^k = \{\overline{\mu}_j\}_{j=1}^k, \{\lambda_j\}_{j=1}^k = \{\overline{\lambda}_j\}_{j=1}^k)$, then the complex ROM (11) can be transformed into a real ROM with the same transfer function, as shown in [3, p. 360].

An alternative view of the above Loewner framework, which is important for the extension of this framework, is obtained by considering the generalized controllability and observability matrices. Define

$$\boldsymbol{\Phi}(s) = (s\mathbf{E} - \mathbf{A})^{-1}. \tag{12}$$

Given the left and right driving frequencies $\mu_1, \ldots, \mu_k, \lambda_1, \ldots, \lambda_k$, the generalized controllability matrix is

$$\mathcal{R} = [\boldsymbol{\Phi}(\lambda_1)\mathbf{b},\ \boldsymbol{\Phi}(\lambda_2)\mathbf{b},\ \ldots,\ \boldsymbol{\Phi}(\lambda_k)\mathbf{b}] \in \mathbb{C}^{n\times k} \tag{13a}$$

and the generalized observability matrix is

$$\mathcal{O}^T = \left[\boldsymbol{\Phi}(\mu_1)^T\mathbf{c},\ \boldsymbol{\Phi}(\mu_2)\mathbf{c},\ \ldots,\ \boldsymbol{\Phi}(\mu_k)^T\mathbf{c}\right] \in \mathbb{C}^{n\times k}. \tag{13b}$$

The Loewner matrices can be factored in terms of the generalized controllability/observability matrices as

$$\mathbb{L} = -\mathcal{O}\mathbf{E}\mathcal{R}, \quad \mathbb{L}_s = -\mathcal{O}\mathbf{A}\mathcal{R}. \tag{14}$$

Note also that the following relations hold

$$\big(\mathbf{H}(\lambda_1), \ldots, \mathbf{H}(\lambda_k)\big) = \mathbf{c}^T\mathcal{R}, \quad \big(\mathbf{H}(\mu_1), \ldots, \mathbf{H}(\mu_k)\big)^T = \mathcal{O}\mathbf{b}. \tag{15}$$

If we define

$$\mathbf{V} = \mathcal{R}\mathbf{X} \in \mathbb{C}^{n\times r}, \quad \mathbf{W} = \mathcal{O}^*\mathbf{Y} \in \mathbb{C}^{n\times r}, \tag{16}$$

then (11), (14) and (15) show that the Loewner ROM (11) can also be written as

$$\widehat{\mathbf{E}} = \mathbf{W}^*\mathbf{E}\mathbf{V}, \quad \widehat{\mathbf{A}} = \mathbf{W}^*\mathbf{A}\mathbf{V}, \quad \widehat{\mathbf{b}} = \mathbf{W}^*\mathbf{b}, \quad \widehat{\mathbf{c}} = \mathbf{V}^*\mathbf{c}. \tag{17}$$

Thus while the Loewner ROM (11) can be obtained directly from data via the Loewner and shifted Loewner matrices (9), it could also be obtained via Petrov-Galerkin projection.

2.2 Quadratic Bilinear Systems

A starting point to extend the Loewner framework and other ROM approaches to the quadratic bilinear case is an expansion of the original system (6) using Carleman linearization or the so-called variational equation approach. See, e.g., Rugh [16, Chap. 3]. We use the latter approach. This expansion is justified for so-called linear-analytic state equations [16, Theorem 3.1], which include (6).

Consider inputs of the form $a\mathbf{u}(t)$, for a positive scalar a, and assume that the solution of (6a) can be written as a power series $\mathbf{x}(t) = \sum_{\ell=1}^{\infty} a^\ell \mathbf{x}_\ell(t)$. Inserting this representation into (6a) and equating a^ℓ terms leads to

$$\mathbf{E}\tfrac{d}{dt}\mathbf{x}_1(t) = \mathbf{A}\mathbf{x}_1(t) + \mathbf{b}\mathbf{u}(t), \qquad \ell = 1, \tag{18a}$$

$$\mathbf{E}\tfrac{d}{dt}\mathbf{x}_2(t) = \mathbf{A}\mathbf{x}_2(t) + \mathbf{Q}\,\mathbf{x}_1(t) \otimes \mathbf{x}_1(t) + \mathbf{N}\mathbf{x}_1(t)\mathbf{u}(t), \qquad \ell = 2, \tag{18b}$$

$$\vdots$$

with homogeneous initial data. We temporarily assume that $\mathbf{E} = \mathbf{I}$ is the identity to simplify notation. Since $\mathbf{E}$ is nonsingular we can alway achieve this by multiplying (18) by $\mathbf{E}^{-1}$. The solutions of (18) are $\mathbf{x}_1(t) = \int_0^t e^{\mathbf{A}\tau_1}\mathbf{b}\mathbf{u}(t-\tau_1)d\tau_1$, $\mathbf{x}_2(t) = \int_0^t e^{\mathbf{A}\tau_2}[\mathbf{Q}\,\mathbf{x}_1(t-\tau_2) \otimes \mathbf{x}_1(t-\tau_2) + \mathbf{N}\mathbf{x}_1(t-\tau_2)\mathbf{u}(t-\tau_2)]d\tau_2$, etc. The expression for $\mathbf{x}_1$ is inserted to obtain a representation of $\mathbf{x}_2$ that depends only in $\mathbf{u}$. Expressions for other solution components can be obtained analogously.

Given the expansion for $\mathbf{x}$, the output (6b) (recall $d = 0$) is $\mathbf{y}(t) = \sum_{\ell=1}^{\infty} \mathbf{c}^T \mathbf{x}_\ell(t)$. We truncate after $\ell = 2$ and insert the expressions for $\mathbf{x}_1$ and $\mathbf{x}_2$ to obtain the approximate output

$$\begin{aligned}\mathbf{y}(t) = &\int_0^t h_1(\tau_1)\mathbf{u}(t-\tau_1)d\tau_1 + \int_0^t \int_0^{t-\tau_2} h_2(\tau_1,\tau_2)\mathbf{u}(t-\tau_1-\tau_2)\mathbf{u}(t-\tau_2)d\tau_1 d\tau_2 \\ &+ \int_0^t \int_0^{t-\tau_3} \int_0^{t-\tau_3} h_3(\tau_1,\tau_2,\tau_3)\mathbf{u}(t-\tau_1-\tau_3)\mathbf{u}(t-\tau_2-\tau_3)d\tau_1 d\tau_2 d\tau_3,\end{aligned}$$

where $h_1(\tau_1) = \mathbf{c}^T e^{\mathbf{A}\tau_1}\mathbf{b}$, $h_2(\tau_1,\tau_2) = \mathbf{c}^T e^{\mathbf{A}\tau_2}\mathbf{N}e^{\mathbf{A}\tau_1}\mathbf{b}$, and $h_3(\tau_1,\tau_2,\tau_3) = \mathbf{c}^T e^{\mathbf{A}\tau_3}\mathbf{Q}\left(e^{\mathbf{A}\tau_2}\mathbf{b} \otimes e^{\mathbf{A}\tau_1}\mathbf{b}\right)$. Application of the multivariate Laplace transform leads to the generalized rational transfer functions, which correspond to the kernels h_1, h_2, h_3. The Loewner framework constructs ROMs that approximate these transfer functions at certain frequencies. We sketch the approach and refer to [8] for details.

As in the linear case, the interpolation points are partitioned into two disjoint sets of left and right interpolation points. Since the transfer functions corresponding to (18) depend on multiple frequencies (with our truncation on up to three frequencies), the interpolation points need to be arranged in a suitable way. To simplify the presentation, assume that we have $k = 3\bar{k}$ left and right interpolation points, which are renamed as follows:

$$\mu_1^{(1)}, \mu_2^{(1)}, \mu_3^{(1)}, \ldots, \mu_1^{(\bar{k})}, \mu_2^{(\bar{k})}, \mu_3^{(\bar{k})}, \qquad \lambda_1^{(1)}, \lambda_2^{(1)}, \lambda_3^{(1)}, \ldots, \lambda_1^{(\bar{k})}, \lambda_2^{(\bar{k})}, \lambda_3^{(\bar{k})}. \tag{19}$$

Next, the left and right interpolation points are grouped in multi-tuples

$$\boldsymbol{\mu}^{(j)} = \{(\mu_1^{(j)}),\ (\mu_1^{(j)},\ \mu_2^{(j)}),\ (\mu_1^{(j)},\ \lambda_1^{(j)},\ \mu_3^{(j)})\},\quad j = 1, \ldots, \bar{k}, \tag{20a}$$

$$\boldsymbol{\lambda}^{(j)} = \{(\lambda_1^{(j)}),\ (\lambda_2^{(j)},\ \lambda_1^{(j)}),\ (\lambda_3^{(j)},\ \lambda_1^{(j)},\ \lambda_1^{(j)})\},\quad j = 1, \ldots, \bar{k}. \tag{20b}$$

The generalized controllability matrix $\mathcal{R} \in \mathbb{C}^{n\times k}$ associated with the right multi-tuples $\boldsymbol{\lambda}^{(1)}, \boldsymbol{\lambda}^{(2)}, \ldots, \boldsymbol{\lambda}^{(\bar{k})}$ is

$$\mathcal{R} = \left[\mathcal{R}^{(1)},\ \mathcal{R}^{(2)},\ \cdots,\ \mathcal{R}^{(\bar{k})}\right], \tag{21}$$

where the matrices $\mathcal{R}^{(j)} \in \mathbb{C}^{n\times 3}$, $j = 1, \ldots, \bar{k}$, are associated with the j-th multi-tuple $\boldsymbol{\lambda}^{(j)}$ in (20) are given by

$$\mathcal{R}^{(j)} = \left[\boldsymbol{\Phi}(\lambda_1^{(j)})\,\mathbf{b},\ \boldsymbol{\Phi}(\lambda_2^{(j)})\,\mathbf{N}\,\boldsymbol{\Phi}(\lambda_1^{(j)})\,\mathbf{b},\ \boldsymbol{\Phi}(\lambda_3^{(j)})\mathbf{Q}\big(\boldsymbol{\Phi}(\lambda_1^{(j)})\mathbf{b} \otimes \boldsymbol{\Phi}(\lambda_1^{(j)})\mathbf{b}\big)\right]. \tag{22}$$

Similarly, the generalized observability matrix $\mathcal{O} \in \mathbb{C}^{k\times n}$ associated with the left multi-tuples $\boldsymbol{\mu}^{(1)}, \boldsymbol{\mu}^{(2)}, \ldots, \boldsymbol{\mu}^{(\bar{k})}$ is given by

$$\mathcal{O} = \left[(\mathcal{O}^{(1)})^T,\ (\mathcal{O}^{(2)})^T,\ \ldots (\mathcal{O}^{(\bar{k})})^T\right]^T \in \mathbb{C}^{k\times n}, \tag{23}$$

where the matrices $\mathcal{O}^{(j)} \in \mathbb{C}^{3\times n}$, $j = 1, \ldots, \bar{k}$, correspond to the j-th multi-tuple $\boldsymbol{\mu}^{(j)}$ in (20) and are given by

$$\mathcal{O}^{(j)} = \begin{bmatrix} \mathbf{c}^T\,\boldsymbol{\Phi}(\mu_1^{(j)}) \\ \mathbf{c}^T\,\boldsymbol{\Phi}(\mu_1^{(j)})\,\mathbf{N}\,\boldsymbol{\Phi}(\mu_2^{(j)}) \\ \mathbf{c}^T\,\boldsymbol{\Phi}(\mu_1^{(j)})\,\mathbf{Q}\left(\boldsymbol{\Phi}(\lambda_1^{(j)})\mathbf{b} \otimes \boldsymbol{\Phi}(\mu_3^{(j)})\right) \end{bmatrix}. \tag{24}$$

Now the Loewner matrix $\mathbb{L}$ and the shifted Loewner matrix $\mathbb{L}_s$ are defined using the generalized controllability (21) and observability (23) matrices as

$$\mathbb{L} = -\mathcal{O}\,\mathbf{E}\,\mathcal{R},\quad \mathbb{L}_s = -\mathcal{O}\,\mathbf{A}\,\mathcal{R}. \tag{25}$$

The fact that the Loewner matrices are factorized in terms of the pairs of matrices $(\mathbf{E}, \mathbf{A})$ and $(\mathcal{O}, \mathcal{R})$ is an inherent property of the Loewner framework which holds true for both the bilinear and quadratic-bilinear extensions of the method.

As in Sect. 2.1 we focus in the case where a redundant amount of data is provided. We use the SVD to extract the relevant data. Consider the (short) SVDs (10) of the Loewner matrices (25). The matrices $\mathbf{Y}, \mathbf{X} \in \mathbb{C}^{k\times r}$ are obtained by selecting the first r columns of the matrices $\mathbf{Y}_1$ and $\mathbf{X}_2$. We define

$$\mathbf{V} = \mathcal{R}\mathbf{X} \in \mathbb{C}^{n\times r},\quad \mathbf{W} = \mathcal{O}^*\mathbf{Y} \in \mathbb{C}^{n\times r}. \tag{26}$$

The Loewner ROM is

$$\widehat{\mathbf{E}} = -\mathbf{Y}^*\mathbb{L}\mathbf{X} = \mathbf{W}^*\mathbf{E}\mathbf{V}, \qquad \widehat{\mathbf{A}} = -\mathbf{Y}^*\mathbb{L}_s\mathbf{X}^* = \mathbf{W}^*\mathbf{E}\mathbf{V}, \tag{27a}$$

$$\widehat{\mathbf{Q}} = \mathbf{W}^*\mathbf{O}(\mathbf{V}\otimes\mathbf{V}), \qquad \widehat{\mathbf{N}} = \mathbf{W}^*\mathbf{N}\mathbf{V}, \tag{27b}$$

$$\widehat{\mathbf{b}} = \mathbf{W}^*\,\mathbf{b}, \qquad \widehat{\mathbf{c}} = \mathbf{V}^*\,\mathbf{c}. \tag{27c}$$

As in the linear case the projection matrices $\mathbf{V}$, $\mathbf{W}$ and the ROM (17) are complex, but we can obtain real projection matrices $\mathbf{V}$, $\mathbf{W}$ and corresponding ROMs if the sets of left and of right interpolation points contain also the conjugate complex data. See [3, p. 360].

The definition of the Kronecker product and (5b) imply

$$\widehat{\mathbf{Q}}(\widehat{\mathbf{x}}\otimes\widehat{\mathbf{z}}) = \mathbf{W}^*\mathbf{O}(\mathbf{V}\otimes\mathbf{V})(\widehat{\mathbf{x}}\otimes\widehat{\mathbf{z}}) = \mathbf{W}^*\mathbf{O}(\mathbf{V}\widehat{\mathbf{x}}\otimes\mathbf{V}\widehat{\mathbf{z}}) = \mathbf{W}^*\mathbf{G}(\mathbf{V}\widehat{\mathbf{x}}, \mathbf{V}\widehat{\mathbf{z}}).$$

Since the nonlinearity is bilinear, the small $r \times r^2$ matrix $\widehat{\mathbf{Q}}$ can be precomputed for efficient evaluations of the ROM.

We have introduced the Loewner ROM (27) as a Petrov-Galerkin projection. This obviously requires access to the original system (6) matrices $\mathbf{E}$, $\mathbf{A}$, etc. Just as in the linear case, it is possible to generate the same Loewner ROM directly from measurements of the generalized transfer functions. See [8, Sect. 3.4] for details.

3 Implementation of the Loewner Framework

So far we have used the Kronecker product representation (5b) of the quadratic term. As we have mentioned before, the Kronecker product is always possible but not convenient from a practical point of view. In this section we show how computations involving expressions of the form $\mathbf{Q}\,\mathbf{x}\otimes\mathbf{z}$ can actually be implemented using $\mathbf{G}$. We focus on computations that arise in the Loewner framework, but similar operations also arise in other ROM approaches, see, e.g., [4, 9]. Thus the following discussions are also relevant for those ROM approaches.

The partial Jacobians of the bilinear map $(\mathbf{x}, \mathbf{z}) \mapsto \mathbf{G}(\mathbf{x}, \mathbf{z})$ are $\mathbf{G}_1(\mathbf{z}) = D_{\mathbf{x}}\mathbf{G}(\mathbf{x}, \mathbf{z}) \in \mathbb{R}^{n\times n}$ and $\mathbf{G}_2(\mathbf{x}) = D_{\mathbf{z}}\mathbf{G}(\mathbf{x}, \mathbf{z}) \in \mathbb{R}^{n\times n}$. Because $\mathbf{G}$ is bilinear

$$\mathbf{G}(\mathbf{x}, \mathbf{z}) = \mathbf{G}_1(\mathbf{z})\mathbf{x} = \mathbf{G}_2(\mathbf{x})\mathbf{z} \qquad \forall \mathbf{x}, \mathbf{z} \in \mathbb{R}^n. \tag{28}$$

Recall the definition (12) of $\boldsymbol{\Phi}(s) = (s\mathbf{E} - \mathbf{A})^{-1}$ and that $\mathbf{b}, \mathbf{c} \in \mathbb{R}^n$. Given scalars $\lambda_1, \lambda_2, \lambda_3, \mu_1, \mu_2, \mu_3 \in \mathbb{C}$, the Loewner framework requires evaluation of quantities like

$$\boldsymbol{\Phi}(\lambda_1)\,\mathbf{Q}\,\big(\boldsymbol{\Phi}(\lambda_2)\mathbf{b}\otimes\boldsymbol{\Phi}(\lambda_3)\mathbf{b}\big) \in \mathbb{C}^n, \tag{29a}$$

$$\mathbf{c}^T\,\boldsymbol{\Phi}(\mu_1)\,\mathbf{Q}\,\big(\boldsymbol{\Phi}(\mu_2)\mathbf{b}\otimes\boldsymbol{\Phi}(\mu_3)\big) \in \mathbb{C}^{1\times n}, \tag{29b}$$

see (22) and (23). Next we discuss how to express these quantities in terms of $\mathbf{G}$. We consider general scalars $\lambda_1, \lambda_2, \lambda_3, \mu_1, \mu_2, \mu_3$. In (22) and (23) some of them are related which can be exploited to reduce some of the computations below in a straight forward way.

The vector (29a) is

$$\boldsymbol{\Phi}(\lambda_1)\ \mathbf{Q}\ \big(\boldsymbol{\Phi}(\lambda_2)\mathbf{b} \otimes \boldsymbol{\Phi}(\lambda_3)\mathbf{b}\big) = \boldsymbol{\Phi}(\lambda_1)\ \mathbf{G}\big(\boldsymbol{\Phi}(\lambda_2)\mathbf{b}, \boldsymbol{\Phi}(\lambda_3)\mathbf{b}\big)$$

and is evaluated as follows.

1. Solve $(\lambda_2\mathbf{E} - \mathbf{A})\mathbf{z}_2 = \mathbf{b}$ to get $\mathbf{z}_2 = \boldsymbol{\Phi}(\lambda_2)\mathbf{b} \in \mathbb{C}^n$.
2. Solve $(\lambda_3\mathbf{E} - \mathbf{A})\mathbf{z}_3 = \mathbf{b}$ to get $\mathbf{z}_3 = \boldsymbol{\Phi}(\lambda_3)\mathbf{b} \in \mathbb{C}^n$.
3. Evaluate $\mathbf{y} = \mathbf{G}(\mathbf{z}_2, \mathbf{z}_3) \in \mathbb{C}^n$.
4. Solve $(\lambda_1\mathbf{E} - \mathbf{A})\mathbf{x} = \mathbf{y}$ to get $\mathbf{x} = \boldsymbol{\Phi}(\lambda_1)\ \mathbf{G}\big(\boldsymbol{\Phi}(\lambda_2)\mathbf{b}, \boldsymbol{\Phi}(\lambda_3)\mathbf{b}\big)$.

The vector (29b) can be expressed in terms of $\mathbf{G}$ using (28). If $\mathbf{e}_j$ denotes the j-th unit vector, the j-th entry of the vector (29b) is

$$\begin{aligned}\mathbf{c}^T\ \boldsymbol{\Phi}(\mu_1)\ \mathbf{Q}\ \big(\boldsymbol{\Phi}(\mu_2)\mathbf{b} \otimes \boldsymbol{\Phi}(\mu_3)\big)\mathbf{e}_j &= \mathbf{c}^T\ \boldsymbol{\Phi}(\mu_1)\ \mathbf{Q}\ \big(\boldsymbol{\Phi}(\mu_2)\mathbf{b} \otimes \boldsymbol{\Phi}(\mu_3)\mathbf{e}_j\big)\\ = \mathbf{c}^T\ \boldsymbol{\Phi}(\mu_1)\ \mathbf{G}\big(\boldsymbol{\Phi}(\mu_2)\mathbf{b},\ \boldsymbol{\Phi}(\mu_3)\mathbf{e}_j\big) &= \mathbf{c}^T\ \boldsymbol{\Phi}(\mu_1)\ \mathbf{G}_2\big(\boldsymbol{\Phi}(\lambda_1)\mathbf{b}\big)\ \boldsymbol{\Phi}(\mu_3)\mathbf{e}_j\end{aligned}$$

so that

$$\mathbf{c}^T\ \boldsymbol{\Phi}(\mu_1)\ \mathbf{Q}\ \big(\boldsymbol{\Phi}(\mu_2)\mathbf{b} \otimes \boldsymbol{\Phi}(\mu_3)\big) = \mathbf{c}^T\ \boldsymbol{\Phi}(\mu_1)\ \mathbf{G}_2\big(\boldsymbol{\Phi}(\mu_2)\mathbf{b}\big)\ \boldsymbol{\Phi}(\mu_3). \tag{30}$$

Computationally, (30) is evaluated as follows.

1. Solve $(\mu_1\mathbf{E} - \mathbf{A})^T\mathbf{z} = \mathbf{c}$ to get $\mathbf{z}^T = \mathbf{c}^T\ \boldsymbol{\Phi}(\mu_1) \in \mathbb{C}^{1\times n}$.
2. Solve $(\mu_2\mathbf{E} - \mathbf{A})\mathbf{w} = \mathbf{b}$ to get $\mathbf{w} = \boldsymbol{\Phi}(\mu_2)\mathbf{b} \in \mathbb{C}^{n\times 1}$.
3. Evaluate $\mathbf{y}^T = \mathbf{z}^T\mathbf{G}_2(\mathbf{w}) \in \mathbb{C}^{1\times n}$.
4. Solve $(\mu_3\mathbf{E} - \mathbf{A})^T\mathbf{x} = \mathbf{y}$ to get $\mathbf{x}^T = \mathbf{c}^T\ \boldsymbol{\Phi}(\mu_1)\ \mathbf{G}_2\big(\boldsymbol{\Phi}(\mu_2)\mathbf{b}\big)\ \boldsymbol{\Phi}(\mu_3) \in \mathbb{C}^{1\times n}$.

4 Application to Burgers' Equation

4.1 *Burgers' Equation*

As an example we consider Burgers' equation with Robin boundary conditions. Given viscosity $\nu > 0$ and parameters $\sigma_0 \le 0$, $\sigma_1 \ge 0$, consider

$$\frac{\partial}{\partial t}v(x,t) - \nu\frac{\partial^2}{\partial x^2}v(x,t) + v(x,t)\frac{\partial}{\partial x}v(x,t) = 0, \quad x \in (0,1), t \in (0,T), \tag{31a}$$

$$\nu\frac{\partial}{\partial x}v(x,0) + \sigma_0 v(0,t) = u_0(t), \quad t \in (0,T), \tag{31b}$$

$$\nu\frac{\partial}{\partial x}v(x,1) + \sigma_1 v(1,t) = u_1(t), \quad t \in (0,T), \tag{31c}$$

$$v(x,0) = 0, \quad x \in (0,1). \tag{31d}$$

The functions u_0, u_1 are the inputs. Since we have restricted our presentation to the case of one input, we will later set $u_1 \equiv 0$ and consider u_0 as the only input. Our output is

$$y(t) = \int_0^1 v(x,t)\,dx.$$

We seek weak solutions of (31) in the function space

$$W(0,T) = \big\{\phi \in L^2(0,T;H^1(0,1)) \;:\; \frac{\partial}{\partial t}\phi \in L^2(0,T;(H^1(0,1))')\big\},$$

where $(H^1(0,1))'$ is the dual of $H^1(0,1)$. The function $v \in W(0,T)$ with $v(x,0) = 0$ a.e. is a weak solution of (31) if

$$\begin{aligned}
&\frac{d}{dt}\int_0^1 v(x,t)\varphi(x)dx + \nu\int_0^1 \frac{\partial}{\partial x}v(x,t)\frac{\partial}{\partial x}\varphi(x)dx \\
&\quad - \sigma_0 v(0,t)\varphi(0) + \sigma_1 v(1,t)\varphi(1) + \int_0^1 \frac{\partial}{\partial x}v(x,t)v(x,t)\varphi(x)dx \\
&= u_0(t)\varphi(0) - u_1(t)\varphi(1), \qquad \text{for all } \varphi \in H^1(0,1) \text{ and a.a. } t \in (0,T).
\end{aligned} \tag{32}$$

An existence and uniqueness proof for (31) is given, e.g., by Volkwein [17]. In particular, [17, Theorem 2.3] proves that for every $u_0, u_1 \in L^2(0,T)$ there exists a unique solution $v \in W(0,T) \cap L^\infty((0,1)\times(0,T))$ that obeys

$$\|v\|_{W(0,T)} + \|v\|_{L^\infty} \le C\big(1 + \|u_0\|_{L^2(0,T)} + \|u_1\|_{L^2(0,T)}\big) \tag{33}$$

for a constant $C > 0$ that depends on $T, \nu > 0$, but not on u_0, u_1.

Since the Loewner framework depends only on transfer function information, it can in theory be applied directly in a function space setting. However, transfer function information is only analytically available for special linear examples. Therefore, we use a fixed finite element semi-discretization to generate transfer function information numerically. Dependence of the Loewner ROM on the mesh size is part of future research.

We discretize (32) in space using linear finite elements on a uniform grid $x_i = ih$, $i = 0,\ldots,n-1$, $h = 1/(n-1)$. The weak solution of Burgers' equation (32) is approximated by $v_h(x,t) = \sum_{j=0}^{n-1} v_j(t)\varphi_j(x)$, where $\varphi_i \in H^1(0,1)$, $i = 0,\ldots,n-1$, are the usual piecewise linear 'hat' functions. We set $u_1 \equiv 0$ and consider u_0 as the only input to arrive at a system (2) with $\mathbf{N} = \mathbf{0}$, and $\mathbf{D} = \mathbf{0}$. For given vectors $\mathbf{v} = (v_0, \ldots, v_{n-1})^T, \mathbf{z} = (z_0, \ldots, z_{n-1})^T$, the i-th component of the bilinear map $\mathbf{G}$ is

$$\mathbf{G}_i(\mathbf{v}, \mathbf{z}) = -\int_0^1 \Big(\frac{d}{dx}\sum_{j=0}^{n-1} v_k\varphi_k(x)\Big)\Big(\sum_{l=0}^{n-1} z_l\varphi_l(x)\Big)\,\varphi_i(x)\,dx.$$

4.2 Numerical Results

We use the problem data $\nu = 0.01$, $\sigma_0 = 0$, $\sigma_1 = 0.1$. The FOM is the linear finite element semi-discretization with $n = 257$. The semi-discretized system (2) is approximately solved using backward Euler over varying time intervals $[0, T]$ specified below with time step size $\Delta t = 1/128$. In all simulations we use the input $u_0(t) = 0.1\sin(4\pi t)$ and $u_1 \equiv 0$.

The interpolation points to construct the Loewner ROM are chosen as follows. First we create 300 logarithmically spaced points ξ_j, $j = 1, \ldots, 300$, between 1 and 10^3 (in Matlab `logspace(0,3,300)`), and then we select the left interpolation points $\mu_{2j-1} = \xi_{2j-1}i$, $\mu_{2j} = -\xi_{2j-1}i$, $j = 1, \ldots, 150$, and the right interpolation points $\lambda_{2j-1} = \xi_{2j}i$, $\lambda_{2j} = -\xi_{2j}i$, $j = 1, \ldots, 150$. The choice of interpolation points clearly has an impact on the quality of the ROM approximation and how to choose 'good' interpolation points is still an open question. Thus, the above choice is somewhat arbitrary.

The singular values of the matrices in (25) are shown in Fig. 1. Let σ_j denote the singular values of $[\mathbb{L}\ \mathbb{L}_s]$. The size r of the ROM is chosen to the smallest r with $\sigma_r/\sigma_1 > 10^{-11}$. This leads to a Loewner ROM of size $r = 22$.

The outputs of the FOM and of the Loewner ROM are shown in Fig. 2. For approximately $t \leq 1.5$ the agreement between the FOM and the Loewner ROM output is good, but there are larger differences between both outputs for approximately $t > 1.5$. Moreover, the Loewner ROM exhibits instabilities starting around $t = 1.5$,

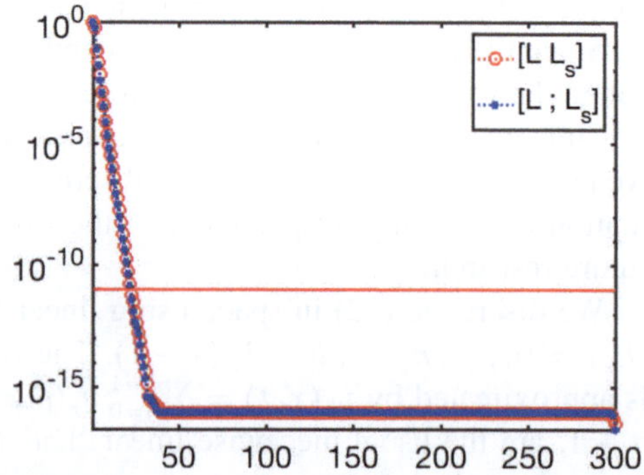

Fig. 1 The normalized singular values of the Loewner matrices (25). The size $r = 22$ of the ROM is chosen to the smallest r with $\sigma_r/\sigma_1 > 10^{-11}$

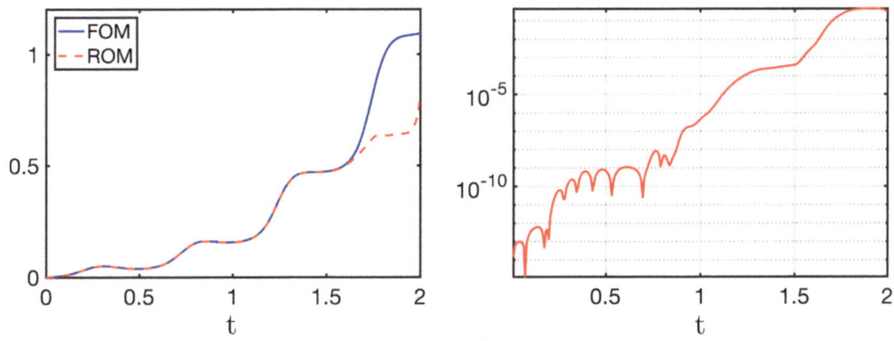

Fig. 2 Left plot: output of the FOM (solid blue line) and of the Loewner ROM with Petrov-Galerkin projection matrices $\mathbf{V} \neq \mathbf{W}$ (dashed red line). Right plot: error between outputs of the FOM and the Loewner ROM

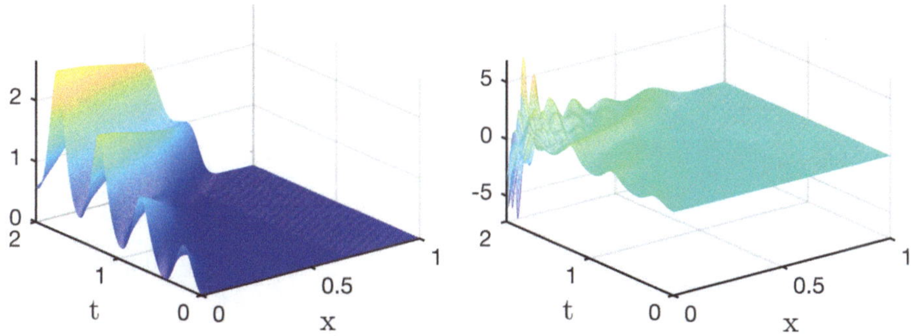

Fig. 3 Solution of the FOM (left) and of the Loewner ROM with Petrov-Galerkin projection matrices $\mathbf{V} \neq \mathbf{W}$ (right). The Loewner ROM exhibits instabilities

as can be seen from the states $\mathbf{x}$ generated by the FOM and the state $\mathbf{V}\widehat{\mathbf{x}}$ generated by the Loewner ROM, which are shown in Fig. 3.

Stability results for the Burgers' equation like (33) are based on the weak form (32) and can be carried over to Galerkin approximations of (32), such as the finite element discretization or Galerkin projection based ROMs with $\mathbf{V} = \mathbf{W}$. In the standard Loewner approach the projection matrices $\mathbf{V} \neq \mathbf{W}$, and it is not clear in the general case how to construct a Loewner ROM with $\mathbf{V} = \mathbf{W}$. We merge the Loewner projection matrices $\mathbf{V}, \mathbf{W} \in \mathbb{R}^{n \times r}$ into one larger matrix $[\mathbf{V}, \mathbf{W}] \in \mathbb{R}^{n \times 2r}$ (we actually compute an orthonormal basis of the columns of $[\mathbf{V}, \mathbf{W}]$ to ensue that the resulting matrix is full rank), and we use a Galerkin projection with this matrix $[\mathbf{V}, \mathbf{W}] \in \mathbb{R}^{n \times 2r}$. We refer to the resulting ROM as a Loewner Galerkin ROM.

The outputs of the FOM and of the Loewner Galerkin ROM are shown in Fig. 4. We can now simulate the Loewner Galerkin ROM at least until $T = 6$ and there is good agreement between the outputs of the FOM and of the Loewner Galerkin ROM. The states $\mathbf{x}$ and $\mathbf{V}\widehat{\mathbf{x}}$ generated by the FOM and the Loewner Galerkin ROM

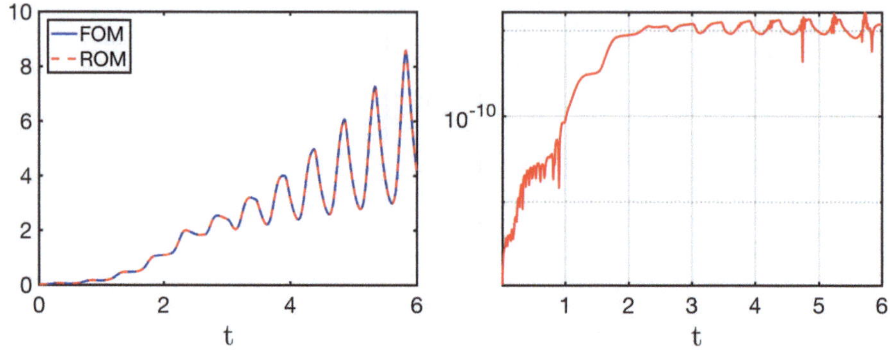

Fig. 4 Left plot: output of the FOM (solid blue line) and of the Loewner ROM with Galerkin projection matrix [**V**, **W**] (dashed red line). Right plot: error between outputs of the FOM and the Loewner ROM

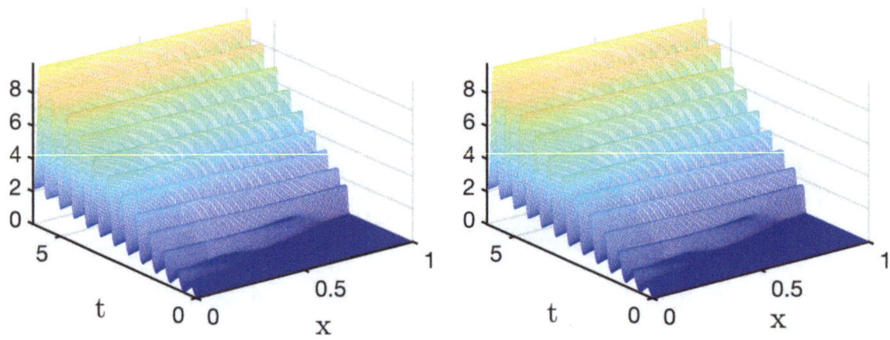

Fig. 5 Solution of the FOM (left) and of the Loewner ROM with Galerkin projection matrix [**V**, **W**] (right)

are shown in Fig. 5. The FOM solution in Fig. 5 restricted to the time interval [0, 2] is identical to the FOM solution shown in Fig. 3.

Of course, since our Loewner Galerkin ROM is twice the size of the standard Loewner Petrov-Galerkin ROM, it is not entirely clear whether this improvement in results is due to the increased ROM size, or the switch from a Petrov-Galerkin projection to a Galerkin projection. We did change the accuracy and corresponding ROM size r of the standard Loewner Petrov-Galerkin ROM slightly and still observed instabilities in the resulting ROMs. Thus it seems more accurate Loewner Petrov-Galerkin ROMs alone do not restore stability, but this issue is still under investigation.

In our current implementation of the Loewner approach, we generate $\mathbf{V}, \mathbf{W} \in \mathbb{R}^{n \times r}$ and compute the ROM explicitly as a Petrov-Galerkin projection ROM (27). As mentioned at the end of Sect. 2.2, the same Loewner ROM can be computed directly from measurements of the generalized transfer functions. In this case our approach to enforce stability via the use of Loewner Galerkin ROM with $[\mathbf{V}, \mathbf{W}] \in \mathbb{R}^{n \times 2r}$

is no longer possible. Extension of the Loewner approach to generate ROM that correspond to a Galerkin projection ROM is an interesting research question.

5 Conclusions and Future Work

We have presented an extension of the Loewner framework to compute ROMs of quadratic-bilinear systems. Specifically, we have used the Kronecker product representation of quadratic-bilinear systems to present the algorithm, but then used the bilinear maps that naturally arise in semi-discretizations of fluid flow problems, such as Burgers' equation or the Navier-Stokes equations to express the actual computations. This makes it possible to apply the Loewner framework to large-scale problems. In this paper we have applied to the viscous Burgers' equation. Application to the Navier-Stokes equations is ongoing work.

The application to Burgers' equation showed the potential of the Loewner framework, but also raises some questions that still need to be addressed. Generally, the selection of interpolation points (the μ_j's and λ_j's) is an issue. Current numerical experiments indicate that the more interpolation points can be used, the better given a constant ROM size r. Recall that the data gets assembled in the Loewner and shifted Loewner matrices and then is compressed via the SVD. Thus more data does not necessarily mean larger ROMs. We consider SISO systems. The extension to multiple input and multiple output systems is possible, using so-called tangential interpolation. For linear systems this is described in the tutorial paper [3].

An important issue is stability. Our numerics have shown that the standard Loewner ROM may not be stable. Currently, our Loewner ROM is equivalent to a Petrov-Galerkin projection, $\mathbf{W} \neq \mathbf{V}$. At the same time, stability results like (33) for Burgers' equation are based on the week form and Galerkin projection. Thus if we can modify the Loewner framework to enforce $\mathbf{W} = \mathbf{V}$, then the resulting ROM inherits the stability properties of the underlying original system. In the linear case stability issues can be treated by postprocessing, see, e.g., Gosea and Antoulas [7]. If we explicitly compute $\mathbf{V}, \mathbf{W} \in \mathbb{R}^{n \times r}$ we can enforce stability via the use of Loewner Galerkin ROM with $[\mathbf{V}, \mathbf{W}] \in \mathbb{R}^{n \times 2r}$ as demonstrated in Sect. 4.2. However, this is not possible if the Loewner ROM is computed directly from data.

Finally, the Loewner framework starts from system representations in frequency domain and is based on measurements in frequency domain. This is inconvenient for many applications where only time domain measurements or simulations are accessible. Initial work towards time-domain Loewner ROMs is presented by Peherstorfer et al. [12].

Acknowledgements The authors gratefully acknowledge support by NSF grants CNS-1701292 and DMS-1522798. We thank the two referees for their comments, which have lead to improvements in the presentation.

References

1. Antoulas, A.C.: Approximation of large-scale dynamical systems. In: Advances in Design and Control, vol. 6. Society for Industrial and Applied Mathematics (SIAM), Philadelphia, PA (2005). https://doi.org/10.1137/1.9780898718713
2. Antoulas, A.C., Gosea, I.V., Ionita, A.C.: Model reduction of bilinear systems in the Loewner framework. SIAM J. Sci. Comput. **38**(5), B889–B916 (2016). https://doi.org/10.1137/15M1041432
3. Antoulas, A.C., Lefteriu, S., Ionita, A.C.: Chapter 8: A tutorial introduction to the Loewner framework for model reduction. In: P. Benner, A. Cohen, M. Ohlberger, K. Willcox (eds.) Model Reduction and Approximation: Theory and Algorithms, pp. 335–376. SIAM, Philadelphia (2017). https://doi.org/10.1137/1.9781611974829.ch8
4. Benner, P., Breiten, T.: Two-sided projection methods for nonlinear model order reduction. SIAM J. Sci. Comput. **37**(2), B239–B260 (2015). https://doi.org/10.1137/14097255X
5. Breiten, T., Damm, T.: Krylov subspace methods for model order reduction of bilinear control systems. Syst. Control Lett. **59**(8), 443–450 (2010). https://doi.org/10.1016/j.sysconle.2010.06.003
6. Elman, H.C., Silvester, D.J., Wathen, A.J.: Finite Elements and Fast Iterative Solvers with Applications in Incompressible Fluid Dynamics, 2nd edn. Numerical Mathematics and Scientific Computation. Oxford University Press, Oxford (2014). https://doi.org/10.1093/acprof:oso/9780199678792.001.0001
7. Gosea, I.V., Antoulas, A.C.: Stability preserving post-processing methods applied in the Loewner framework. In: IEEE 20th Workshop on Signal and Power Integrity (SPI), pp. 1–4 (2016). https://doi.org/10.1109/SaPIW.2016.7496283
8. Gosea, I.V., Antoulas, A.C.: Data-driven model order reduction of quadratic-bilinear systems. Numer. Linear Algebra Appl. (2018). Under review
9. Gu, C.: QLMOR: a projection-based nonlinear model order reduction approach using quadratic-linear representation of nonlinear systems. IEEE Trans. Comput. Aided Des. Integr. Circuits Syst. **30**(9), 1307–1320 (2011). https://doi.org/10.1109/TCAD.2011.2142184
10. Hesthaven, J.S., Rozza, G., Stamm, B.: Certified Reduced Basis Methods for Parametrized Partial Differential Equations. Springer Briefs in Mathematics. Springer, New York (2015). https://doi.org/10.1007/978-3-319-22470-1
11. Layton, W.: Introduction to the numerical analysis of incompressible viscous flows. Computational Science and Engineering, vol. 6. Society for Industrial and Applied Mathematics (SIAM), Philadelphia, PA (2008). https://doi.org/10.1137/1.9780898718904
12. Peherstorfer, B., Gugercin, S., Willcox, K.: Data-driven reduced model construction with time-domain loewner models. SIAM J. Sci. Comput. **39**(5), A2152–A2178 (2017). https://doi.org/10.1137/16M1094750
13. Quarteroni, A., Manzoni, A., Negri, F.: Reduced Basis Methods for Partial Differential Equations. An Introduction. Unitext, vol. 92. Springer, Cham (2016). https://doi.org/10.1007/978-3-319-15431-2
14. Rowley, C.W., Dawson, S.T.M.: Model reduction for flow analysis and control. Ann. Rev. Fluid Mech. **49**(1), 387–417 (2017). https://doi.org/10.1146/annurev-fluid-010816-060042
15. Rozza, G., Huynh, D.B.P., Patera, A.T.: Reduced basis approximation and a posteriori error estimation for affinely parametrized elliptic coercive partial differential equations: application to transport and continuum mechanics. Arch. Comput. Methods Eng. **15**(3), 229–275 (2008). https://doi.org/10.1007/s11831-008-9019-9
16. Rugh, W.J.: Nonlinear System Theory. The Volterra/Wiener Approach. Johns Hopkins University Press, Baltimore, Md. (1981). https://sites.google.com/site/wilsonjrugh. Accessed 22 Feb 2018
17. Volkwein, S.: Second order conditions for boundary control problems of the Burgers equation. Control Cybernet **30**(3), 249–278 (2001). http://www.oxygene.ibspan.waw.pl:3000/contents/export?filename=2001-3-02_volkwein.pdf. Accessed 22 Feb 2018

Model Reduction for a Pulsed Detonation Combuster via Shifted Proper Orthogonal Decomposition

Philipp Schulze, Julius Reiss and Volker Mehrmann

Abstract We consider the problem of finding an optimal data-driven modal decomposition of flows with multiple convection velocities. To this end, we apply the shifted proper orthogonal decomposition (sPOD) which is a recently proposed mode decomposition technique. It overcomes the poor performance of classical methods like the proper orthogonal decomposition (POD) for a class of transport-dominated phenomena with large gradients. This is achieved by identifying the transport directions and velocities and by shifting the modes in space to track the transports. We propose a new algorithm for computing an sPOD which carries out a residual minimization in which the main cost arises from solving a nonlinear optimization problem scaling with the snapshot dimension. We apply the algorithm to snapshot data from the simulation of a pulsed detonation combuster and observe that very few sPOD modes are sufficient to obtain a good approximation. For the same accuracy, the common POD needs ten times as many modes and, in contrast to the sPOD modes, the POD modes do not reflect the moving front profiles properly.

Keywords Transport-dominated phenomena
Shifted proper orthogonal decomposition · Mode decomposition
Pulsed detonation combuster

P. Schulze (✉) · V. Mehrmann
Institute of Mathematics, Technische Universität Berlin, 10623 Berlin, Germany
e-mail: pschulze@math.tu-berlin.de
URL: http://www.tu-berlin.de/?id=66282

V. Mehrmann
e-mail: mehrmann@math.tu-berlin.de

J. Reiss
Institute of Fluid Dynamics and Technical Acoustics, Technische Universität Berlin, 10623 Berlin, Germany
e-mail: reiss@tnt.tu-berlin.de
URL: http://www.cfd.tu-berlin.de/reiss/

R. King (ed.), *Active Flow and Combustion Control 2018*,
Notes on Numerical Fluid Mechanics and Multidisciplinary Design 141,
https://doi.org/10.1007/978-3-319-98177-2_17

1 Introduction

Model reduction, see e.g. [2, 3, 11], is an essential requirement in almost all areas of science and technology to obtain efficient multi-parameter simulations and, in particular, optimization and control methods. Often the full-order model (FOM) arises from a semi-discretization in space of a partial differential equation (PDE) and the state dimension scales with the number of grid points which is typically large. However, one is usually not interested in a detailed description of the complete dynamics but often only in a low-dimensional manifold where the solution of interest approximately evolves.

Model reduction for nonlinear dynamical systems is often based on mode decomposition techniques as the proper orthogonal decomposition (POD) [3, 4, 24] or the dynamic mode decomposition [14, 22]. Standard mode decomposition techniques are based on the concept of representing the unknown solution as a linear combination of modes. More precisely, let q be a function in space x and time t representing the state of the dynamical system, then a common model reduction ansatz is an approximation

$$q\left(x,t\right) \approx \sum_{k=1}^{r} \alpha_k\left(t\right) \psi_k\left(x\right) \tag{1}$$

with space-dependent modes ψ_k, time-dependent coefficients, or amplitudes, α_k, and r is the number of modes.

While the amplitudes typically become the unknowns of the reduced-order model, the modes have to be determined in advance. To determine the modes, one typically simulates the system and computes space- and time-discrete snapshots of a numerical approximation q_m which are stored in a snapshot matrix $X \in \mathbb{R}^{m\times n}$, i.e., $[X]_{ij} = q_m(x_i, t_j) \approx q(x_i, t_j)$ for $i = 1, \ldots, m$ and $j = 1, \ldots, n$. With the coefficients of the snapshot matrix one obtains a discrete analogue of (1) as

$$[X]_j \approx \sum_{k=1}^{r} a_{k,j} w_k \tag{2}$$

for $j = 1, \ldots, n$, where $[X]_j$ denotes the jth column of X, $w_k \in \mathbb{R}^m$ are coefficient vector representations of the modes ψ_k, and $a_{k,j}$ are the corresponding amplitudes at time point t_j.

A classical way to obtain modes and amplitudes is the POD which is based on a singular value decomposition (SVD) of the snapshot matrix X. The POD representation is optimal in the sense that it minimizes the residual in the discrete representation (2). The resulting reduced-order model is obtained as projection onto the span of the so obtained modes.

In many applications the assumption that POD delivers a good approximation of the form (1) or (2) with a small number r is valid and model reduction schemes like POD lead to models with dimensions that are orders of magnitude smaller than those of the full-order model [12].

However, when the dynamics of the system features structures with high gradients that are propagating through the domain, then schemes of the form (1) typically need a large number of modes to approximate the dynamics well, and hence model order reduction is not very effective. To overcome this difficulty, recently, there have been several suggestions for model reduction of such transport-dominated systems. In [19] the authors use ideas of symmetry reduction to decompose the solution into a frozen profile and a translation group accounting for the transport. The advantages over standard model reduction schemes are demonstrated by means of the Burgers' equation. In [21] the authors present a method which is able to decompose multiple transport phenomena. The main ingredients are SVDs of several shifted snapshot matrices combined with a greedy algorithm. The method is cheap to apply but it often needs more shifted modes than necessary, as illustrated with results for the linear wave equation. For further references on model reduction for transport-dominated problems, see [1, 5, 9, 13, 16, 23]. Most of these approaches consider transport-dominated systems with only one transport velocity and assume periodic boundary conditions. However, in many applications, multiple transport velocities are encountered, e.g., by different waves propagating through the domain. To deal with such phenomena, in [20] the shifted POD (sPOD) method has been proposed to obtain mode decompositions suitable for multiple transport phenomena. This new technique differs from (1) by shifting the modes in space into different reference frames according to the different transports of the system, i.e.,

$$q(x,t) \approx \sum_{\ell=1}^{N_s} \mathcal{T}_{\text{per}}\left(\Delta^{\ell}(t)\right) \sum_{k=1}^{r_\ell} \alpha_k^{\ell}(t)\psi_k^{\ell}(x) \tag{3}$$

where $\mathcal{T}_{\text{per}}(\cdot)$ is a shift operator defined on a periodic domain $[0, L]$ via

$$\mathcal{T}_{\text{per}}\left(\Delta(t)\right) f\left(x,t\right) := f\left((x+\Delta(t)) \bmod L, t\right),$$

N_s denotes the number of shifted reference frames, mod denotes the *modulo operator* reflecting the periodicity of the domain, and $\Delta^{\ell}(t)$ are time-dependent shifts which track the locations of, e.g., different wave profiles over time. Similar to POD one obtains a discrete analogue of (3) via

$$[X]_j \approx \sum_{\ell=1}^{N_s} T_{\text{per}}\left(d_j^{\ell}\right) \sum_{k=1}^{r_\ell} a_{k,j}^{\ell} w_k^{\ell} \tag{4}$$

for $j = 1, \ldots, n$, where T_{per} is a discrete approximation of $\mathcal{T}_{\text{per}}$ and d_j^{ℓ} are shifts at discrete time points t_j. In [20] a heuristic algorithm is proposed to compute a decomposition of the form (4) in an iterative procedure, and it has been demonstrated that this approach is very successful for several examples including two separating vortex pairs and the linear wave equation. In the latter case the method needs less modes than other methods such as e.g. [21] and also retrieves the known analytic solution. It should be mentioned that while we expect the shifted POD to perform well for systems with a few dominant moving coherent structures, it may not be effective in its current form for turbulent flows where also higher modes significantly contribute to the dynamics.

In this paper, we propose an optimization procedure to compute an *optimal* decomposition of the form (4). To this end, we generalize the cost functional that is used to obtain the optimality of the POD method to the sPOD setting. We first consider the optimization on the infinite-dimensional level, see Sect. 2, and then present an algorithm which computes the decomposition in the fully discrete setting, see Sects. 3 and 4. The computational cost is higher than for the method in [20] but the obtained approximations are locally optimal in the sense that a residual is minimized.

The focus of our work is on obtaining an optimal mode decomposition which then can be used for the construction of a reduced-order model, e.g., by a Galerkin projection. A rigorous treatment of non-periodic boundary conditions is also discussed elsewhere.

To demonstrate the efficiency of the new approach, we present results for a pulsed detonation combuster (PDC). The snapshot data originate from a data assimilation, cf. [10], and exhibit multiple transport phenomena which interact nonlinearly with each other and with the boundary.

2 Optimal sPOD Approximation

As a model problem for a partial differential equation whose solution features multiple transport velocities we consider the linear acoustic wave equation

$$\begin{aligned} \partial_t \rho + \rho_{\mathrm{ref}} \partial_x u &= 0, \\ \partial_t u + c^2/\rho_{\mathrm{ref}} \partial_x \rho &= 0, \end{aligned} \tag{5}$$

on a one-dimensional spatial domain $\Omega = (0, 1)$ with periodic boundary conditions. Here, u is the velocity, ρ the density, ρ_{ref} a reference density, and c the speed of sound. The analytic solution of (5) can be expressed as

$$\begin{bmatrix} \rho(x,t) \\ u(x,t) \end{bmatrix} = q_-(x+ct) \begin{bmatrix} \rho_{\mathrm{ref}} \\ -c \end{bmatrix} + q_+(x-ct) \begin{bmatrix} \rho_{\mathrm{ref}} \\ c \end{bmatrix}, \tag{6}$$

where q_- and q_+ are the Riemann invariants which are uniquely determined by the initial conditions. In the following, we use $\rho_{\mathrm{ref}} = 1$ and $c = 1$ and we consider the initial conditions

$$\rho(x,0) = \rho_0(x) = \exp\left(-\left(\frac{x-0.5}{0.01}\right)^2\right), \qquad u(x,0) = u_0(x) \equiv 0,$$

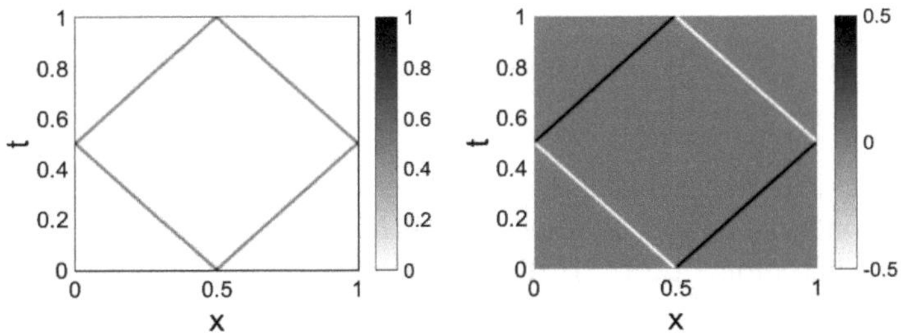

Fig. 1 Linear wave equation: snapshots of the full-order solution for the density (left) and the velocity (right)

Fig. 2 Linear wave equation: singular value decay of the snapshot matrix

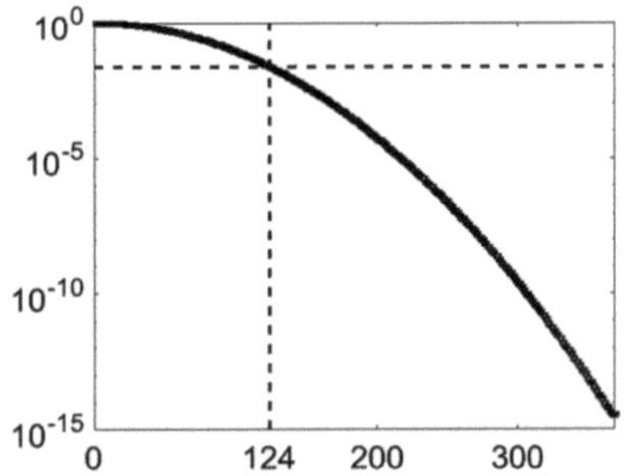

which represent a pressure pulse with large gradients. The analytic solution, see Fig. 1, is hard to approximate by a classical POD approach, since the singular values of the snapshot matrix, that is obtained by sampling the analytic solution, are decaying very slowly, cf. Fig. 2. To demonstrate the difficulties that POD has for this problem consider the *relative approximation error*

$$\left(\sum_{j=1}^{n} \left\| [X]_j - \left[\tilde{X}\right]_j \right\|^2\right) \Big/ \left(\sum_{j=1}^{n} \left\| [X]_j \right\|^2\right) \tag{7}$$

of an approximation $\tilde{X}$ of the snapshot matrix X, $\|\cdot\|$ being the Euclidean norm.

In this model problem, to obtain a relative error of less than 1%, the POD needs 124 modes (cf. dashed lines in Fig. 2) although the analytic solution is simply represented by the sum of two shifted functions. Indeed, the analytic solution (6) can be formulated within the more general representation (3) with only two modes and

$$N_s = 2, \quad r_1 = r_2 = 1, \quad \Delta^1(t) = -\Delta^2(t) = t, \quad \alpha_1^1(t) = \alpha_1^2(t) \equiv 0.5,$$
$$\psi_1^1(x) = \rho_0(x) \begin{bmatrix} 1 & -1 \end{bmatrix}^T, \; \psi_1^2(x) = \rho_0(x) \begin{bmatrix} 1 & 1 \end{bmatrix}^T.$$

However, the question arises how to compute such a decomposition when only snapshot data are available. In this case the POD is optimal in the sense that it minimizes the residual, i.e., it solves the optimization problem

$$\min_{\psi,\alpha} \int_0^T \int_\Omega \left(q(x,t) - \sum_{k=1}^{r} \alpha_k(t)\psi_k(x) \right)^2 \mathrm{d}x\mathrm{d}t \quad \text{s. t.} \ \langle \psi_i(x), \psi_j(x) \rangle_{L^2(\Omega)} = \delta_{ij} \quad (8)$$

for $i, j = 1, \ldots, r$, where δ denotes the Kronecker delta. In this way the modes ψ_j form an orthonormal basis with respect to the L^2 inner product in Ω.

To extend this optimality of (8) to the more general decomposition (3), we consider the optimization problem

$$\min_{\psi,\alpha} \int_0^T \int_\Omega \left(q(x,t) - \sum_{\ell=1}^{N_\mathrm{s}} \mathcal{T}_\mathrm{per}\left(\Delta^\ell(t)\right) \sum_{k=1}^{r_\ell} \alpha_k^\ell(t)\psi_k^\ell(x) \right)^2 \mathrm{d}x\mathrm{d}t, \qquad (9)$$

where for the moment we assume that the shift frames Δ are available or can be approximated before the optimization for the modes ψ and their time amplitudes α is carried out. Methods to estimate these shifts based on given snapshot data have been discussed in [20].

In contrast to (8) and (9) is an unconstrained optimization problem without the orthonormality restriction for the modes ψ_j. The reason why we drop this orthonormality requirement is that in a decomposition of the form (3) even linearly dependent modes may lead to optimal approximations.

To illustrate the necessity to allow linearly dependent modes, consider again the linear wave equation but this time only the density, i.e., take $q(x,t) = \rho(x,t)$. In this case a solution of the optimization problem (9) is obtained with

$$N_\mathrm{s} = 2, \quad r_1 = r_2 = 1, \ \Delta^1(t) = -\Delta^2(t) = t,$$
$$\alpha_1^1(t) = \alpha_1^2(t) \equiv 0.5, \quad \psi_1^1(x) = \psi_1^2(x) = \rho_0(x),$$

i. e., there is an optimal approximation with linearly dependent modes $\psi_1^1 = \psi_1^2$.

Thus, we omit orthogonality constraints on the modes in (9), at least when there is more than one transport velocity.

3 Residual Minimization

In this section we discuss the optimization problem (9) with a general linear shift operator $\mathcal{T}$, i.e., we consider

$$\min_{\psi,\alpha} \int_0^T \int_\Omega \left(q(x,t) - \sum_{\ell=1}^{N_\mathrm{s}} \mathcal{T}\left(\Delta^\ell(t)\right) \sum_{k=1}^{r_\ell} \alpha_k^\ell(t)\psi_k^\ell(x) \right)^2 \mathrm{d}x\mathrm{d}t. \qquad (10)$$

The solution of the POD optimization problem (8) can be obtained by solving an operator eigenvalue problem, which in the discrete setting corresponds to computing an SVD. Since in the setting of (10), the modes may be linearly dependent, we have to solve a nonlinear optimization problem instead. To this end, we apply numerical optimization techniques on the discrete level but, prior to that, we analyze some properties of (10).

First, it should be noted that the solution is in general not unique. This can be seen by taking for instance the simple case where $\mathcal{T} = \mathcal{T}_{\text{per}}$ and $q(x,t) = q_1(x+t) + q_2(x-t) + \cos(t)q_3(x)$ with some arbitrary functions q_i for $i = 1, \dots, 3$. Then, a solution of (10) is given by

$$\begin{aligned} &N_s = 3, \quad r_1 = r_2 = r_3 = 1, \; \Delta^1(t) = -\Delta^2(t) = t, \quad \Delta^3(t) \equiv 0, \\ &\alpha_1^1(t) = \alpha_1^2(t) \equiv 1, \qquad\qquad \alpha_1^3(t) = \cos(t), \quad \psi_1^i(x) = q_i(x), \quad \text{for } i = 1, \dots, 3. \end{aligned}$$

On the other hand, by making use of the trigonometric identities $\sin(x \pm t) = \sin(x)\cos(t) \pm \cos(x)\sin(t)$, another solution is

$$\begin{aligned} &N_s = 3, \quad r_1 = r_2 = r_3 = 1, \quad \Delta^1(t) = -\Delta^2(t) = t, \quad \Delta^3(t) \equiv 0 \\ &\alpha_1^1(t) = \alpha_1^2(t) \equiv 1, \quad \alpha_1^3(t) = \cos(t), \\ &\psi_1^i(x) = q_i(x) + \sin(x), \quad \text{for } i = 1, 2, \quad \psi_1^3(x) = q_3(x) - 2\sin(x). \end{aligned}$$

Both these solutions are optimal, since the cost functional is zero.

As discussed in Sect. 1, many of the currently discussed model reduction approaches for transport-dominated phenomena consider the case of only one transport velocity ($N_s = 1$) and periodic boundary conditions. In this special case the cost functional takes the form

$$\min_{\psi,\alpha} \int_0^T \int_\Omega \left(q(x,t) - \mathcal{T}_{\text{per}}(\Delta(t)) \sum_{k=1}^{r} \alpha_k(t)\psi_k(x) \right)^2 \mathrm{d}x\mathrm{d}t \tag{11}$$

and one can enforce the modes to form an orthonormal basis, since orthogonality is preserved under the action of the periodic shift operator, i.e.,

$$\left\langle \mathcal{T}_{\text{per}}(\Delta(t))\,\psi_i(x), \mathcal{T}_{\text{per}}(\Delta(t))\,\psi_j(x) \right\rangle_{L^2(\Omega)} = \left\langle \psi_i(x), \psi_j(x) \right\rangle_{L^2(\Omega)} = \delta_{ij} \tag{12}$$

for $i, j = 1, \dots, r$. This follows, since $\mathcal{T}_{\text{per}}(\cdot)$ is a unitary operator, cf. [7].

Since the adjoint operator of $\mathcal{T}_{\text{per}}(\Delta)$ is given by $\mathcal{T}_{\text{per}}{}^*(\Delta) = \mathcal{T}_{\text{per}}(-\Delta)$, the optimization problem (11) associated with the constraints (12) is equivalent to

$$\min_{\psi,\alpha} \int_0^T \int_\Omega \left(\mathcal{T}_{\text{per}}(-\Delta(t))\, q(x,t) - \sum_{k=1}^{r} \alpha_k(t)\psi_k(x) \right)^2 \mathrm{d}x\mathrm{d}t, \quad \text{s. t. (12)}.$$

Thus in this special case, the optimization problem leads to a POD of the transformed function $\mathcal{T}_{\text{per}}(-\Delta(t))\, q\,(x,t)$, which has been used, e.g., in [5].

In the general case of more than one transport velocity ($N_s > 1$), we have to solve the optimization problem (10) numerically. Carrying out a discretization, we have to solve the optimization problem

$$\min_{w,a} \underbrace{\sum_{j=1}^{n} \left\| [X]_j - \sum_{\ell=1}^{N_s} T\left(d_j^\ell\right) \sum_{k=1}^{r_\ell} a_{k,j}^\ell w_k^\ell \right\|^2}_{=:J}. \tag{13}$$

where n is the number of snapshots. Introducing the notation

$$\begin{aligned} a_j &:= [a_{1,j}^1 \quad \dots \quad a_{r_1,j}^1 \;\; a_{1,j}^2 \quad \dots \quad a_{r_{N_s},j}^{N_s}]^T, \\ K_j &:= \left[T\left(d_j^1\right) w_1^1 \dots T\left(d_j^1\right) w_{r_1}^1 \; T\left(d_j^2\right) w_1^2 \dots T\left(d_j^{N_s}\right) w_{r_{N_s}}^{N_s}\right], \end{aligned}$$

the cost functional in (13) can be expressed as the least squares problem

$$J = \sum_{j=1}^{n} \left\| [X]_j - K_j a_j \right\|_2^2. \tag{14}$$

Considering the dependency of J with respect to the amplitudes a_j for fixed modes w, the necessary optimality condition is given by

$$\nabla_{a_j} J = -2K_j^T\left([X]_j - K_j a_j\right) = 0,$$

or equivalently

$$K_j^T K_j a_j = K_j^T\, [X]_j \tag{15}$$

for $j = 1, \dots, n$. The general solution of (15) is given by

$$a_j = V_{j,1} \Sigma_{j,1}^{-1} U_{j,1}^T\, [X]_j + V_{j,2} \beta_j, \tag{16}$$

where β_j is an arbitrary vector of suitable dimension, and the matrices $V_{j,1}$, $\Sigma_{j,1}$, $U_{j,1}$, and $V_{j,2}$ are defined via the SVD of K_j

$$K_j = \left[U_{j,1}\; U_{j,2}\right] \begin{bmatrix} \Sigma_{j,1} & 0 \\ 0 & 0 \end{bmatrix} \begin{bmatrix} V_{j,1}^T \\ V_{j,2}^T \end{bmatrix},$$

where $\Sigma_{j,1}$ contains the non-zero singular values of K_j [8]. If the shifted modes are linearly independent at a time point t, then $V_{j,2}$ is void and the solution (16) is unique, otherwise (15) has infinitely many solutions.

Substituting (16) into (14), the cost functional takes the form

$$J = \sum_{j=1}^{n} \left\| [X]_j - U_{j,1} U_{j,1}^T [X]_j \right\|_2^2,$$

which only depends on the modes w hidden in the matrices $U_{j,1}$. Simple calculations show that minimizing J is equivalent to the minimization problem

$$\min_{w} \tilde{J} = - \sum_{j=1}^{n} \left\| U_{j,1}^T [X]_j \right\|_2^2. \tag{17}$$

The gradient of $\tilde{J}$ with respect to a mode w_k^ℓ is given by

$$\nabla_{w_k^\ell} \tilde{J} = \sum_{j=1}^{n} a_{k,j}^\ell T \left(d_j^\ell\right)^T \left(I_m - U_{j,1} U_{j,1}^T\right) [X]_j .$$

An algorithm to compute an optimal solution is presented in Sect. 4.

4 Algorithm and Implementation

Since it is a priori unclear how many modes are necessary to achieve a certain error tolerance, we propose to solve the optimization problem (17) starting with a small number of modes and iteratively adding modes in a greedy fashion, cf. Algorithm 1. To initiate the algorithm we choose a vector $\boldsymbol{r}^0 \in \mathbb{N}^{N_s}$ containing the initial mode numbers for each velocity frame and prescribed shifts d_j^ℓ for each velocity frame and discrete time step. The algorithm starts with computing a mode decomposition with mode numbers $\boldsymbol{r}^0$. For this, the optimization problem (17) is solved with a nonlinear optimization solver of choice, e.g., Newton's method or quasi-Newton methods, see e.g. [18]. Since the optimization problem scales with the full dimension, we recommend an inexact Newton method or a limited-memory quasi-Newton method which is more efficient [18]. Motivated by the case with one velocity frame and a periodic shift operator discussed in Sect. 3, we choose the first $[\boldsymbol{r}^0]_\ell$ singular vectors of the transformed snapshot matrix

$$\left[T\left(-d_1^\ell\right)[X]_1 \cdots T\left(-d_n^\ell\right)[X]_n\right]$$

as starting values for the modes of the ℓth velocity frame. Following this, in line 5 the relative error is compared with the tolerance and if the tolerance is not achieved, then the algorithm continues by adding modes in a greedy manner. More precisely, in the **for** loop in lines 7–11, we add one mode to each frame at a time, solve the optimization problem (17), construct $\tilde{X}$, and compute the error. Subsequently, the errors corresponding to the different mode number vectors $\boldsymbol{r}^{p-1} + e_i$ are compared, where $e_i \in \mathbb{R}^{N_s}$ denotes the ith unit vector, and only that mode is kept which results in the smallest error. This **while** loop continues until the error is below the tolerance or the maximum iteration number is reached.

Algorithm 1 sPOD algorithm based on residual minimization

Input: snapshot matrix X; initial mode numbers $\boldsymbol{r}^0$; shifts d_j^ℓ for $j = 1, \ldots, n$ and $\ell = 1, \ldots, N_s$; routine for the calculation of $T(\cdot)$; error tolerance tol; maximum iteration number $p_{\max}$

Output: modes w_k^ℓ; amplitudes $a_{k,j}^\ell$ for $j = 1, \ldots, n$, $\ell = 1, \ldots, N_s$, and $k = 1, \ldots, r_\ell$

1: Solve (17) with mode numbers $\boldsymbol{r}^0$ for the modes w
2: Compute the amplitudes a from (16)
3: Reconstruct $\tilde{X}$ as in (4) and compute the error as in (7)
4: $p = 0$
5: **while** (error $>$ tol) and ($p < p_{\max}$) **do**
6:     $p \leftarrow p + 1$
7:     **for** $i = 1 : N_s$ **do**
8:         Solve (17) with mode numbers $\boldsymbol{r}^{p-1} + e_i$ for the modes w
9:         Compute the amplitudes a, $\tilde{X}$, and the error as in lines 2 and 3
10:         tempError(i) $\leftarrow$ error
11:     **end for**
12:     Find the index q for which tempError is minimal
13:     error $\leftarrow$ tempError(q)
14:     $\boldsymbol{r}^p \leftarrow \boldsymbol{r}^{p-1} + e_q$
15: **end while**

The major computational cost of Algorithm 1 arises from the solution of the optimization problems in lines 1 and 8 and depends on the chosen solver. The computation time can be decreased significantly by performing the **for** loop in lines 7–11 in parallel. Another opportunity for a speedup is to use multigrid methods for the optimization, see e.g. [17].

Most parts of Sect. 3, as well as Algorithm 1 are valid for general matrix functions T which do not necessarily have to be associated with a shift operation. Thus, the use of matrix functions which simulate other effects like rotation or dilation is possible, however, this topic is not within the scope of this paper.

Instead, in Sect. 5 we use a shift operator with constant extrapolation, i.e.,

$$\mathcal{T}_c\left(\Delta(t)\right) f(x,t) := \begin{cases} f\left(x - \Delta(t), t\right) & \text{for } 0 \leq x - \Delta(t) \leq L, \\ f(0) & \text{for } x - \Delta(t) < 0, \\ f(L) & \text{for } x - \Delta(t) > L. \end{cases}$$

Such a shift operator has proven to be well-suited for moving shock waves, cf. [20]. For the discrete analogue T_c on a uniform grid with mesh width h, we distinguish between two cases: If the shift is a multiple of h, then $T_c(\cdot)$ is defined as

$$T_c(kh) = \begin{bmatrix} \begin{array}{c|c} 1\,0\ldots0 & 0 \\ \hline & 0 \\ I_{m-1} & \vdots \\ & 0 \end{array} \end{bmatrix}^k, \quad T_c(-kh) = \begin{bmatrix} \begin{array}{c|c} 0 & \\ \vdots & I_{m-1} \\ 0 & \\ \hline 0 & 0\cdots0\,1 \end{array} \end{bmatrix}^k \tag{18}$$

with $k \in \mathbb{N}$. If the shift is not a multiple of h we use an interpolation scheme, i.e., for instance, a linear interpolation like $T_c(0.5h) = 0.5(T_c(0) + T_c(h))$. Similarly, a shift matrix function for the periodic case has been introduced in [21].

5 Test Case: Pulsed Detonation Combuster

As a realistic test example, we consider density, velocity, pressure, and effective species snapshot data of a *Pulsed Detonation Combuster* (*PDC*) with a shock-focusing geometry where the effective species ranges from 0 (burned) to 1 (unburned). The data is based on a simulation of the reactive, compressible Navier-Stokes equations where physical parameters have been adjusted by a data assimilation, see [10]. The density and species snapshots are depicted in Fig. 3.

In the snapshots of the species we observe a reaction front propagating through the domain. The density snapshots show initially two transports, the reaction front and a leading shock, slightly diverging before they converge again and interact. This *deflagration to detonation transition (DDT)* is caused by a nozzle at around $x = 0.2$, cf. [10]. Following this, the reaction front and the leading shock continue as a detonation wave moving to the right. At the same time, a reflected wave is moving to the left before being reflected at the boundary. When it reaches the nozzle again, another partial reflection is visible. The velocity and pressure snapshots look similar.

Before we apply Algorithm 1 we need to find good candidates for the shifts corresponding to the transports of the system. Here, we focus on the four most dominant transports: the reaction front, the leading shock, the reflected wave, and the partial reflection at the nozzle which is referred to as *re-reflected wave* in the following. We track these transports based on the snapshot data without any a priori

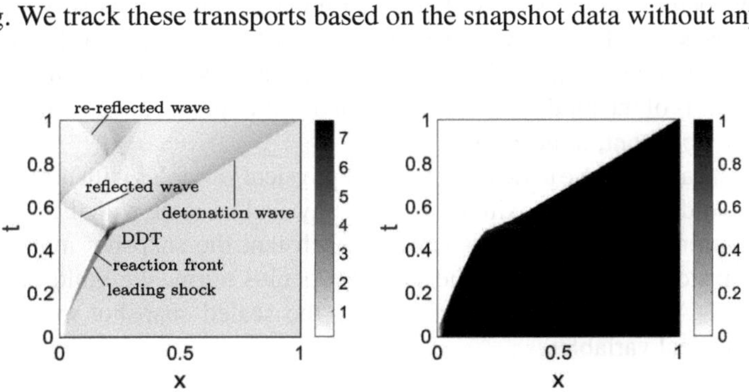

Fig. 3 PDC: snapshots of the full-order solution for density (left) and species (right)

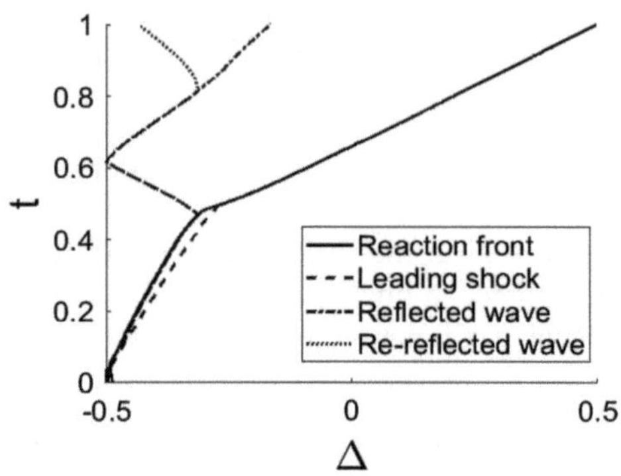

Fig. 4 PDC: tracked shifts for the different transports

knowledge of their velocities. The reaction front is the easiest to detect, since it is clearly visible in the species snapshots as a large gradient. To track it, we determine the location of the maximum in each column of the difference matrix whose jth column is defined as the difference between the $j + 1$st and jth column of the species snapshot matrix. The resulting tracked shift is depicted in Fig. 4, solid line. Here, negative shift values occur since the reaction front is shifted such that it is centered in the middle of the computational domain.

The tracking of the other transports works similarly, but is a little more elaborate since we need to distinguish them from each other. To this end, we restrain the region of the computational domain where the location of the maximum slope is computed. This subregion depends on both the considered transport and time interval. In our tests, this decomposition in subregions has been done manually based on the velocity snapshots. The corresponding tracked shifts are depicted in Fig. 4. In addition, we also add a frame with zero velocity to account for the structures that we cannot capture well by the other velocity frames.

We apply Algorithm 1 with a shift operator with constant extrapolation as in (18) with Lagrange polynomials of degree three for the interpolation. In addition, we specify $\text{tol} = 0.01$, $p_{\max} = n$, and $\boldsymbol{r}^0 = [1\ 1\ 1\ 1\ 0]$, i.e., one mode for each of the non-zero velocity frames. The nonlinear optimization problem is solved using the MATLAB package HANSO which is based on a limited-memory BFGS method [15]. Moreover, to avoid parasitic structures in the approximation of the species, we force those parts of the modes which correspond to the species and to other transports than the reaction front, to be zero.

In this test case we have to deal with data of physical variables with highly different scales. To avoid that the approximation of the physical variable with the highest scale becomes dominant we scale the snapshots such that the snapshot matrices of the different physical variables have the same Frobenius norm. We build the snapshot matrix X for Algorithm 1 by concatenating the scaled snapshot matrices of the different physical variables.

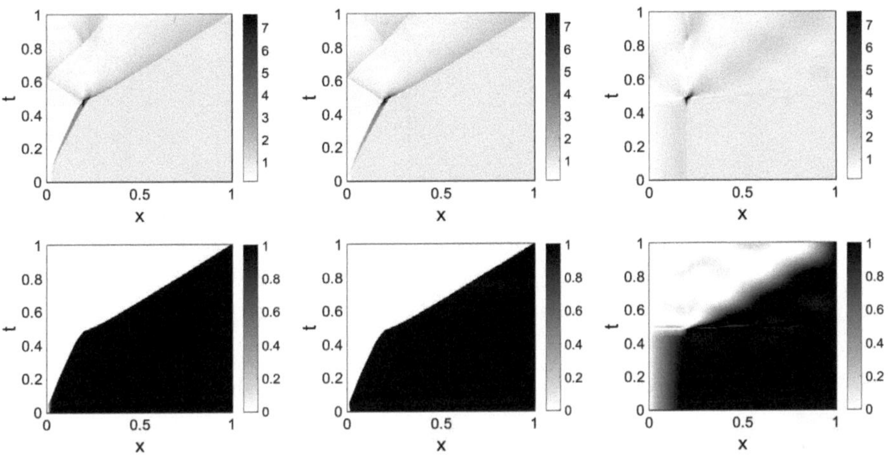

Fig. 5 PDC: comparison between full-order solution (left column), sPOD approximation with 7 modes (middle column), and POD approximation with 7 modes (right column). The top row shows the results for the density, the bottom row for the species

Algorithm 1 terminates after 3 iterations in the **while** loop with an error of 0.71% and $\boldsymbol{r}^3 = [3\ 1\ 1\ 1\ 1]$, i.e., two modes have been added to the reaction front and one mode to the zero velocity frame. This means that we meet the error tolerance with 7 modes in total. The sPOD approximation for the density and the species is depicted in Fig. 5, middle column. Although some deviations to the full-order solution are visible, the sPOD captures the dynamics well and the dominant transports are clearly distinct. This becomes even more striking when comparing it to the POD with the same number of modes which is plotted in Fig. 5, right. As is common in the POD literature, we first subtracted the mean value of each row of the snapshot matrix to center the data around the origin, cf. [6]. The POD approximation of the density features a high peak in the region of the DDT while the other structures are hardly recognizable. For the species, the reaction front is at least indicated, but blurred, and further distortions are visible especially near the DDT. To obtain a POD approximation of the same accuracy as the sPOD with 7 modes, 73 POD modes are needed for this example.

Another advantage of the sPOD becomes clear when looking at the POD and sPOD modes. In Fig. 6 the first sPOD mode for the species in the reaction front frame is depicted and compared to the first POD mode. While the sPOD mode clearly reveals the reaction front as a jump in the middle, the POD is rather smooth and does not show any structure resembling a reaction front.

In Fig. 7 the first sPOD mode for the density is depicted for the reaction front, leading shock, and reflected wave and compared to the first three POD modes. The latter ones mainly focus on the DDT which agrees with Fig. 5, top right, while the moving fronts are not captured. The sPOD modes are not as clear as in Fig. 6 but still each of them features a clear front profile in the middle (marked by dashed lines)

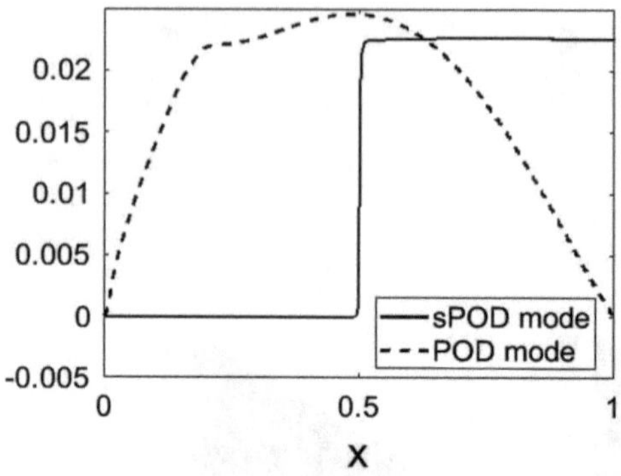

Fig. 6 PDC: comparison of first POD mode and first sPOD mode for the species

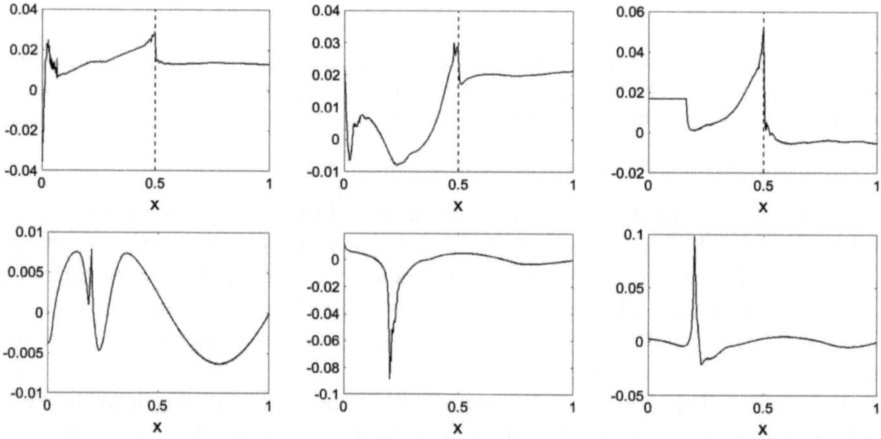

Fig. 7 PDC: comparison of first sPOD modes (top row) for the reaction front, the leading shock, and the reflected wave (from left to right) and first three POD modes (bottom row, from left to right) for the density

corresponding to the sharp fronts visible in Fig. 3. Thus, the sPOD modes capture the principal transport phenomena dominating the PDC dynamics properly. However, especially at the left boundary they differ strongly: The mode for the reflected wave, top right in Fig. 7, reveals a flat profile at the left boundary. This is due to the fact that this part of the mode is not used in the sPOD approximation, since the corresponding shift, depicted in Fig. 4, does not attain values greater than -0.16. The modes for the reaction front and the leading shock reveal some oscillations at the left boundary. A possible reason for this is the use of the shift operator with constant extrapolation which provides the values at the boundaries of the mode with a disproportionate weight.

6 Summary and Outlook

We have presented a new algorithm for computing a shifted proper orthogonal decomposition (sPOD) based on a residual minimization applied to snapshot data. We have applied the algorithm to snapshots determined from a pulsed detonation combuster (PDC) and compared the results with the standard proper orthogonal decomposition (POD). The sPOD yields a reasonable approximation of the snapshots with only very few modes. In contrast, the POD approximation with the same number of modes is blurred and the dynamics is not captured well. Moreover, the sPOD modes clearly reveal the front profiles of the different transports, whereas the POD is not suitable for identifying structures in this test case. In comparison to the heuristic sPOD algorithm proposed in [20], the new algorithm is based on a residual minimization and hence at least locally optimal. A drawback of the new algorithm is that it is more expensive than the POD and the original sPOD approach of [20]. The reason is that a large-scale nonlinear optimization problem has to be solved.

The results of the new sPOD algorithm look promising in terms of both the number of required modes and the physical structures identified by the sPOD modes. The use of the sPOD modes to obtain a reduced-order model via projection is currently under investigation. With this projection framework and the sPOD modes presented in this paper, we aim for constructing dynamic reduced-order models for the PDC for investigating different operating points in an efficient way. Further interesting research directions are a rigorous treatment of non-periodic boundary conditions and an optimization of the shifts together with the modes and the amplitudes.

Acknowledgements The authors gratefully acknowledge support by the Deutsche Forschungsgemeinschaft (DFG) as part of collaborative research center SFB 1029 "Substantial efficiency increase in gas turbines through direct use of coupled unsteady combustion and flow dynamics" on project A02.

References

1. Abgrall, R., Amsallem, D., Crisovan, R.: Robust model reduction by L^1-norm minimization and approximation via dictionaries: application to nonlinear hyperbolic problems. Adv. Model. Simul. Eng. Sci. **3**(1) (2016)
2. Antoulas, A.C.: Approximation of Large-Scale Dynamical Systems. SIAM, Philadelphia (2005)
3. Benner, P., Gugercin, S., Willcox, K.: A survey of projection-based model reduction methods for parametric dynamical systems. SIAM Rev. **57**(4), 483–531 (2015)
4. Berkooz, G., Holmes, P., Lumley, J.L.: The proper orthogonal decomposition in the analysis of turbulent flows. Annu. Rev. Fluid Mech. **25**, 539–575 (1993)
5. Cagniart, N., Maday, Y., Stamm, B.: Model order reduction for problems with large convection effects. Preprint hal-01395571 (2016). https://hal.archives-ouvertes.fr
6. Chatterjee, A.: An introduction to the proper orthogonal decomposition. Current Sci. **78**(7), 808–817 (2000)
7. Cohen, L.: The Weyl Operator and its Generalization. Springer, Basel (2013)

8. Gander, W., Gander, M.J., Kwok, F.: Sci. Comput. Springer International Publishing, Cham (2014)
9. Gerbeau, J.-F., Lombardi, D.: Approximated Lax pairs for the reduced order integration of nonlinear evolution equations. J. Comput. Phys. **265**, 246–269 (2014)
10. Gray, J.A.T., Lemke, M., Reiss, J., Paschereit, C.O., Sesterhenn, J., Moeck, J.P.: A compact shock-focusing geometry for detonation initiation: experiments and adjoint-based variational data assimilation. Combust. Flame **183**, 144–156 (2017)
11. Hesthaven, J.S., Rozza, G., Stamm, B.: Certified Reduced Basis Methods for Parametrized Partial Differential Equations. Springer International Publishing, Cham (2016)
12. Holmes, P., Lumley, J.L., Berkooz, G.: Turbulence, Coherent Structures Dynamical Systems and Symmetry. Cambridge University Press, Cambridge (1996)
13. Iollo, A., Lombardi, D.: Advection modes by optimal mass transfer. Phys. Rev. E **89**(2), 022923 (2014)
14. Kutz, J.N., Brunton, S.L., Brunton, B.W., Proctor, J.L.: Dynamic Mode Decomposition. SIAM, Philadelphia (2016)
15. Lewis, A.S., Overton, M.L.: Nonsmooth optimization via quasi-Newton methods. Math. Program. **141**(1–2), 135–163 (2013)
16. Mojgani, R., Balajewicz, M.: Lagrangian basis method for dimensionality reduction of convection dominated nonlinear flows. Preprint 1701.04343v1 (2017)
17. Nash, S.G.: A multigrid approach to discretized optimization problems. Optim. Methods Softw. **14**(1–2), 99–116 (2000)
18. Nocedal, J., Wright, S.J.: Numerical Optimization, 2nd edn. Springer, New York (2006)
19. Ohlberger, M., Rave, S.: Nonlinear reduced basis approximation of parameterized evolution equations via the method of freezing. C. R. Math. Acad. Sci. Paris **351**(23–24), 901–906 (2013)
20. Reiss, J., Schulze, P., Sesterhenn, J., Mehrmann, V.: The shifted proper orthogonal decomposition: a mode decomposition for multiple transport phenomena. SIAM J. Sci. Comput. (To appear)
21. Rim, D., Moe, S., LeVeque, R.J.: Transport reversal for model reduction of hyperbolic partial differential equations. Preprint 1701.07529v1 (2017)
22. Schmid, P.J., Sesterhenn, J.L.: Dynamic mode decomposition of numerical and experimental data. In: 61st Annual Meeting of the APS Division of Fluid Dynamics, p. 208, San Antonio, USA (2008)
23. Sesterhenn, J., Shahirpour, A.: A Lagrangian dynamic mode decomposition. Preprint 1603.02539v1 (2016)
24. Volkwein, S.: Optimal control of a phase-field model using proper orthogonal decomposition. ZAMM Z. Angew. Math. Mech. **81**(2), 83–97 (2001)

Part V
Numerical Aspects in Combustion

Control of Condensed-Phase Explosive Behaviour by Means of Cavities and Solid Particles

Louisa Michael and Nikolaos Nikiforakis

Abstract Controlling the sensitivity of condensed-phase explosives is a matter of safe handling of the materials and a necessity for efficient blasting. It is known that impurities such as air cavities or solid particles can be used to sensitise the material by reducing the time to ignition. As the ignition of the explosive is a temperature-driven event, analysing the temperature field following the interaction of a shock wave with these impurities gives a measure of the effect of the impurity on the sensitisation of the material. Air cavity collapse in explosives has been extensively studied and recently focus has shifted on the accurate recovery of the temperature field during the collapse process. The interaction of a shock wave with solid particles or with a combination of cavities and particles, has been studied to a lesser extent. In this work, we assess the effect of the different impurities in isolation, in a multi-cavity and a multi-bead configuration and as a combined particle-cavity matrix. Results indicate that the beads have a more subtle effect on the sensitisation of the material, compared to cavities. An informed combination of the two (leading order by cavities and marginal adjustment by particles) could result to a fairly accurate control of the explosive.

Keywords Cavity collapse · Shock-bead interaction · Hot spots · Sensitivity
Condensed-phase explosives · Nitromethane

1 Introduction

Controlling the performance of condensed-phase explosives is of interest to the mining industry. The ability to control the ignition sensitivity of the explosive material is not only a matter of safety during handling and transportation of materials but

L. Michael (✉) · N. Nikiforakis
Laboratory for Scientific Computing, Cavendish Laboratory, Department of Physics, University of Cambridge, Cambridge CB3 0HE, UK
e-mail: lm355@cam.ac.uk
URL: https://www.lsc.phy.cam.ac.uk/

R. King (ed.), *Active Flow and Combustion Control 2018*,
Notes on Numerical Fluid Mechanics and Multidisciplinary Design 141,
https://doi.org/10.1007/978-3-319-98177-2_18

also a necessity for efficient blasting [14]. Common techniques used to sensitise a condensed-phase explosive include the addition of air cavities, glass microbaloons and glass or other solid beads in the body of the explosive. These artificial impurities affect the sensitivity of the material to different extent, but they all result in the generation of regions where the pressure and temperature are locally higher than the rest of the material, known as hot-spots, and lead to earlier ignition than in a neat material. Moreover, the inclusion of different types of impurities will have a different effect in the performance of the explosive compared to a single type of added impurities. Understanding the effect these impurities have (in isolation, in single-impurity type matrices and in multiple-impurity type matrices) on the ignition sensitivity of the material will allow better control of the explosive performance.

To this end, experimental and numerical studies have been performed to identify the process governing the cavity collapse and determine the mechanical effects behind hot spot generation. For an extensive discussion on these see the paper by Michael and Nikiforakis [10] and references therein. However, the interaction of a shock wave with solid particles or with a matrix combining cavities and solid particles has not been studied extensively. To the authors' knowledge, Bourne and Field [3] are the only ones who presented a study of cavity collapse in an inert liquid laden with solid (lead and nylon) particles.

Similar studies were done by numerical means, examining the shock-cavity interaction process in various configurations, mostly considering the collapse in inert gaseous or solid media, or the pressure field and pressure amplification in multi-cavity scenarios. For a detailed discussion of these the reader is refereed to [10, 11]. The two and three-dimensional isolated cavity collapse in inert and reactive nitromethane was presented by Michael and Nikiforakis in [7, 10, 11], focusing on the temperature field induced by the collapse and subsequent ignition of the explosive.

A relatively small number of numerical studies can be found on the shock interaction with deformable particles. Ling et al. [5] studied the shock interaction of an aluminium particle in nitromethane, and Zhang et al. [19] modelled the shock interaction with magnesium, tungsten, beryllium and uranium in nitromethane, in isolation and in clusters, to study the velocities attained by the particles. The acceleration and heating of aluminium particles of several sizes in detonating nitromethane was studied by Ripley et al. [15]; Sridharan et al. [18] computed the transient drag of an aluminium particle in nitromethane, after its interaction with a shock. Menikoff [6] studied the hot spot formation from glass bead shock reflections in inert nitromethane.

In conclusion, although there is some body of evidence towards understanding the generation of hot spots by cavity collapse, more insight is needed to understand the effect of shock-particle interaction on the temperature field of the material and hence on the control of the performance of the explosive. Moreover, besides [3], there are no studies on the effects of the combined effect of particles and cavities.

In this work we present and compare the effects of different types of impurities, namely PMMA particles and gas cavities, in isolation and in matrix configuration (multi-cavity, multi-bead and cavity-bead combination), assessing their effect on the ignition control of the explosive.

A multi-physics methodology is employed to perform these simulations. The hydrodynamic model presented in [8] is used to model the explosive and air cavities and a full elastoplastic model is used to describe the response of PMMA particles. The two models are of the same hyperbolic form and are thus solved with high resolution shock-capturing schemes. Communication between the two materials and the corresponding sets of equations is achieved by means of a variant of the ghost fluid method which uses mixed Riemann solvers, see [9]. This approach overcomes several challenges presented in simulating such a complex physical scenario. The hydrodynamic model used for the explosive and air cavities allows the non-trivial use of complex equations of state for describing the explosive and the cavities, retains at least 1000:1 density difference across the cavity boundary while maintaining oscillation free interfaces (in terms of pressures, velocities and temperatures) and allows the recovery of realistic temperature fields in the explosive matrix.

A major challenge worthy of special attention is the accurate and oscillation-free recovery of temperature fields in the explosive matrix. This is of critical importance as the ignition process of the energetic material is a temperature-driven effect, thus the accurate prediction of ignition relies on physically meaningful temperatures. For more information on this the reader is referred to [10, 11]. This model allows for oscillation-free temperature fields. The elastoplastic model we use to model the solid materials is rendered in hyperbolic form and thus can be solved on an Eulerian mesh, using finite volume methods. This eliminates mesh tangling issues that might occur in Lagrangian approaches and allows the immediate communication between the hyperbolic and elastoplastic materials by means of the ghost fluid method.

2 Mathematical Models

In this section, the distinct mathematical formulations used to describe the fluid and solid elastoplastic materials in the interaction of a shock wave with voids and particles in an explosive are presented.

The explosives (hydrodynamic) model The air cavities immersed in nitromethane are modelled using the MiNi16 formulation [8], which is summarised below. The gas inside the cavities is described as *phase 1*, with density, velocity vector and pressure $(\rho_1, \mathbf{u}_1, p_1)$. The nitromethane is denoted as *phase 2* with density, velocity vector and pressure $(\rho_2, \mathbf{u}_2, p_2)$. We denote by z a colour function, which can be considered to be the volume fraction of the air with respect to the volume of the total mixture of phases 1 and 2, with density ρ. For convenience, we denote z by z_1 and $1 - z$ by z_2. Then, the closure condition $z_1 + z_2 = 1$ holds. Velocity and pressure equilibrium applies between the all phases, such that $u_\alpha = u_\beta = u_1 = u_2 = u$ and $p_\alpha = p_\beta = p_1 = p_2 = p$.

Then, the MiNi16 system is described as in [8] by:

$$\frac{\partial z_1\rho_1}{\partial t} + \nabla \cdot (z_1\rho_1\mathbf{u}) = 0, \tag{1}$$

$$\frac{\partial z_2\rho_2}{\partial t} + \nabla \cdot (z_2\rho_2\mathbf{u}) = 0, \tag{2}$$

$$\frac{\partial}{\partial t}(\rho\mathbf{u_i}) + \nabla \cdot (\rho\mathbf{u_i}\mathbf{u}) + \frac{\partial p}{\partial \mathbf{x_i}} = 0, \tag{3}$$

$$\frac{\partial}{\partial t}(\rho E) + \nabla \cdot (\rho E + p)\mathbf{u} = 0, \tag{4}$$

$$\frac{\partial z_1}{\partial t} + \mathbf{u} \cdot \nabla z_1 = 0, \tag{5}$$

$$\frac{\partial z_2\rho_2\lambda}{\partial t} + \nabla \cdot (z_2\rho_2\mathbf{u}\lambda) = z_2\rho_2 K, \tag{6}$$

where $\mathbf{u} = (u, v, w)$ denotes the total vector velocity, i denotes space dimension, $i = 1, 2, 3$, ρ the total density of the system and E the specific total energy given by $E = e + \frac{1}{2}\sum_i u_i^2$, with e the total specific internal energy of the system. We denote by λ the mass fraction of the explosive, such that $\lambda = 1$ denotes fully unburnt material and $\lambda = 0$ denotes fully burnt material. As this work is restricted to the inert scenario, $\lambda = 0$ everywhere and the equations reduce to those by Allaire et al. [1].

In this work, all fluid components described by the MiNi16 model are assumed to be governed by a Mie-Grüneisen equation of state, of the form:

$$p = p_{ref_i} + \rho_i \Gamma_i (e_i - e_{ref_i}), \text{ for } i = 1, 2. \tag{7}$$

Material interfaces between the phases are described by a diffused interface technique. Hence, mixture rules need to be defined for the diffusion zone, relating the thermodynamic properties of the mixture with those of the individual phases. The mixture rules for the specific internal energy, density and adiabatic index ($\gamma = 1 + \frac{1}{\xi}$) are:

$$\rho e = z_1\rho_1 e_1 + z_2\rho_2 e_2, \quad \rho = z_1\rho_1 + z_2\rho_2, \text{ and } \xi = z_1\xi_1 + z_2\xi_2, \tag{8}$$

where e_1, e_2 denote the specific internal energies of phases *1* and *2*.

The sound speed also follows a mixture rule given as:

$$\xi c^2 = \sum_i y_i \xi_i c_i^2, \tag{9}$$

where c_i is the individual sound speed of phase i and $y_i = \frac{\rho_i z_i}{\rho}$ its mass fraction. For more information on this as well as for the numerical evaluation of the total equation of state the reader is referred to [8]. Validation of the hydrodynamic mathematical model can be found in [10].

Equations of state for nitromethane and air To close the hydrodynamic system, the Cochran-Chan equation of state [16] is employed to describe the liquid nitromethane. This is an equation of state of Mie-Grüneisen form given by Eq. 7 with reference pressure given by

$$p_{\text{ref}}(\rho) = \mathcal{A}\left(\frac{\rho_0}{\rho}\right)^{-\mathcal{E}_1} - \mathcal{B}\left(\frac{\rho_0}{\rho}\right)^{-\mathcal{E}_2}, \tag{10}$$

reference energy given by

$$e_{\text{ref}}(\rho) = \frac{-\mathcal{A}}{\rho_0(1-\mathcal{E}_1)}\left[\left(\frac{\rho_0}{\rho}\right)^{1-\mathcal{E}_1} - 1\right] + \frac{\mathcal{B}}{\rho_0(1-\mathcal{E}_2)}\left[\left(\frac{\rho_0}{\rho}\right)^{1-\mathcal{E}_2} - 1\right] \tag{11}$$

and Grüneisen coefficient $\Gamma(\rho) = \Gamma_0$. The gas inside the cavity (where applicable) is modelled by the ideal gas equation of state, which is of Mie-Grüneisen form as well, with $p_{\text{ref}} = 0$ and $e_{\text{ref}} = 0$. The parameters for the equations of state of the two materials are given in Table 1.

Recovery of temperature The multi-phase nature of the model allows for separate temperature fields to be computed for each material as

$$T_i = \frac{p - p_{\text{ref}_i}(\rho)}{\rho_i \Gamma_i c_{v_i}}, \quad \text{for} \quad i = 1, 2. \tag{12}$$

As a result, the nitromethane temperature ($T_{\text{NM}} = T_2$) is computed explicitly from the equation of state and can be used directly in the reaction rate law.

Computing the temperature of a general condensed phase explosive ($T_{\text{CF}} = T_2$) can involve completing the equation of state starting from the basic thermodynamic law $T\frac{\mathrm{d}S}{\mathrm{d}v} = c_v\frac{\mathrm{d}T}{\mathrm{d}v} + c_v\frac{\Gamma T}{v}$ and by integrating to obtain a reference temperature ($T_{\text{ref}_{\text{CF}}}$) such that:

$$T_{\text{CF}} = \frac{p - p_{\text{ref}_{\text{CF}}}(\rho)}{\rho_{\text{CF}}\Gamma_{\text{CF}}c_{v\text{CF}}} - T_{\text{ref}_{\text{CF}}}. \tag{13}$$

When the reference curve is an isentrope, $\mathrm{d}S/\mathrm{d}v = 0$ and hence we can simply compute $T_{\text{ref}_{\text{CF}}} = T_0(\frac{\rho}{\rho_0})^{\Gamma}$. When the reference curve is a Hugoniot curve, the basic thermodynamic law cannot be integrated directly and often the Walsh Christian technique and numerical ODE-integration techniques are used to compute the reference Hugoniot temperature.

Table 1 Equation of state parameters for nitromethane and air

Equation of state parameters	Γ_0 [–]	$\mathcal{A}$ [GPa]	$\mathcal{B}$ [GPa]	$\mathcal{E}_1$ [–]	$\mathcal{E}_2$ [–]	ρ_0 [kg m^{-3}]	c_v [J kg^{-1} k^{-1}]
Nitromethane [16]	1.19	0.819	1.51	4.53	1.42	1134	1714
Air	0.4	–	–	–	–	–	718

For this work, substitution of the parameters of the equation of state for nitromethane and imposing an initial temperature of 298 K for $\rho = \rho_0$ gives $T_0 = 0$. This is in line with other work using the Cochran-Chan equation of state, where T_0 takes zero or very small values. The form (12) gives temperatures that match experiments as demonstrated in [10], but for other materials or other equations of state (e.g., shock Mie-Grüneisen) care should be taken as a different reference curve (as per Eq. 13) would be necessary.

The ideal gas equation of state results in the overheating of the gas inside the cavity since high pressures are reached within the cavity during the collapse process. However, the very short timescale of the process means that heat transfer does not take place and thus the cavity temperature does not affect the ambient nitromethane temperature. Since temperatures inside the cavity are not of interest for this work, they are not presented henceforth.

Note that the two-phase nature of the model allows for large (>1000:1) density gradients to be sustained across material boundaries and both the density and temperature fields are maintained oscillation-free.

The elastoplastic model In this work, we use the elastic solid model described by Schoch et al. [17] and Barton et al. [2], based on the formulation by Godunov and Romenskii [4] to describe the physical behaviour of the solid particles. Plasticity of the material is included, following the work of Miller and Colella [13].

In an Eulerian frame employed in this work, there is no mesh distortion that can be used to describe the solid material deformation. Thus the material distortion needs to be accounted for in a different way. Here, this is done by defining the elastic deformation gradient as $F^e_{ij} = \frac{\partial x_i}{\partial X_j}$, which maps the coordinate $\mathbf{X}$ in the initial configuration to the coordinate $\mathbf{x}$ in the deformed configuration.

The state of the solid is characterised by the elastic deformation gradient, velocity u_i and entropy S. Following the work by Barton et al. [2], the complete three-dimensional system forms a hyperbolic system of conservation laws for momentum, strain and energy:

$$\frac{\partial \rho u_i}{\partial t} + \frac{\partial(\rho u_i u_m - \sigma_{im})}{\partial x_m} = 0, \tag{14}$$

$$\frac{\partial \rho E}{\partial t} + \frac{\partial(\rho u_m E - u_i \sigma_{im})}{\partial x_m} = 0, \tag{15}$$

$$\frac{\partial \rho F^e_{ij}}{\partial t} + \frac{\partial(\rho F^e_{ij} u_m - \rho F^e_{mj} u_i)}{\partial x_m} = -u_i \frac{\partial \rho F^e_{mj}}{\partial x_m} + P_{ij}, \tag{16}$$

$$\frac{\partial \rho \kappa}{\partial t} + \frac{\partial(\rho u_m \kappa)}{\partial x_m} = \rho \dot{\kappa}, \tag{17}$$

with the vector components $\cdot_i$ and tensor components $\cdot_{ij}$. The first two equations along with the density-deformation gradient relation $\rho = \rho_0/\det\mathbf{F}^e$, where ρ_0 is the density of the initial unstressed medium, essentially evolve the solid material hydrodynamically. Here, σ is the stress, E the total energy such that $E = \frac{1}{2}|u|^2 + e$, with

e the specific internal energy and κ the scalar material history that tracks the work hardening of the material through plastic deformation. We denote the source terms associated with the plastic update as P_{ij}.

The system is closed by an analytic constitutive model relating the specific internal energy to the deformation gradient, entropy and material history parameter (if applicable) $e = e(\mathbf{F}^e, S, \kappa)$. For more information the reader is referred to [9].

The deformation is purely elastic until the physical state is evolved beyond the yield surface ($f > 0$), which in this work is taken to be:

$$f(\boldsymbol{\sigma}) = ||\mathrm{dev}\boldsymbol{\sigma}|| - \sqrt{\frac{2}{3}}\sigma_Y = 0, \text{ with } \mathrm{dev}\boldsymbol{\sigma} = \boldsymbol{\sigma} - \frac{1}{3}(\mathrm{tr}\boldsymbol{\sigma})I, \tag{18}$$

where σ_Y is the yield stress and the matrix norm $||.||$ the Shur norm ($||\boldsymbol{\sigma}||^2 = \mathrm{tr}(\boldsymbol{\sigma}^T\boldsymbol{\sigma})$).

As this identifies the maximum yield allowed to be reached by an elastic-only step, a predictor-corrector method is followed to re-map the solid state onto the yield surface. Assuming that the simulation timestep is small, this is taken to be a straight line, using the associative flow rate ($\dot{\epsilon}^p = \eta\frac{\partial F}{\partial \sigma}$), satisfying the maximum plastic dissipation principle (i.e. the steepest path). In general, this re-mapping procedure is governed by the dissipation law $\psi_{plast} = \boldsymbol{\Sigma} : ((\mathbf{F}^p)^{-1}\dot{\mathbf{F}}^p)$, where $\boldsymbol{\Sigma} = \boldsymbol{G}\boldsymbol{\sigma}\boldsymbol{F}$ and $:$ is the double contraction of tensors (e.g. $\boldsymbol{\sigma} : \boldsymbol{\sigma} = \mathrm{tr}(\boldsymbol{\sigma}^T\boldsymbol{\sigma})$). The initial prediction is $\mathbf{F} = \mathbf{F}^e$ and $\mathbf{F}^p = \mathbf{I}$, where $\mathbf{F}$ is the specific total deformation tensor and $\mathbf{F}^p$ the plastic deformation tensor that contains the contribution from plastic deformation. This is then relaxed to the yield surface according to the procedure of Miller and Colella [13].

The explosive and solid mathematical formulations described in this section are solved numerically using high-resolution, shock-capturing, Riemann-problem based methods and structured, hierarchical adaptive mesh refinement, as described in previous work [7, 8, 12, 17].

Equation of state for PMMA To close the elastoplastic system, the PMMA is described by an energy-independent shock Mie-Grüneisen (Hugoniot) equation of state where the parameters for PMMA for ρ_0, the reference density for identity deformation, s the linear shock speed-particle speed ratio, c_0 the unshocked sound speed and T_0 the reference temperature are given in Table 2. Validation of the elastoplastic mathematical model can be found in [9].

Table 2 Shock Mie-Grüneisen equation of state parameters the elastoplastic (with perfect plasticity) PMMA

Hyperelastic, shear and plasticity parameters	ρ_0 [kg m^{-3}]	c_0 [m s^{-1}]	T_0 [K]	s [–]	G [MPa]	σ_Y [MPa]
PMMA	1180	2260	300	1.82	1148	85

The multi-material approach In this work, we use level set methods to track the solid-explosive[1] interface. The behaviour of the material components at the interface is modelled by the implementation of dynamical boundary conditions with the aid of the Riemann ghost fluid method and devised mixed-material Riemann solvers to solve the interfacial Riemann problems between materials. For more details on the method the reader is referred to [9].

3 Results

In this section we study the interaction of an isolated PMMA particle, a 2×2 PMMA particles matrix, an air cavity, a 2×2 air cavity matrix and a 2×2 matrix of 2 cavities and 2 particles with a 10.98 GPa shock wave in non-reactive liquid nitromethane. Nitromethane is modelled by the Cochran-Chan equation of state given by Eqs. (7)–(11) and PMMA using a shock Mie-Grüneisen with constant shear and perfect plasticity both described in Sect. 2. As the ignition and thermal runaway in an explosive are attributed to the complex interaction between non-linear gas dynamics and chemistry, it is intuitive to consider in the first instance the induced temperature field in the explosive in the absence of chemical reactions. This will allow the purely gas-dynamical effects to be elucidated. In effect, by controlling the induced temperature field in the explosive material (by the judiciary inclusion of voids and beads) we can control its ignition time. The simulations are performed in two dimensions, with effective grid size $dx = dy = 0.625\ \mu m$. The initial conditions for the simulations in this work are given in Table 3.

3.1 *Single PMMA Bead*

In Fig. 1 we present the temperature field generated in the nitromethane upon the interaction of the incident shock wave (S_0) with the PMMA bead originally centred at $(x, y) = (0.18, 0.2)$ mm, of radius of 0.08 mm. The temperatures in the bead are omitted as the timescales are small for heat transfer to occur between the two materials. Instead, we display a mock-schlieren plot inside the bead. As the bead is symmetric about the horizontal axis we present only the upper half of the configuration. The interaction of the incident shock wave with the bead generates two new shock waves, one travelling upstream into the nitromethane (S_1) and one downstream into the bead (S_2). The upstream travelling shock compresses again the nitromethane, which reaches temperatures of 1300 K, only ~30 K higher than those generated by the original incident shock wave (Fig. 1a). The angle of interaction of the shock wave and the bead continuously changes, resulting to a transition from a regular

[1]Note that by explosive we refer to any hydrodynamic system modelled by MiNi16 or its reduced systems, including the simultaneous modelling of the nitromethane and the air-cavities.

Table 3 Initial conditions for the shock-bead interaction and shock-induced cavity collapse in inert nitromethane considered in this work

	Material	ρ_1 [kg m^{-3}]	ρ_2 [kg m^{-3}]	u [m s^{-1}]	v [m s^{-1}]	p [Pa]	z_1
$x < 100\,\mu$m	Shocked nitromethane	2.4	1934.0	2000.0	0.0	10.98×10^9	10^{-6}
$x \geq 100\,\mu$m	Ambient nitromethane	1.2	1134.0	0.0	0.0	1×10^5	10^{-6}
Bubble	Air	1.2	1134.0	0.0	0.0	1×10^5	$1 - 10^{-6}$
		u [m s^{-1}]		v [m s^{-1}]		p [Pa]	
Bead	PMMA	0.0		0.0		1×10^5	

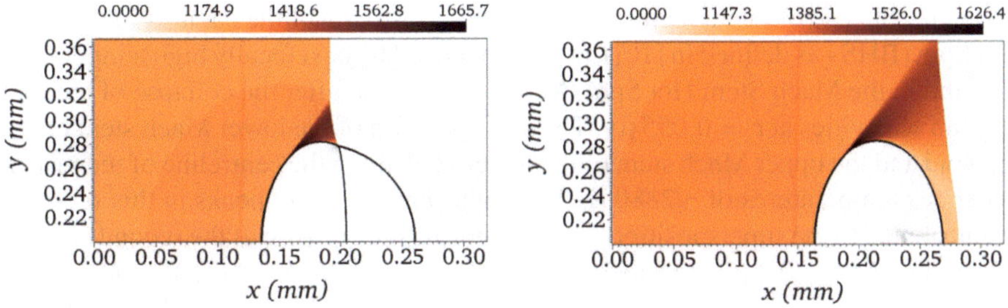

(a) $t = 0.02\mu$s (b) $t = 0.04\mu$s

Fig. 1 Temperature field in nitromethane

shock reflection to a Mach reflection. A pair of Mach stems is generated at the top and bottom of the bead; the top one is seen in Fig. 1a. In fact, the largest temperature increase in this configuration is attributed to the Mach stem, leading to temperatures of $\sim$400 – 500 K higher than the post incident-shock temperature (Fig. 1a). The Mach stem grows and the Mach stem triple point moves away from the bead, along the incident shock wave (Fig. 1b) forming a band of high temperatures. Finally the shock wave traversing the bead exits into the nitromethane and continues travelling with the incident shock wave. The higher impedance of the bead compared to the nitromethane contributes (along with the Mach reflection) to the curvature of the wave front ($S_{0,2}$), travelling now downstream the bead. Upon exiting the beam, the shock wave (S_2) is weaker than the incident shock wave, leading to temperatures of only $\sim$1000 – 1050 K (Fig. 1b), which are lower than the original post-shock temperatures induced by the incident shock wave (S_0). Another interesting observation is that the temperature along the final downstream-travelling shock wave ($S_{0,2}$) is not uniform.

3.2 2 × 2 Matrix of Air Cavities

A common way of controlling the generation of higher temperatures in the explosive material is the inclusion of cavities. The authors have extensively studied single-cavity collapse in nitromethane in [10, 11] thus in this work we do not repeat these results. In multi-cavity configurations, the waves generated upon the collapse process of each cavity interact in the regions in between the voids leading to elevations of temperatures higher than in single-cavity configurations [14]. The authors have studied this scenario before for cavities collapsing in water [12] generating the same wave patterns so we will limit here the discussion to the wave interaction and its effect on the nitromethane temperature field.

The first locally high temperature ($T \sim 2960$ K) in this scenario is encountered upon the collapse of the first column of cavities, upstream of the cavities in the Back Hot Spot (BHS - as defined in [10]), at $t = 0.04\,\mu$s. The next locally high temperature is found in the Mach Stem Hot Spot (MSHS) generated after the collapse of the first column of cavities at $t = 0.055\,\mu$s. The superposition of the lower Mach stem of the top void and the upper Mach stem of the lower void along the centreline of the matrix generates temperatures of ~2880 K. The highest temperature peaks in this scenario are attributed to the supersposition of waves during the collapse of the second column of cavities. At $t = 0.1\,\mu$s the superposition of waves upstream of the second column gives temperatures of ~5275 K in between the cavities' lobes. Similarly, in between the lobes of the first column's cavities highs of $T \sim 4660$ K are seen at $t = 0.115\,\mu$s. It is concluded that wave superposition plays the most important role in temperature increase in this multi-cavity scenario.

3.3 2 × 2 Matrix of PMMA Beads

In this section we investigate the effect of a 2 × 2 matrix of PMMA beads on the nitromethane temperature field. In the mock schlieren plots of Fig. 2a we see the first interaction of the S_1 from the first two beads along the centreline of the matrix, perpendicular to the incident shock. Subsequently, S_{1B}, the S_1 wave from the bottom bead impacts onto the top bead and similarly S_{1T}, the S_1 wave from the top bead impacts onto the bottom bead. This leads to new shock waves inside and outside the beads. The shocks outside the beads interact with the Mach stems and the two Mach stems eventually intersect. After the exit of the shocks from the beads, the new lead shock (S_0'), along which different temperature ranges can be found, interacts with the next two beads and the shock-bead interaction as well as the wave superpositions are repeated (Fig. 2b).

The interaction of the shock with each bead leads to the formation of a high-temperature band on each side of each bead. This can be seen, for the outer parts of the beads, in Fig. 2b. The temperature in the bands, however, is lower than in the original Mach stem. The bands on the inner sides of the beads, as seen in the

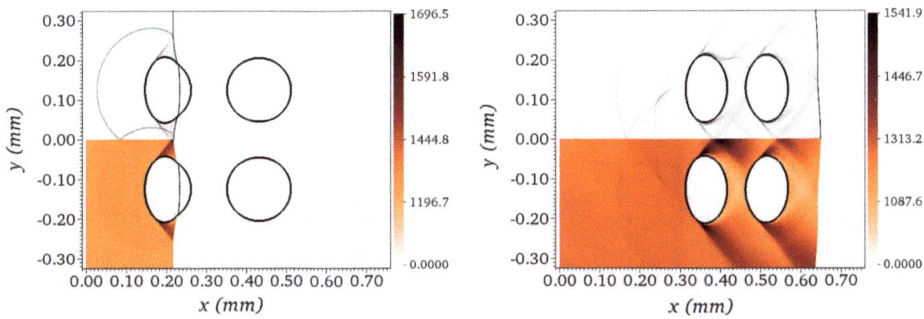

(a) $t = 0.07\mu s$ (b) $t = 0.12\mu s$

Fig. 2 Mock-schlieren plots (top half) illustrating the interaction of waves and temperature field in nitromethane (bottom half) in a 2 × 2 PMMA bead configuration

same figure, are superimposed in the region in between the beads, leading to new high temperature regions. Consequently, the new lead shock (S_0') has variable temperature ranges along its front and a higher temperature along its middle compared to the isolated shock-bead interaction scenario. Moreover, the part of the new lead shock that is now directly in front of the first column of beads, in the region directly in front of the beads it is actually weaker then the original incident shock wave. Thus, the subsequent beads that are in the shadow of the first column beads will feel a weaker shock, leading to lower temperatures compared to the temperatures produced by the first column.

In this configuration, the first high temperature peak is seen when the Mach stem is generated at the top and bottom of the beads of the first column ($T \sim 1770$ K) at $t = 0.015\,\mu$s. The second high temperature peak is seen when the Mach stem is generated at the sides of the beads of the second column ($T \sim 1750$ K) at $t = 0.065\,\mu$s.

3.4 Combination of 2 Cavities and 2 Beads

In this section, we combine two air cavities and two PMMA beads in a 2 × 2 array, with a clockwise ordering of cavity-bead-bead-cavity. The clockwise ordering of bead-cavity-cavity-bead is the same scenario reflected about the horizontal axis and it is thus not discussed separately. In this configuration, the highest temperatures are observed during the cavity collapse and not the shock-bead interaction (Fig. 3b, d). Looking at the interaction of the incident shock wave with the first column of impurities we observe that the shock wave S_1 generated upon the interaction of the incident shock and the bead is superimposed with the rarefaction wave (R_1) generated upon the interaction of the incident shock wave and the air cavity (Fig. 3a). As a result this shock wave weakens and when it interacts with the cavity it does not lead to its

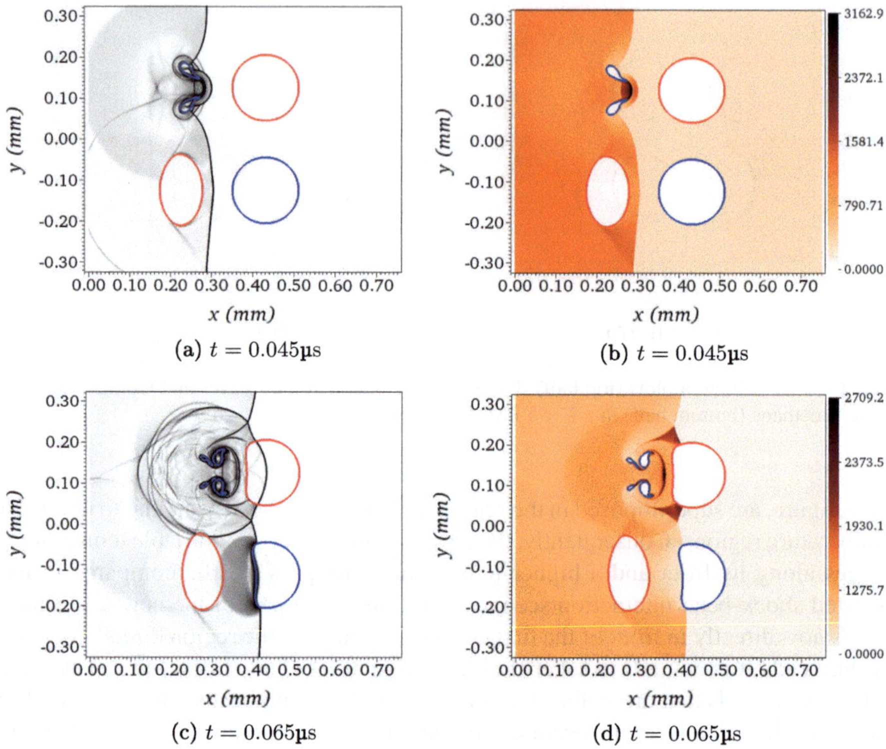

Fig. 3 Mock-schlieren plots (left) illustrating the interaction of waves and the induced temperature field in nitromethane (right) in a clockwise cavity-bead-bead-cavity configuration

asymmetric collapse (Fig. 3a). The shock waves generated upon the collapse of the top cavity, however have a significant effect on the deformation of the lower bead (Figs. 3c, 4a, c). The interaction of the shock wave emanating at the collapse of the top cavity, as well as the Mach stem generated by the interaction of the top bead with the incident shock (S_0' in this case) leads, however to the asymmetric collapse of the lower cavity. The jet deviation can be seen in Fig. 4a and the earlier generation of the upper Mach stem (compared to the lower one) around the lower cavity is seen in Fig. 4c. This has as a result a higher temperature in this upper Mach stem of the lower cavity compared to the Mach stems of the upper cavity (3295 K compared to 2300 K). The localised maxima of high temperatures in this scenario correspond to the MSHS of the top bead ($T = 1645$ K) at $t = 0.02\,\mu$s, the BHS of the lower void, ($T = 2940$ K) at $t = 0.04\,\mu$s, and lower cavity top MSHS ($T = 3240$ K) at $t = 0.105\,\mu$s.

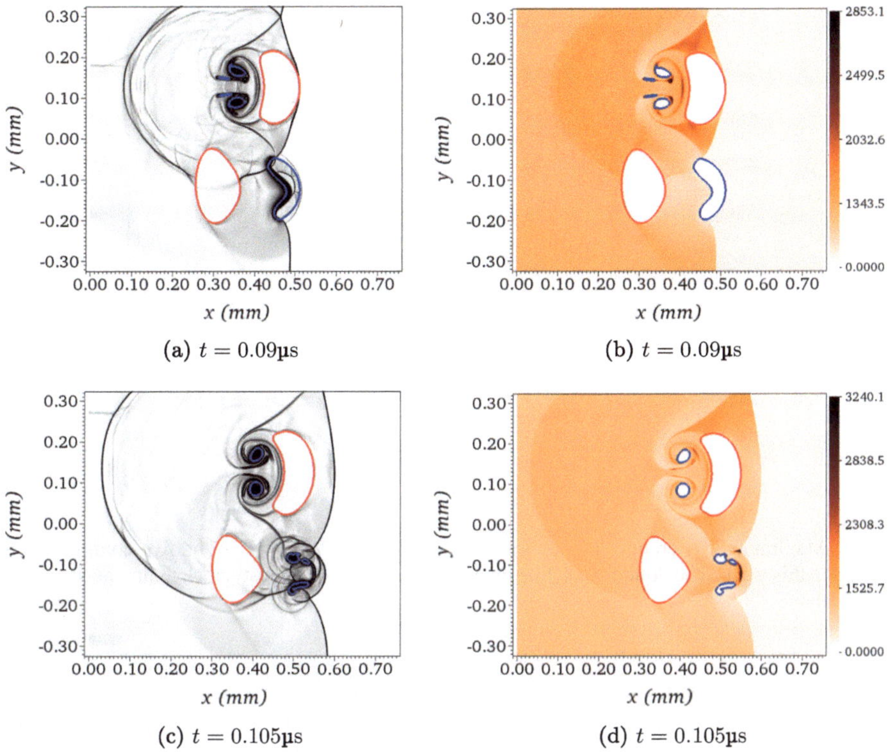

Fig. 4 Mock-schlieren plots (left) illustrating the interaction of waves and the induced temperature field in nitromethane (right) in a clockwise cavity-bead-bead-cavity configuration (continued from Fig. 3)

3.5 Analysis of the Temperature Field

In order to infer the effect of the impurities on the shocked material, we need to consider the maximum temperature of the explosive for any given combination of impurities. To this end we compare in Fig. 5 the maximum nitromethane temperature as a function of time, for five different impurity configurations. These include: an isolated cavity, an isolated bead, a 2×2 matrix of cavities, a 2×2 matrix of beads and a 2×2 matrix combining 2 cavities and 2 beads. We also include as a dashed black line the post-shock temperature of neat nitromethane; i.e., the temperature that the shocked material would reach if no impurities were present.

We observe that the smallest temperature increase occurs by the single bead scenario and slightly higher temperatures in the multi-bead example, which indicates that the inclusion of beads is suitable for subtle adjustment of temperature. The inclusion of voids should be preferred when higher temperature elevations are needed, leading to a more abrupt sensitivity increase of the material. In practice, the desired temperature rise can be achieved to a leading order by means of cavities, while marginal adjustment can be done by solid particles.

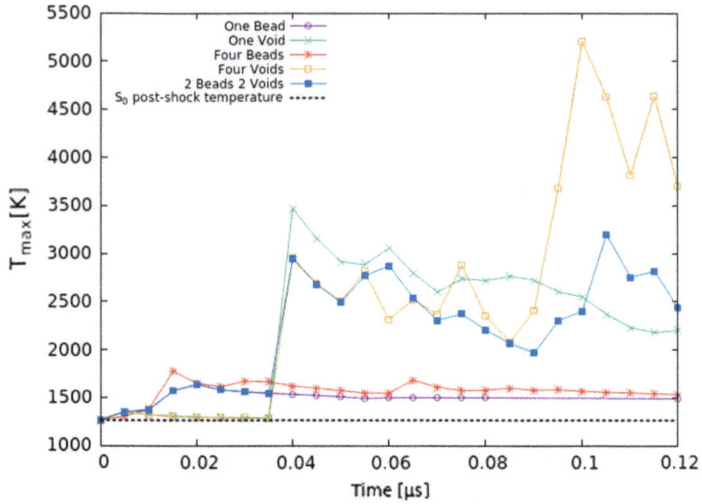

Fig. 5 Maximum temperature distribution in nitromethane over time, in the five configurations studied in this work. The dotted black line gives the reference of the post-shock temperature in neat nitromethane

These initial results have to be verified by large-scale computations, where a statistically-significant number of cavities and particles are included in a larger sample of the explosive. We also anticipate that dimensionality effects are important, as could potentially be the non-uniform distribution of the impurities and the material of the particles.

4 Conclusion

In this work we employ a multi-physics computational framework to study the effect of air cavities and PMMA particles on the sensitisation of condensed-phase explosives as a means of controlling their performance. The framework simultaneously solves a multi-phase hydrodynamic model for the explosive and for the air cavities, and an elastoplastic model for the solid particles. Communication between the different states of matter (the solid and the two fluids) is achieved by means of a variant of the ghost fluid method. We study the effect of five configurations of impurities (isolated cavity, isolated bead, a multi-cavity and a multi-bead configuration and a combined cavity-bead matrix) in nitromethane and determine their relative effect on the temperature field. Initial results indicate that cavities have a more profound effect on sensitisation, compared to PMMA particles; more extended studies are necessary in order to assess the effect of dimensionality, distribution of the impurities and of the material of the particles. This knowledge can be used to accurately control the sensitivity and the performance of nonideal mining explosives.

Acknowledgements The authors gratefully thank Alan Minchinton from Orica—Research and Innovation for useful discussions.

References

1. Allaire, G., Clerc, S., Kokh, S.: A five-equation model for the simulation of interfaces between compressible fluids. J. Comput. Phys. **181**(2), 577–616 (2002)
2. Barton, P.T., Drikakis, D., Romenski, E., Titarev, V.A.: Exact and approximate solutions of riemann problems in non-linear elasticity. J. Comput. Phys. **228**(18), 7046–7068 (2009)
3. Bourne, N., Field, J.: Cavity collapse in a liquid with solid particles. J. Fluid Mech. **259**, 149–165 (1994)
4. Godunov, S.K., Romenskii, E.: Elements of continuum mechanics and conservation laws. Springer Science & Business Media (2013)
5. Ling, Y., Haselbacher, A., Balachandar, S., Najjar, F., Stewart, D.: Shock interaction with a deformable particle: direct numerical simulation and point-particle modeling. J. Appl. Phys. **113**(1), 013, 504 (2013)
6. Menikoff, R.: Hot spot formation from shock reflections. Shock Waves **21**(2), 141–148 (2011)
7. Michael, L., Nikiforakis, N.: The temperature field around collapsing cavities in condensed phase explosives. In: 15th International Detonation Symposium, pp. 60–70 (2014)
8. Michael, L., Nikiforakis, N.: A hybrid formulation for the numerical simulation of condensed phase explosives. J. Comput. Phys. **316**, 193–217 (2016)
9. Michael, L., Nikiforakis, N.: A multi-physics methodology for the simulation of the two-way interaction of reactive flow and elastoplastic structural response. J. Comput. Phys. **367**, 1–27 (2018)
10. Michael, L., Nikiforakis, N.: The evolution of the temperature field during cavity collapse in liquid nitromethane. Part I: inert case. Shock Waves (in print, 2018)
11. Michael, L., Nikiforakis, N.: The evolution of the temperature field during cavity collapse in liquid nitromethane. Part II: reactive case. Shock Waves (in print, 2018)
12. Michael, L., Nikiforakis, N., Bates, K.: Numerical simulations of shock-induced void collapse in liquid explosives. In: 14th International Detonation Symposium (2010)
13. Miller, G., Colella, P.: A conservative three-dimensional Eulerian method for coupled solid-fluid shock capturing. J. Comput. Phys. **183**(1), 26–82 (2002)
14. Minchinton, A.: On the influence of fundamental detonics on blasting practice. In: 11th International Symposium on Rock Fragmentation by Blasting, Sydney, pp. 41–53 (2015)
15. Ripley, R., Zhang, F., Lien, F.: Detonation interaction with metal particles in explosives. In: 13th International Detonation Symposium (2006)
16. Saurel, R., Petitpas, F., Berry, R.A.: Simple and efficient relaxation methods for interfaces separating compressible fluids, cavitating flows and shocks in multiphase mixtures. J. Comput. Phys. **228**(5), 1678–1712 (2009)
17. Schoch, S., Nordin-Bates, K., Nikiforakis, N.: An Eulerian algorithm for coupled simulations of elastoplastic-solids and condensed-phase explosives. J. Comput. Phys. **252**, 163–194 (2013)
18. Sridharan, P., Jackson, T., Zhang, J., Balachandar, S., Thakur, S.: Shock interaction with deformable particles using a constrained interface reinitialization scheme. J. Appl. Phys. **119**(6), 064, 904 (2016)
19. Zhang, F., Thibault, P.A., Link, R.: Shock interaction with solid particles in condensed matter and related momentum transfer. In: Proceedings of the Royal Society of London A, vol. 459, pp. 705–726 (2003)

An Open and Parallel Multiresolution Framework Using Block-Based Adaptive Grids

Mario Sroka, Thomas Engels, Philipp Krah, Sophie Mutzel, Kai Schneider and Julius Reiss

Abstract A numerical approach for solving evolutionary partial differential equations in two and three space dimensions on block-based adaptive grids is presented. The numerical discretization is based on high-order, central finite-differences and explicit time integration. Grid refinement and coarsening are triggered by multiresolution analysis, i.e. thresholding of wavelet coefficients, which allow controlling the precision of the adaptive approximation of the solution with respect to uniform grid computations. The implementation of the scheme is fully parallel using MPI with a hybrid data structure. Load balancing relies on space filling curves techniques. Validation tests for 2D advection equations allow to assess the precision and performance of the developed code. Computations of the compressible Navier-Stokes equations for a temporally developing 2D mixing layer illustrate the properties of the code for nonlinear multi-scale problems. The code is open source.

Keywords Adaptive block-structured mesh · Multiresolution · Wavelets
Parallel computing · Open source · Linear advection · Compressible navier-stokes

1 Introduction

For many applications in computational fluid dynamics, adaptive grids are more advantageous than uniform grids, because computational efforts are put at locations required by the solution. Since small-scale flow structures may travel, emerge and

M. Sroka · P. Krah · S. Mutzel · J. Reiss (✉)
Technische Universität Berlin, Müller-Breslau-Strasse 15, 10623 Berlin, Germany
e-mail: reiss@tnt.tu-berlin.de

T. Engels
École normale supérieure, LMD (UMR 8539), 24, Rue Lhomond,
75231 Paris Cedex 05, France
e-mail: thomas.engels@ens.fr

K. Schneider
Aix–Marseille Université, CNRS, Centrale Marseille, I2M UMR 7373,
39 rue Joliot-Curie, 13451 Marseille Cedex 20, France

R. King (ed.), *Active Flow and Combustion Control 2018*,
Notes on Numerical Fluid Mechanics and Multidisciplinary Design 141,
https://doi.org/10.1007/978-3-319-98177-2_19

disappear, the required local resolution is time-dependent. Therefore dynamic gridding, which tracks the evolution of the solution, is more efficient than static grids. However, suitable grid adaptation techniques are necessary to dynamically track the solution. These techniques can increase the computational cost, therefore their efficiency is problem dependent and related to the sparsity of the adaptive grid.

Examples where adaptivity is beneficial are reactive flows with localized flame fronts, detonations and shock waves [1, 23], coherent structures in turbulence [24] and flapping insect flight [12]. For the latter the time-varying geometry generates localized turbulent flow structures. These applications motivate and trigger the development of a novel multiresolution framework, which can be used for many mixed parabolic/hyperbolic partial differential equations (PDE).

The idea of adaptivity is to refine the grid where required and to coarsen it where possible, while controlling the precision of the solution.

Such approaches have a long tradition and can be traced back to the late seventies [5]. Adaptive mesh refinement and multiresolution concepts developed by Berger et al. [2] and Harten [14, 15], respectively, are meanwhile widely used for large scale computations (e.g. [10, 18, 20]).

Berger suggested a flexible refinement strategy by overlaying different grids of various orientation and size, in the following referred to as *adaptive mesh refinement* (AMR). Harten instead discusses a mathematical more rigorous wavelet based method, termed *multiresolution* (MR). For AMR methods, the decision where to adapt the grid is based on error indicators, such as gradients of the solution or derived quantities. In contrast in MR, the multiresolution transform allows efficient compression of data fields by thresholding detail coefficients. This multiresolution transform is equivalent to biorthogonal wavelets, see e.g. [15]. An important feature of MR is the reliable error estimator of the solution on the adaptive grid, as the error introduced by removing grid points can be directly controlled.

In wavelet-based approaches the governing equations are discretized, either by using wavelets in a Galerkin or collocation approach [24], or using a classical discretization, e.g. finite volumes or differences, where the grid is adapted locally using MR analysis [4, 10].

MR methods typically keep only the information which is dictated by a threshold criterion, which is refereed to as sparse point representation (SPR), introduced in [16]. AMR methods often utilize blocks and refine complete areas, by which the maximal sparsity is typically abandoned in favor of a simpler code structure. An example of this approach is the AMROC code [8], where blocks of arbitrary size and shape are refined. A detailed comparison of MR with AMR techniques has been carried out in [9].

For practical applications both the data compression and the speed-up of the calculation are crucial. The latter is reduced by the computational overhead to handle the adaptive grid and corresponding datastructures. This effort differs substantially between different approaches [19]. It can be reduced by refining complete blocks, thereby reducing the elements to manage, and by exploiting simple grid structures.

A MR method using a quad- or octtree representation to simplify the grid structure is reported, e.g., in [10, 11] and has later also been used in [22].

For detailed reviews on the subject of multiresolution methods we refer the reader to [7, 10, 18, 24, 24]. Implementation issues have been discussed in [6].

Our aim is to provide a multiresolution framework, which can be easily adapted to different two- and three-dimensional simulations encountered in CFD, and which can be efficiently used on fully parallel machines.

To this end the chosen framework is block based, with nested blocks on quad- or octree grids. The individual blocks define structured grids with a fixed number of points. Refinement and coarsening are controlled by a threshold criterion applied to the wavelet coefficients. The software, termed "wavelet adaptive block-based solver for interactions with turbulence" (`WABBIT`), is open-source and freely available[1] in order to maximize its utility for the scientific community and for reproducible science.

The purpose of this paper is to introduce the code, present its main features and explain structural and implementation details. It is organized as follows. In Sect. 2 we give an overview of implementation and structure details. Numerics will only be shortly described, but special issues of our data structure, interpolation, and the `MPI` coding will be explained in detail. Section 3 considers a classical validation test case, including a discussion on the adaptivity and convergence order of `WABBIT`. In Sect. 4 we present computations for a temporally developing double shear layer, governed by the compressible Navier-Stokes equations. Section 5 draws conclusions and gives perspectives for future work.

2 Code Structure

In this section we present a detailed description of the data and code structure. One of the main concepts in WABBIT is the encapsulation and separation of the set of PDE from the rest of the code, thus the PDE implementation is not significantly different from that in a single domain code and can easily be exchanged. The code solves evolutionary PDE of the type $\partial_t \phi = N(\phi)$. The spatial part $N(\phi)$ is referred to as *right hand side* in this report. A primary directive for the code is its "explicit simplicity", which means avoiding complex programming structures to improve maintainability. WABBIT is written in Fortran 95 and aims at reaching high performance on massively parallel machines with distributed memory architecture. We use the MPI library to parallelize all subroutines, while parallel I/O is handled through the HDF5 library.

2.1 *Multiresolution Algorithm*

The main structure of the code is defined by the multiresolution algorithm. After the initialization phase, the general process to advance the numerical solution $\varphi(t^n, x)$ on the grid $\mathcal{G}^n$ to the new time level t^{n+1} can be outlined as follows.

[1] Available on https://github.com/adaptive-cfd/WABBIT.

1. Refinement. We assume that the grid $\mathcal{G}^n$ is sufficient to adequately represent the solution $\varphi(t^n, x)$, but we cannot suppose this will be true at the new time level. Non-linearities may create scales that cannot be resolved on $\mathcal{G}^n$, and transport can advect existing fine structures. Therefore, we have to extend $\mathcal{G}^n$ to $\widetilde{\mathcal{G}}^n$ by adding a "safety zone" [24] to ensure that the new solution $\varphi\left(t^{n+1}, x\right)$ can be represented on $\widetilde{\mathcal{G}}^n$. To this end, all blocks are refined by one level, which ensures that quadratic non-linearities cannot produce unresolved scales.
2. Evolution. On the new grid $\widetilde{\mathcal{G}}^n$, we first synchronize the layer of ghost nodes (Sect. 2.5) and then solve the PDE using finite differences and explicit time-marching methods.
3. Coarsening. We now have the new solution $\varphi(t^{n+1}, x)$ on the grid $\widetilde{\mathcal{G}}^n$. The grid $\widetilde{\mathcal{G}}^n$ is a worst-case scenario and guarantees resolving $\varphi(t^{n+1}, x)$ using *a priori* knowledge on the non-linearity. It can now be coarsened to obtain the new grid $\mathcal{G}^{n+1}$, removing, in part, blocks created during the refinement stage. Section 2.3 explains this process in more detail.
4. Load balancing. The remaining blocks are, if necessary, redistributed among MPI processes using a space-filling curve [25], such that all processes compute approximately the same number of blocks. The space-filling curve allows preservation of locality and reduces interprocessor communication cost.

2.2 Block- and Grid Definition

Block Definition. The decomposition of the computational domain builds on blocks as smallest elements, as used for example in [11]. The approach thus builds on a hybrid datastructure, combining the advantages of structured and unstructured data types. The structured blocks have a high CPU caching efficiency. Using blocks instead of single points reduces neighbor search operations. A drawback of the block based approach is the reduced compression rate.

A block is illustrated in Fig. 1. Its definition (in 2D) is

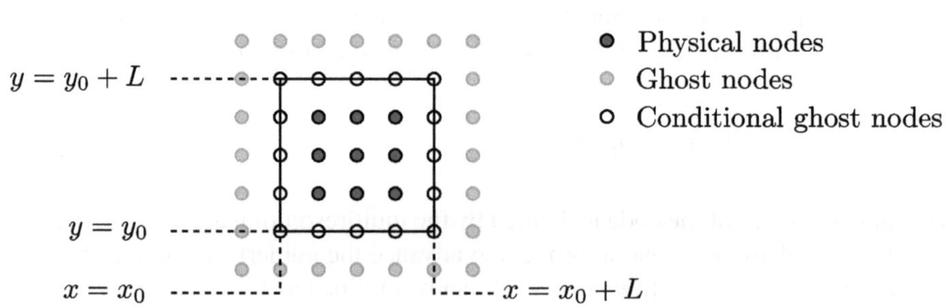

Fig. 1 Definition of a block with $B_s = 5$ and $n_g = 1$

$$\mathcal{B}^{\ell} = \left\{ \underline{x} = \underline{x}_0 + \left(i \cdot \Delta x^{\ell}, j \cdot \Delta x^{\ell}\right)^T, \; 0 \le i, j \le B_s \right\}$$

where $\underline{x}_0$ is the blocks origin, $\Delta x^{\ell} = 2^{-\ell} L/(B_s - 1)$ is the lattice spacing at level ℓ, and L the size of the entire computational domain. The mesh level encodes the refinement from 1 as coarsest to the user defined value $J_{\max}$ as finest. Blocks have B_s points in each direction, where B_s is odd, which is a requirement of the grid definition we use. We add a layer of n_g ghost points that are synchronized with neighboring blocks (see Sect. 2.5). The first layer of physical points is called conditional ghost nodes, and they are defined as follows:

1. If the adjacent block is on the same level, then the conditional ghost nodes are part of both blocks and thus redundant in memory; their values are identical.
2. If the levels differ, the conditional ghost nodes belong to the block on the finer level, i.e., their values will be overwritten by those on the finer block.

Grid Definition. A complete grid consisting of $N_b = 7$ blocks is shown in Fig. 2. We force the grid to be graded, i.e., we limit the maximum level difference between two blocks to one. Blocks are addressed by a quadtree-code (or an octtree in 3D), as introduced in [13], and also shown in Fig. 2. Each digit of the treecode represents one mesh level, thus its length indicates the level ℓ of the block. If a block is coarsened, the last digit is removed, while for refinement refinement, one digit is added. The function of the treecode is to allow quick neighbor search, which is essential for high performance. For a given treecode the adjacent treecodes can easily be calculated [13]. A list of the treecodes of all existing blocks allows us to find the data of the neighboring block, see Sect. 2.4. To ensure unique and invertible neighbor relations, we define them not only containing the direction but also encode if a block covers

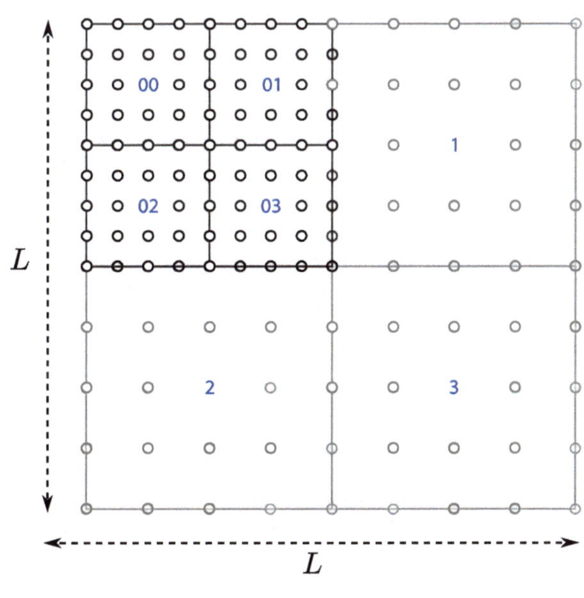

Fig. 2 Example grid with $N_b = 7$ blocks. Three blocks on mesh level 1 (gray) and four on level 2 (black), together with their treecodes. Note that the mesh level is equal to the length of the treecode. Points at the coarse/fine interface belong to finer blocks

only part a border. This situation occurs if two neighboring blocks differ in level. We also account for diagonal neighborhoods. In two space dimensions 16 different relations defined (74 in 3D). This simplifies the ghost nodes synchronization step, since all required information, the neighbor location and interpolation operation are available.

Right Hand Side Evaluation. The PDE subroutine purely acts on single blocks. Therefore efficient, single block finite difference schemes can be used allowing to combine existing codes with the WABBIT framework. Adapting the block size to the CPU cache offers near optimal performance on modern hardware. The size of the ghost node layer can be chosen freely, to match numerical schemes with different stencil sizes.

2.3 Refinement/Coarsening of Blocks

If a block is flagged for refinement by some criteria (see blow) this refinement is executed as illustrated in Fig. 3. The block, with synchronized ghost points, is first uniformly upsampled by midpoint insertion, i.e., missing values on the grid

$$\widetilde{\mathcal{B}}^{\ell} = \left\{ \underline{x} = \underline{x}_0 + \left(i \cdot \Delta x^{\ell}/2, j \cdot \Delta x^{\ell}/2\right)^T, \quad -2n_g \leq i, j \leq 2B_s - 1 + 2n_g \right\}$$

are interpolated (gray points in Fig. 3 center). In other words, a prediction operator $\mathcal{P}_{\ell\to\ell+1}$ is applied [14]. The data is then distributed to four new blocks $\mathcal{B}_i^{\ell+1}$, where one digit is added to the treecode, which are created on the `MPI` process holding the initial ("mother") block. The blocks are nested, i.e. all nodes of a coarser block also exist in the finer one. The reverse process is coarsening, where four sister blocks on the same level are merged into one coarser block by applying the restriction operator

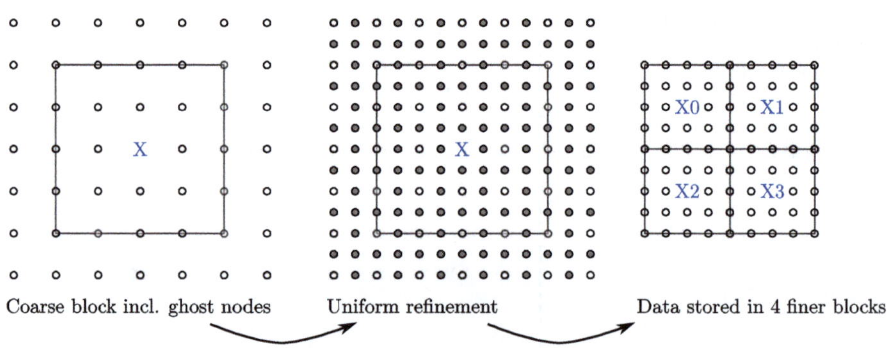

Fig. 3 Process of refining block with treecode X. First, the block is upsampled, including the ghost nodes layer. Then, four new blocks are created, where one digit is added to the treecode

$\mathcal{R}_{\ell\to\ell-1}$, which simply removes every second point. For coarsening, no ghost node synchronization is required, but all four blocks need to be gathered on one `MPI` rank.

The refinement operator uses central interpolation schemes. Using one-sided schemes close to the boundary would not require ghost points and would thus reduce the number of communications. They yield errors only of the order of the threshold ε. However, the small, but non-smooth structures of these errors force very fine meshes, which can increase the number of blocks. This *fill-up* can lead to prohibitively expensive calculations.

Computation of Detail Coefficients. The decision whether a block can be coarsened or not is made by calculating its detail coefficients [24]. The are computed by first applying the restriction operator, followed by the prediction operator. After this round trip of restriction and prediction, the original resolution is recovered, but the values of the data differ slightly. The difference

$$\mathcal{D} = \{d(\underline{x})\} = \mathcal{B}^{\ell} - \mathcal{P}_{\ell-1\to\ell}(\mathcal{R}_{\ell\to\ell-1}(\mathcal{B}^{\ell}))$$

is called details. If details are small, the field is smooth on the current grid level. Therefore, the details act as indicator for a possible coarsening [14]. Non-zero details are obtained at odd indices only (gray points in Fig. 3, center) because of the nested grid definition and the fact that restriction and prediction do not change these values. The refinement flag for a block is then

$$r = \begin{cases} -1 & \text{if } \left\|d(\underline{x})\right\|_{\infty} < \varepsilon \\ 0 & \text{otherwise} \end{cases}$$

where -1 indicates coarsening and 0 no change. In other words, the largest detail sets the status of the block. Note, that WABBIT technically provides the possibility to flag -1 for coarsening, 0 for unaltered and $+1$ for refinement, it can hence be used with arbitrary indicators. Since a block cannot be coarsened if its sister blocks on the same root do not share the $+1$ refinement status, WABBIT assigns the -2 status for blocks that can indeed be coarsened, after checking for completeness and gradedness.

2.4 Data Structure

The data are split into two kinds of data, first, the field data (the flow fields) required to calculate the PDE and, second, the data to administer the block decomposition and the parallel distribution.

Data which are held only on one specific `MPI` process are called *heavy data*. This is the (typically large) field data and the neighbor relations for the blocks held by the MPI process. The field data (`hvy_block`) is a five dimensional array where the first three indices describe the note within a block (3D notation is always used in the code), the fourth index the index of the physical variables and the last one the block index identifying it within the MPI process.

The *light data* (`lgt_block`) are data which are kept synchronous between all processes. They describe the global topology of the adapted grid and change during the computation. The light data consist of the block treecode, the block mesh level and the refinement flag. Additionally, we encode the `MPI` process rank i_{process} and the block index on this process j_{block} by the position I of the data within the *light data* array, $I = (i_{\text{process}} - 1) \cdot N_{\max} + j_{\text{block}}$, where $N_{\max}$ is the maximal number of blocks per process. The *light data* enable each process to determine the process holding neighboring blocks, by looking for the index I corresponding to the adjacent treecode. The number of blocks required during the computation is unknown before running the simulation. To avoid time consuming memory allocation, $N_{\max}$ is typically determined by the available memory. This sets the index range of the last index of the heavy data and determines the size of the light data. Hence, many blocks are typically unused; they are marked by setting the treecode in the light data list to -1. To accelerate the search within the *light data*, we keep a second list of indices holding active entries.

2.5 Parallel Implementation

Data Synchronization. For parallel computing, an efficient data synchronization strategy is essential for good performance. There are two different tasks in WABBIT, namely light and heavy data synchronisation. *Light data* synchronization is an MPI all-to-all operation, where we communicate active entries of the light data only. *Heavy data* synchronization, i.e. filling the ghost nodes layer of each block, is much more complicated. We have to balance a small number of MPI calls and a small amount of communicated data, and additionally we have to ensure that no idle time occurs due to blocking of a process by a communication in which this process is not involved. To this end, we use `MPI` point-to-point communication, namely non-blocking non-buffered send/receive calls. To reduce the number of communications, the ghost point data of all blocks belonging to one process are gathered and send as one chunk. After the `MPI` communications, all processes store received data in the ghost point layers.

The conditional ghost nodes require special attention during the synchronization. To ensure that neighboring blocks always have the same values at these nodes, the redundant nodes are sent, when required, to the neighboring process. Blocks on higher mesh levels (finer grids) always overwrite the redundant nodes to neighbors on lower mesh level (coarser grid). It is assumed that two blocks on the same mesh level never differ at a redundant node, because any numerical scheme should always produce the same values.

Load Balancing. The external neighborhood consists of ghost nodes, which may be located on other processes and therefore have to be sent/received in the *heavy data* synchronization step. Internal ghost nodes can simply be copied within the process memory, which is much faster than `MPI` communication. It is, thus, desired to reduce inter-process neighborhood. We use space filling curves [25] to redistribute the blocks among the processes for their good localization. The computation of the space filling curve is simple, because we can use the treecode to calculate the index on the curve.

3 Advection Test Case

As a validation case, we now consider a benchmarking problem for the 2D advection equation, $\partial_t \varphi + u \cdot \nabla \varphi = 0$, where $\varphi(x, y, t)$ is a scalar and $0 \leq x, y < 1$. The spatially-periodic setup considers time-periodic mixing of a Gaussian blob,

$$\varphi(x, y, 0) = e^{-\left((x-c)^2+(y-d)^2\right)/\beta}$$

where $c = 0.5$, $d = 0.75$ and $\beta = 0.01$. The time-dependent velocity field is given by

$$u(x, y, t) = \cos\left(\frac{\pi t}{t_a}\right)\begin{pmatrix} \sin^2(\pi x)\sin(2\pi y) \\ \sin^2(\pi y)(-\sin(2\pi x)) \end{pmatrix} \tag{1}$$

and swirls the initial distribution, but reverses to the initial state at $t = t_a$. The swirling motion produces increasingly fine structures until $t = t_a/2$, where t_a controls also the size of structures. The larger t_a, the more challenging is the test.

Spatial derivatives are discretized with a 4th-order, central finite-difference scheme and we use a 4th-order Runge–Kutta time integration. Interpolation for the refinement operator is also 4th order. We compute the solution for $t_a = 5$, for various maximal mesh levels $J_{\max}$. The computational domain is a unit square and we use a block size of 33×33.

Figure 4 illustrates φ at the initial time, $t = 0$, and the instant of maximal distortion at $t = 2.5 = t_a/2$. At $t = 2.5$ the grid is strongly refined in regions of fine structures, while the remaining part of the domain features a coarser resolution, e.g., in the center of the domain. Further the distribution among the MPI processes is shown by different colors, revealing the locality of the space filling curve.

In the following we compare soloutions with the finest strutures at $t = t_a/2$ with a reference solution, to investigate the quality and performance. The reference solution is obtained with a pseudo-spectral code on a sufficiently fine mesh to have a negligible error compared with the current results.

Figure 5a illustrates the relative error, computed as the ∞-norm of the difference $\varphi - \varphi_{\text{ex}}$, normalized by $||\varphi_{\text{ex}}||_\infty$. All quantities are evaluated on the terminal grid. A linear least squares fit exhibits convergence orders close to one for the large maximal

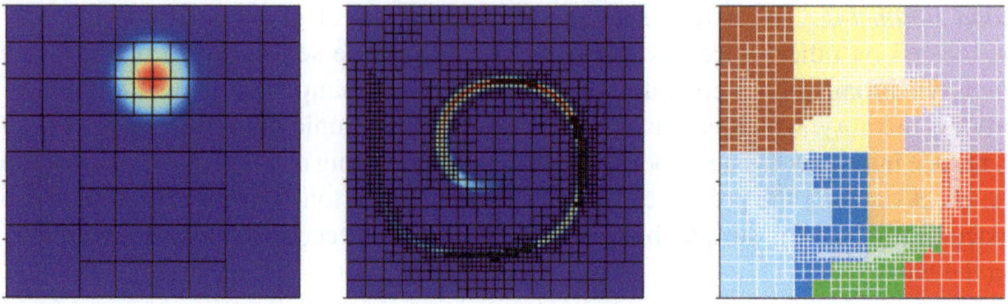

Fig. 4 Shown is a pseudocolor-plot of φ at times $t = 0$, $t = t_a/2 = 2.5$ and the distribution of the blocks among the MPI processes by different colors at $t = 2.5$ (from left to right). Each block covers 33×33 points

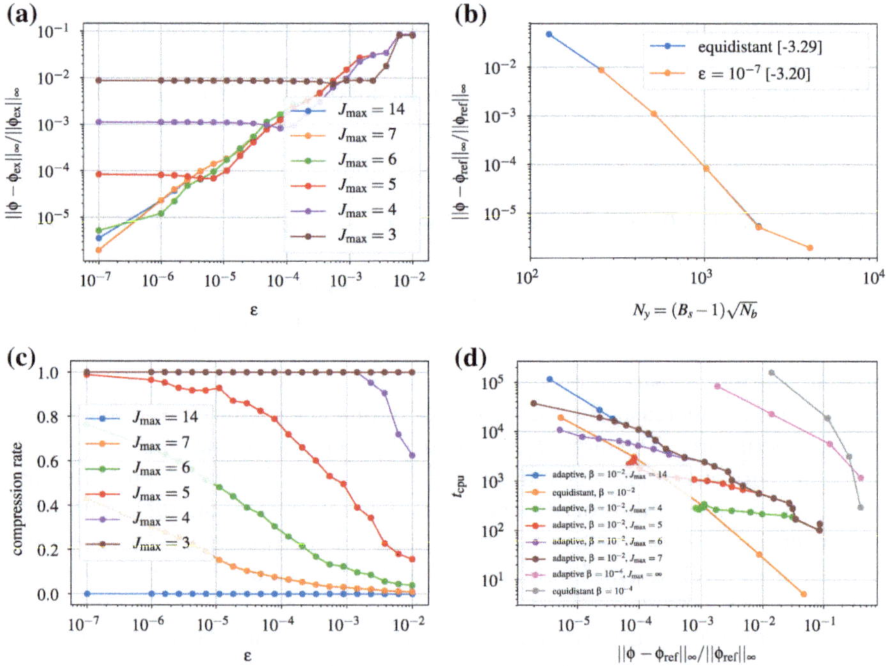

Fig. 5 Swirl test for varying $J_{\max}$ and ε. **a** For different maximal refinement levels a saturation of the error is seen at different values of ε, showing the cross over form threshold- to discretization-error. **b** Error decay for fixed $\varepsilon = 10^{-7}$ and varying $J_{\max}$ (i.e. the rightmost data points in A) as a function of the number of points in one direction. The adaptive computation preserves roughly the 4th order accuracy of the discretization scheme. **c** Compression rate defined as block of the adaptive mesh compared with a equidistant mesh constantly on the same $J_{\max}$. **d** The CPU time as a function of discretization error for two different initial conditions. For the broad pulse ($\beta = 10^{-2}$) the adaptive solution is faster for an appropriate choice of $J_{\max}(\varepsilon)$ for the finer pules ($\beta = 10^{-2}$) it is faster even for a constant $J_{\max} = \infty$ for relevant errors

refinements. In this case the error decays, as expected, linearly in ε. For smaller $J_{\max}$ we find a saturation of the error, which is determined by the highest allowed resolution. This different levels are plotted in Fig. 5b, where a convergence order close to four, as expected by the space and time discretization is found. Thus, the points where the saturation sets in are turnover points form an threshold to and cut-off dominated error. For the sake of efficiency one aims to be close to this turnover point where both errors are of similar size. In Fig. 5c the compression rate, i.e. the number of blocks relative to an equidistant grid constantly on the level of the same $J_{\max}$ is depicted. As expected the compression becomes close to one for small ε. In Fig. 5d the error is shown as a function of the computational time for two initial conditions, the broad pulse with $\beta = 10^{-2}$ and a narrower one with $\beta = 10^{-4}$. For the broad pulse ($\beta = 10^{-2}$) the curves for different $J_{\max}$ are below the equidistant curve only for carefully chosen values of ε. This is explained by the wide area of refinement at the final time, see Fig. 4. Here a multi-resolution method cannot win much. Even for $J_{\max} = 14$, which in practice means deactivating the level restriction, a similar scaling as for the equidistant grid is found with a factor approaching about four. Thus, even without tuning $J_{\max}(\varepsilon)$ accordingly, and given the low cost of the right hand side, the computational complexity of the adaptive code scales reasonably compared to the equidistant solution. For a finer initial condition ($\beta = 10^{-4}$), even without the level restriction ($J_{\max} = \infty$), the adaptive code produces better run-times for practical relevant errors.

4 Navier-Stokes Test Case

In this section we present the results of a second test case, governed by the ideal-gas, constant heat capacity compressible Navier-Stokes equations in the skew-symmetric formulation [21]. A double shear-layer in a periodic domain is perturbed so that the growing instabilities end up with small scale structures, similar to [17]. The size of the computational domain is $L = L_x = L_y = 8$ and the shear layer is initially located at $\frac{L}{2} \pm 0.25$. The density and y-velocity is $\rho_1 = 2$ and $v_1 = 1$ between the shear layers and $\rho_0 = 1$ and $v_0 = -1$ otherwise. At the jumps it is smoothed by $\tanh((y - y_{\text{jump}})/\lambda_w)$ with a width $\lambda_w = L/240$. The initial pressure is uniformly $p = 2.5$. The x-velocity is disturbed to induce the instability in a controlled manner by $u = \lambda \sin(2\pi(y - L/2))$ with $\lambda = 0.1$. The dynamic viscosity is given by $\mu = 10^{-6}$. The adiabatic index is $\gamma = 1.4$ and the Prandtl number is $Pr = 0.71$ and the specific gas constant $R_s = 287.05$.

We discretize spatial derivatives with standard 4th-order central differencing scheme, use the standard 4th-order Runge-Kutta time integration and for interpolation a 4th-order scheme. We use global time stepping so that the time step is (usually) determined by the time step at the highest mesh level. We apply a shock capturing filter as described in [3] with a threshold value of $r_{\text{th}} = 10^{-5}$ in every time step. Filtering, as any procedure to suppress high wave numbers (e.g. flux limiter, slope limiter or numerical damping), interacts with the MR. No special modification beyond the previously described [21] smoothed detector, was necessary for the use with the multi resolution framework. The investigation of the interplay between filtering and MR is left for future work.

In Fig. 6 the density field for adaptive computations with a threshold $\varepsilon = 10^{-3}$ at $t = 4$ is shown. In both density and vorticity field one can observe small scale structures created by the shear layer instability. The size and form of the structures are in agreement with [17].

In the right of Fig. 7 the compression rate of the shear layer is plotted over time. We start with a low number of blocks (i.e. low values of the compression rate), the grid fills up to the maximal refinement with simulation time. This is explained by short wavelength acoustic waves emitted by the shear layer. Depending on the investigation target a modified threshold criterion, e.g., applying it only to certain variables might be beneficial. For this the error estimation must be reviewed and it is left for future work.

In Fig. 7 we show the kinetic energy spectra for these computations compared to the result for a fixed grid. To calculate the energy spectra we refine the mesh after the computation to a fixed mesh level, if needed. They agree well on the resolved scales.

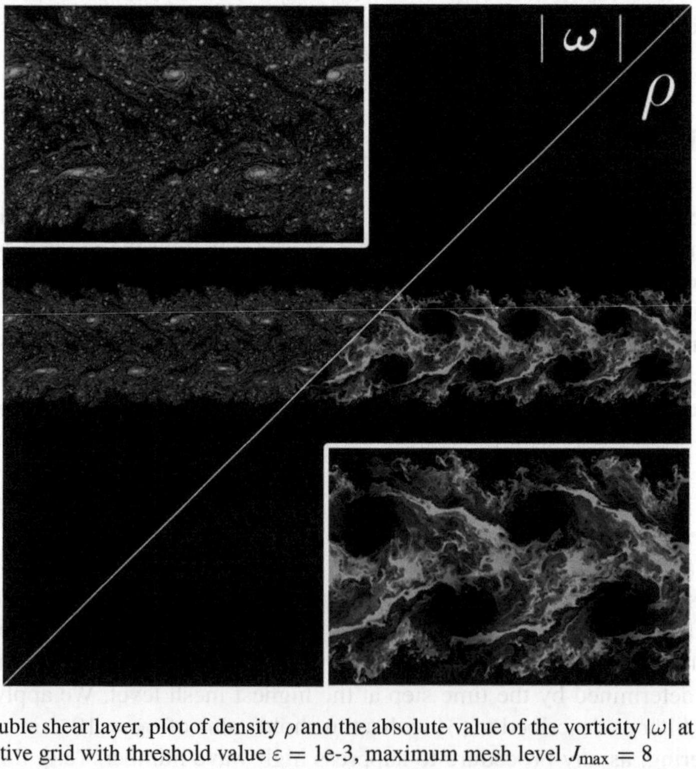

Fig. 6 Double shear layer, plot of density ρ and the absolute value of the vorticity $|\omega|$ at time $t = 4$ on an adaptive grid with threshold value ε = 1e-3, maximum mesh level $J_{\max} = 8$

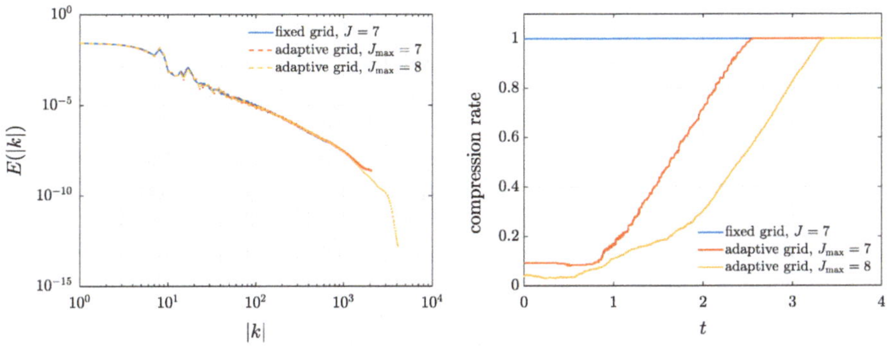

Fig. 7 Left: Energy spectrum of the double shear layer. The computations were performed on a fixed grid with mesh level $J = 7$, and on adaptive grids with threshold value $\varepsilon = 10^{-3}$, maximum mesh level $J_{\max} = 7$, $J_{\max} = 8$, $t = 4$. Right: The compression rate. After high initial copressions the grid fills up due to high wavenumber acoustic waves

Fig. 8 From the strong scaling a parallel fraction of 99% can be estimated

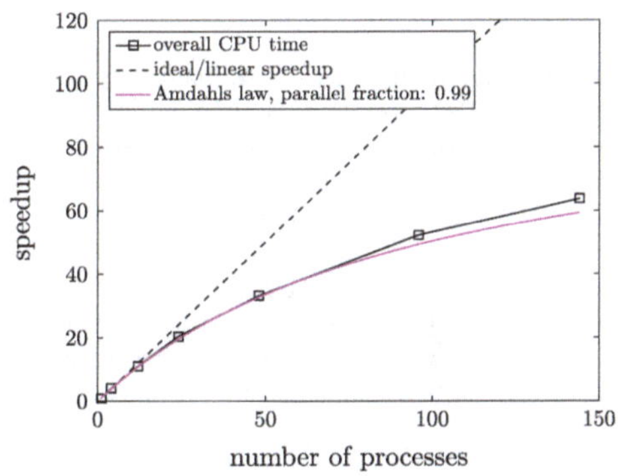

For the higher maximum mesh level $J_{\max}$ we observe a better resolution of the small scale structures. Summarized, if we compare adaptive and fixed mesh computation, we can observe a good resolution of the small scales within the double shear layer.

Figure 8 shows the strong scaling behavior for the adaptive double shear layer computation with $J_{\max} = 7$. We observe a scaling which is predicted by Amdahl's law with a parallel fraction of 0.99. The observed strong scaling is reasonable and we anticipate that code optimization will yield further improvements.

5 Conclusions and Perspectives

The novel framework `WABBIT` with its main structures and concepts has been described. `WABBIT` uses a multiresolution algorithm to adapt the mesh to capture small localized structures. Within the framework different equation sets can be used.

We showed that the error due to the thresholding is controlled and scales nearly linear. In the Gaussian pulse test case we found that the maximum number of blocks is reached at the largest deformation of the pulse and after that the mesh is coarsen with several orders of magnitude. We observed that the fill-up was strongly reduced by using a symmetric interpolation stencil, which will be investigated in future work.

In the second test case we showed an application of the compressible Navier-Stokes equations. Here we saw a good resolution of small scale structures and observed the impact of discarding wavelet coefficient on the physics of the shear layer. In our simulations we observe a reasonable strong scaling. Scaling will be assessed in more detail when foreseen improvements are implemented.

In the near future we will extend the physical situation by using reactive Navier-Stokes equations to simulate turbulent flames. Validation for 3D problems and further improvement of the performance is currently worked on. For this an additional parallelization with `openMP` is in preparation, which should reduce the communication effort further in typical cluster architecture. Further a generic boundary handling within the frame work and an interface to connect other `MPI` programs is under way.

Acknowledgements MS and JR thankfully acknowledge funding by the Deutsche Forschungsgemeinschaft (DFG) (grant SFB-1029, project A4). TE and KS acknowledge financial support from the Agence nationale de la recherche (ANR Grant 15-CE40-0019) and DFG (Grant SE 824/26-1), project AIFIT. This work was granted access to the HPC resources of IDRIS under the allocation 2018-91664 attributed by GENCI (Grand Équipement National de Calcul Intensif). For this work we were also granted access to the HPC resources of Aix-Marseille Université financed by the project Equip@Meso (ANR-10-EQPX- 29-01). TE and KS thankfully acknowledge financial support granted by the ministères des Affaires étrangères et du développement International (MAEDI) et de l'Education national et l'enseignement supérieur, de la recherche et de l'innovation (MENESRI), and the Deutscher Akademischer Austauschdienst (DAAD) within the French-German Procope project FIFIT.

References

1. Bengoechea, S., Gray, J.A.T., Moeck, J.P., Paschereit, C.O., Sesterhenn, J.: Detonation initiation in pipes with a single obstacle for hydrogen-enriched air mixtures. Submitted to Combustion and Flame (2018)
2. Berger, M.J., Oliger, J.: Adaptive mesh refinement for hyperbolic partial differential equations. J. Comp. Phys. **53**(3), 484–512 (1984)
3. Bogey, C., De Cacqueray, N., Bailly, C.: A shock-capturing methodology based on adaptative spatial filtering for high-order non-linear computations. J. Comp. Phys. **228**(5), 1447–1465 (2009)
4. Bramkamp, F., Lamby, P., Müller, S.: An adaptive multiscale finite volume solver for unsteady and steady state flow computations. J. Comp. Phys. **197**(2), 460–490 (2004)
5. Brandt, A.: Multi-level adaptive solutions to boundary-value problems. Math. Comp. **31**(138), 333–390 (1977)
6. Brix, K., Melian, S., Müller, S., Bachmann, M.: Adaptive multiresolution methods: practical issues on data structures, implementation and parallelization. ESAIM: Proc. **34**, 151–183 (2011)
7. Coquel, F., Maday, Y., Müller, S., Postel, M., Tran, Q.H.: New trends in multiresolution and adaptive methods for convection-dominated problems. ESAIM: Proc. **29**, 1–7 (2009)

8. Deiterding, R., Domingues, M.O., Gomes, S.M., Roussel, O., Schneider, K.: Adaptive multiresolution or adaptive mesh refinement? a case study for 2d euler equations. ESAIM: Proc. **29**, 28–42 (2009)
9. Deiterding, R., Domingues, M.O., Gomes, S.M., Schneider, K.: Comparison of adaptive multiresolution and adaptive mesh refinement applied to simulations of the compressible euler equations. SIAM J. Sci. Comp. **38**(5), S173–S193 (2016)
10. Domingues, M.O., Gomes, S.M., Roussel, O., Schneider, K.: Adaptive multiresolution methods. ESAIM: Proc. **34**, 1–96 (2011)
11. Domingues, M.O., Gomes, S.M., Diaz, L.M.A.: Diaz. Adaptive wavelet representation and differentiation on block-structured grids. Appl. Numer. Math. **47**(3), 421–437 (2003)
12. Engels, T., Kolomenskiy, D., Schneider, K., Sesterhenn, J.: Flusi: A novel parallel simulation tool for flapping insect flight using a fourier method with volume penalization. SIAM J. Sci. Comp. **38**(5), S3–S24 (2016)
13. Gargantini, I.: An effective way to represent quadtrees. Commun. ACM **25**(12), 905–910 (1982)
14. Harten, A.: Discrete multi-resolution analysis and generalized wavelets. Appl. Numer. Math. **12**(1), 153–192 (1993). special issue
15. Harten, A.: Multiresolution representation of data: a general framework. SIAM J. Numer. Anal. **33**(3), 1205–1256 (1996)
16. Holmström, M.: Solving hyperbolic pdes using interpolating wavelets. SIAM J. Sci. Comp. **21**(2), 405–420 (1999)
17. Maulik, R., San, O.: Resolution and energy dissipation characteristics of implicit les and explicit filtering models for compressible turbulence. Fluids **2**(2), 14 (2017)
18. Müller, S.: Adaptive Multiscale Schemes for Conservation Laws. Springer (2003)
19. Müller, S.: Multiresolution schemes for conservation laws. In: DeVore, R., Kunoth, A. (eds.), Multiscale, Nonlinear and Adaptive Approximation, pp. 379–408, Berlin, Heidelberg (2009). Springer Berlin Heidelberg
20. Deiterding, R: Block-structured adaptive mesh refinement—theory, implementation and application. ESAIM: Proc. **34**, 97–150 (2011)
21. Reiss, J., Sesterhenn, J.: A conservative, skew-symmetric finite difference scheme for the compressible navier-stokes equations. Comput. Fluids **101**, 208–219 (2014)
22. Rossinelli, D., Hejazialhosseini, B., Spampinato, D.G., Koumoutsakos, P.: Multicore/multi-gpu accelerated simulations of multiphase compressible flows using wavelet adapted grids. SIAM J. Sci. Comp. **33**(2), 512–540 (2011)
23. Roussel, O., Schneider, K.: Adaptive multiresolution computations applied to detonations. Z. Phys. Chem. **229**(6), 931–953 (2015)
24. Schneider, K., Vasilyev, O.V.: Wavelet methods in computational fluid dynamics. Ann. Rev. Fluid Mech. **42**(1), 473–503 (2010)
25. Zumbusch, G.: Parallel multilevel methods: adaptive mesh refinement and loadbalancing. Advances in numerical mathematics. 1 edn (2003)

A 1D Multi-Tube Code for the Shockless Explosion Combustion

Giordana Tornow and Rupert Klein

Abstract Shockless explosion combustion (SEC) has been suggested by Bobusch et al., CST, **186**, 2014, as a new approach towards approximate constant volume combustion for gas turbine applications. The SEC process relies on nearly homogeneous autoignition in a premixed fuel-oxidizer charge and acoustic resonances for cyclic recharge. Operation of a single SEC tube has proven to be rather robust in numerical simulations, provided the flow control assures nearly homogeneous autoignition. Configurations with multiple tubes that fire into a common collector plenum preceding the turbine will be needed, however, to avoid excessive fluctuating thermal and mechanical load on the turbine blades. In such a configuration, the resonating tubes will interact with the volume of the plenum, and proper control of these interactions will be an important part of the engine design process. The present work presents an efficient, rough design tool that simulates the firing of such multi-tube SEC configurations into a torus-shaped turbine plenum. Both the tubes and the plenum are represented by computational quasi-one-dimensional gasdynamics modules implemented in a finite volume code for the reactive Euler equations. Suitable tube-to-plenum coupling conditions based on mass, energy, and plenum-axial momentum conservation represent the gasdynamic interactions of all engine components. First investigations utilising this tool reveal considerable dependence of the SEC-tubes' operating conditions on the tube radius and length, and on the tubes' positioning along the plenum torus. The SEC is especially sensitive to the plenum's radius. Misfiring of one of the tubes does essentially not affect the operation of the others and does not even necessarily lead to a shut-down of the disturbed SEC tube.

Keywords Approximate constant volume combustion · Shockless explosion combustion

G. Tornow (✉) · R. Klein
Department of Mathematics, Geophysical Fluid Dynamics,
Freie Universität Berlin, Arnimallee 6, 14195 Berlin, Germany
e-mail: gtornow@math.fu-berlin.de

R. Klein
e-mail: rupert.klein@math.fu-berlin.de

R. King (ed.), *Active Flow and Combustion Control 2018*,
Notes on Numerical Fluid Mechanics and Multidisciplinary Design 141,
https://doi.org/10.1007/978-3-319-98177-2_20

1 Introduction

The efficiency gain of gas turbines expected by approximate constant volume combustion (aCVC) compared to today's operation mode, the constant pressure deflagration combustion (CPC), is well-known to the community (see for example the analysis of [8]). Various approaches to realising aCVC have been developed and put into practice to different extent and with diverse success. Each approach has its own challenges and drawbacks: the pulsed detonation combustion (PDC) requires a deflagration-to-detonation transition (DDT) in every cycle, the rotating detonation engine (RDE) must be fuelled within a very short time, and the pulse jet engine (PJE) works only in between the ranges of CPC and aCVC, not fully harvesting the potential of constant volume thermodynamic processes. Since 2012, the shockless explosion combustion (SEC) process is under development, [4]. It relies on acoustic resonance for recharge like the PJE, but aims at homogeneous autoignition to approximate constant volume combustion. Thus it avoids the losses associated with the DDT and turbulent deflagration found in the pulsed detonation and pulsed jet engines.

A stringent interpretation of the term "constant volume combustion" demands all parcels of reacting gas to maintain their initial density throughout the process. This is the situation the SEC combustor aims to approximate. During inflow, fresh hot compressed gas enters the SEC tube with a stratified charge that covers about 1/3 of the tube length. The stratification is tuned to induce approximately homogeneous autoignition. With the characteristic time of heat release of realistic fuels being much shorter than the tube's longitudinal acoustic time scale, chemical energy release will take place at approximately constant density, and the pressure will rise substantially within the charge. The ensuing pressure wave transports the released energy down the tube and into the attached turbine plenum. A wave resonance mechanism akin to that utilised in pulse jet engines supports the recharging process.

The conceptual advantage of the SEC over a pulsed detonation combustor (PDC) is that it features much lower peak pressures, shock-related dissipation, and local kinetic energy. This eases the harvesting of the potential efficiency gain from constant volume combustion. Realising nearly homogeneous autoignition in a highly dynamical flow requires very tight control of the fuelling process, however, so that the development of advanced controlling schemes will be crucial for the success of the concept.

A computationally efficient one-dimensional SEC simulation code has been implemented by Berndt [2], that can be used to test and train fuel injection control schemes . It solves the reactive Euler equations by a finite volume method and models the SEC tubes as long-stretched cylinders with axially varying cross-section and a one-dimensional distribution of the state variables along their axis. The software has been employed to study SEC in single-tube operation, e.g., to develop an efficient reduced chemical model designed to probe particular gasdynamic effects of the SEC process [3], or to investigate the sensitivity of the process with respect to various chemical parameters [7].

A full-fledged future gas turbine application will likely feature several SEC tubes firing into an annular intermediate plenum which then connects to the turbine. In a first investigation into such arrangements, multiple copies of the above mentioned quasi-one-dimensional code are coupled here through suitable transition conditions to simulate various arrangements of three SEC tubes coupled to a torus-shaped plenum. The plenum itself is again represented by the same quasi-one-dimensional code, albeit with periodic boundary conditions to model a closed torus. A mass loss from the plenum, determined from a user-defined sub-routine, mimics the turbine mass flow.

The purging and recharge of the SEC tubes relies on pressure wave resonances just as in the pulse jet combustor. As a consequence, a number of interesting questions regarding proper tuning of the interacting non-stationary gasdynamic processes in the coupled tubes and plenum arise. In particular, we study here the influences of the plenum radius and plenum length, the consequences of different arrangements of the SEC tubes along the plenum axis, and the robustness of the cycles in the tubes when one of the fuel supply pipelines is interrupted shortly.

Section 2 briefly summarises the simulation code implemented to investigate these questions. Numerical tests addressing the questions mentioned above are documented in Sect. 3. More possible applications and more complex issues for further code development as well as opportunities for improvement and extension are discussed in Sect. 4.

2 Implementation

In the following the code utilised for the simulation experiments in Sect. 3 will be described briefly with focus on the main features for the present usage: the quasi-one-dimensionality with possibility of lateral in- and outflow and the simplified chemistry model.

2.1 *One-Dimensional Model and Single-Tube Reference Operation*

The basic code that serves as our starting point has been developed by Berndt [2]. It is optimised to work with good accuracy for realistic thermochemical gas properties and to robustly handle strong shocks including detonation waves. It utilises the Harten-Lax-van Leer (HLL) numerical flux with Einfeldt's correction (HLLEM) to solve the Euler equations for a multi-species ideal gas flow. The MUSCL-Hancock or WENO reconstructions and Strang splitting for chemistry are employed to achieve second order accuracy. See detailed references in [2, 3]. Boundaries are modelled by appropriately assigning ghost cell states as usual in this type of scheme. As the SEC tubes are considered to be cylindrical with small aspect ratio, a quasi-one-dimensional approximation is adopted. Smooth axial variation of the tube radius is allowed for via inclusion of a suitable pressure source term (see, e.g. [6]).

To represent an SEC tube, the upstream boundary simulates a pressure valve that opens when the pressure in the first grid cell drops below a given compressor pressure value. When the valve opens, non-reactive compressed "air", possibly preheated to a given reference temperature, purges the tube for about half a millisecond, which we take to be the ignition delay time of the compressed and preheated gas. Subsequently, as long as the valve is still open, fuelled mixture enters the tube. A stratified charge (combustible mixture) is generated by varying the fuel mixture fraction of inflowing gas in time. This variation of the gas composition is tuned to produce a homogeneous autoignition after 0.7 ms under the conditions of standard cyclic operation of a free SEC tube not attached to a plenum.

In this standard cycle, the fresh charge covers about one third of the SEC tube length which is 0.8 m, with a species from the simplified chemistry mechanism modelling a mixture of dimethyl ether (DME) and air close to stoichiometric conditions (details can be found in [3]).

2.2 *Modelling Lateral Inflow and Outflow*

In previous simulations, e.g., in [2, 3], the downstream boundary condition modelled expansion into open space at atmospheric conditions to represent the test rig setup in related laboratory experiments, or into an infinitely large plenum at elevated pressure. Here we initiate the study of interactions between several SEC tubes and a finite size turbine plenum as depicted in Fig. 1. To this end, the code was first extended to allow for modelling lateral mass flow into (or out of) the modelled tube. These processes are represented as they would be in a multidimensional gasdynamics code

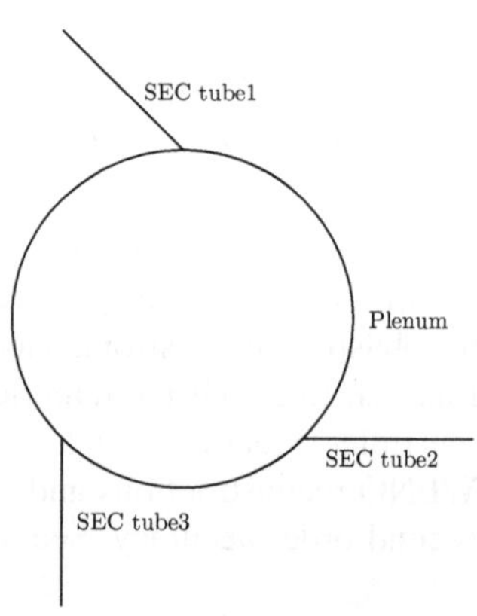

Fig. 1 The SEC tubes–plenum–configuration showing the slanted tubes with the circular plenum and the distribution of one dimensional cells

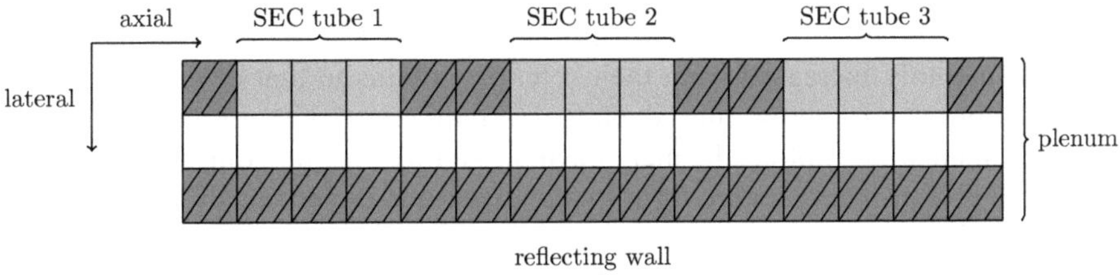

Fig. 2 Modelling the SEC tube–Plenum–Interaction. White cells are the computational domain of the plenum, dark grey cells are solid wall ghost cells and light grey cells are used to couple plenum and SEC tube

using Strang splitting to cover the space dimensions. Thus, for the plenum simulation the code is extended to two space dimensions as indicated in Fig. 2. In the second (lateral) computational direction the computational grid just covers one row of cells representing the main computational domain and two adjacent rows of cells used as dummy cells to impose boundary conditions. The closed tube walls opposite to the exits of the SEC tubes as well as between the SEC tubes on the same side are modelled in the lateral direction by the usual reflecting wall boundary conditions. These guarantee zero mass flux and proper adjustment of the wall pressure. To simulate the exit of the nth SEC tube, the flow state found in the last grid cell of the model simulation for that tube at the same time step is imposed in the corresponding dummy cells (light grey cells in the top row of Fig. 2) before processing the gasdynamic step in the lateral direction for the plenum. In turn, the plenum states averaged over the cells corresponding to the width of the attached SEC tube (three cells in the figure) are imposed in the dummy cells of the SEC tube simulations.

In this process, we allow for non-orthogonal intersection of the SEC tubes with the plenum (45° in Fig. 1). By supplying a directional (unit) vector, the user fixes an angle under which the mass flows from the SEC tubes meet the plenum stream. Consistent with the derivation of quasi-onedimensional gasdynamics models, we assume rapid lateral equilibration of all transport processes. This leads to the present model of immediate dissipation of the lateral momentum and kinetic energy upon entry of the burnt gas into the plenum. This is realised by converting the components of momentum in the last cells of the SEC tube simulations in the plenum's lateral direction (vertical direction in the Fig. 2) into internal energy, while the component of momentum aligned with the plenum axis is maintained. In our code this is done by simply setting the momentum in lateral direction to zero but keeping the energy value. This defines the SEC tube states seen by the pertinent plenum's lateral dummy cells (light grey in Fig. 2). All other cells are treated as a reflecting wall as stated above. To account for the feedback of the plenum to one of the SEC tubes, the states in every plenum grid cell that directly couples to this tube are averaged, and this state serves to impose the boundary condition in the ghost cells of the tube. In this fashion, a two-way interaction between the SEC tubes and the plenum is realised. To model a possible difference in radius between the SEC tubes and the plenum, the tubes' radii are smoothly increased from their reference radius to that of the plenum. In all simulations shown below, this adaptation of radii covered 2% of the tube length. This implementation exploits the mentioned possibility of simulating axial variation of radius via a pressure source term in the quasi-one-dimensional computational implementation.

The mass flow out of the plenum that drives the turbine is modelled in the present first approach in a very rudimentary way. The reference single tube SEC run with an opening into infinite space at given mean exit pressure generates a mean mass flux $\dot{m}$ over many cycles. Now, for n tubes attached to the plenum, a total mass of $\Delta t\, n\, \dot{m}$ is subtracted from the plenum, equidistributed over all the plenum cells. In doing so, we let the mass deducted from each cell carry the local specific momentum and energy.

The overall algorithm proceeds as follows: Every SEC tube and the plenum are distinct computational domains and treated one after the other beginning with the SEC tubes. Manually, a global time step size is fixed but before every solution step in each domain the stability criteria are tested. The chemical kinetics model (from [3]) and the gasdynamics model based on the Euler equations are then advanced via operator splitting. The computation accounts for the mass flow through the turbine by reducing the density by a user supplied amount in every grid cell scaling with spacial and temporal step width and number of SEC tubes as explained above. For the last operational step, the tubes' interactions with the plenum are determined as also explained above.

2.3 Reduced Chemical Model for SEC Simulations

The strongly simplified mechanism for kinetics developed for gasdynamic investigations of the SEC process in [3] is included here. This scheme involves three iconic species: the energy-carrying fuel, an energetically neutral (zero binding enthalpy) "radical" species whose build-up controls the onset of energy release from the fuel component, and a non-reactive product. The model was tuned to mimic the behaviour of a realistic fuel igniting in one stage as far as characteristic time scale ratios of ignition delay and excitation time (heat release rate) are concerned. Reactions are implemented as a sequence of one-step Arrhenius reactions and follow only one path from fuel to radical to product.

3 Results from Numerical Tests

In this section we consider a configuration of three SEC tubes coupled to a torus-shaped plenum as already seen in Fig. 1. The SEC tubes are slanted by 45° relative

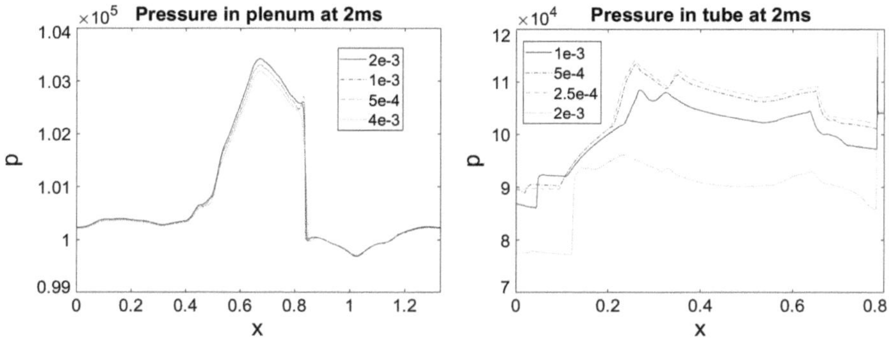

Fig. 3 Pressure in plenum (left) and SEC tube (right) over space, 2 ms after ignition with three different resolutions: fine (dashed lines), used in further investigations (solid lines) and coarse (dash-dot lines)

to the plenum axis. Their length is 0.8 m with a resolution of $dx_T = 8 \times 10^{-4}$ m and the basic radius R_T, except for the radius adjustment towards the plenum, is 0.02 m. The plenum is resolved with a grid cell size of $dx_P = 1.6 \times 10^{-3}$ m. To justify the chosen resolution a series of test runs was conducted with only one SEC tube and a plenum of 1.33 m length with a third, half and double the grid width each. For the plenum resolution Fig. 3 makes clear that even a coarser grid would have been acceptable. As more of the chemistry and dynamics take place in the SEC tubes, these domains are resolved with more grid cells. Figure 3 shows a much larger difference between the coarse and the medium than between the medium and the finer grids, indicating convergence. As the current investigations are designed to qualitatively show the effects of certain parameters on the multi-tube configuration, the medium grid width for the SEC tube seemed to be a good choice. In future works this issue will be tackled by more flexible, non-equidistant meshing.

The fixed time step is chosen to be $dt = 5 \times 10^{-5}$ ms. This value has been extracted from a series of test runs and is found to be a good choice in terms of computational effort.

The right boundary represents the plenum state where the velocity is translated into the SEC tube's coordinate system and other than x-direction velocity components are converted to inner energy just as in the opposite case. Because the SEC tube is connected to more than one plenum cell, all plenum cells which interact with the same tube are averaged to form the boundary ghost cell. The plenum is a torus with periodic boundaries.

For all but the tests in Sect. 3.1, SEC tubes and plenum simulations are started in the middle of a standard cycle of the given configuration. The initial values for the SEC tubes represent a state in the working cycle after purging and fuelling and just before the next ignition when the radical species is at its highest concentration. The fresh charge occupies about 0.3 m. All tubes are set up equally initially, so they would fire simultaneously. The plenum temperature is initialised everywhere by the

Table 1 Parameter setting in simulation experiments. R_p is the SEC tubes radius, L_p the length of the plenum and x_T the axial (mid-point) position of the SEC tubes along the plenum

Case	Param.		
	R_p (m)	L_p (m)	x_T (m)
Reference	0.08	4	1.01
			2.34
			3.68
Sect. 3.1	0.03, 0.06	4	1.01
			2.34
			3.68
Sect. 3.2	0.08	0.5	1.01
			2.34
			3.68
Sects. 3.3	0.08	4	1.93
			2
			2.07

temperature of the rightmost cell of a SEC tube averaged over four standard cycles. The pressure is chosen to be elevated to 1.1 bar.

In the following simulation experiments three variables have been tested for their influence on the SEC cycle: the radius and length of the plenum, and the positioning of the tubes along the plenum. The fourth simulation is a test case with interrupted fuel supply in one of the SEC tubes, which tests the robustness of operation of the tubes. In Table 1 the values of the tested parameters are listed. The configurations with given parameters are compared to the reference case in the respective sections.

3.1 Plenum Radius

The refilling of the SEC tube is realised via a suction wave. This is the reflection of the pressure wave from ignition at the downstream end of the SEC tube. Therefore, the radius of the plenum is expected to be crucial for the cyclic operation. We surmise that if it does not behave sufficiently similarly to an ideally open end the refilling will fail. This could be substantiated by the following simulation experiments: The initial values were selected such that the plenum is filled with compressor "air" at rest at 1 bar and 1000 K. The tube is fuelled within the first 0.32 m with radicals so that ignition is just about to begin. The 0.48 m downstream are also filled with compressor "air". The plenum radius was set to be 1.5, 3 and 4 times the SEC tube's radius, i.e., within 20 grid cells the SEC tube widens to 0.03, 0.06 and 0.08 m. Figure 4 shows that the cyclic recharge and ignition process fails for the two smaller radii. With larger radius the SEC cycle survives somewhat longer (middle panel), but even from

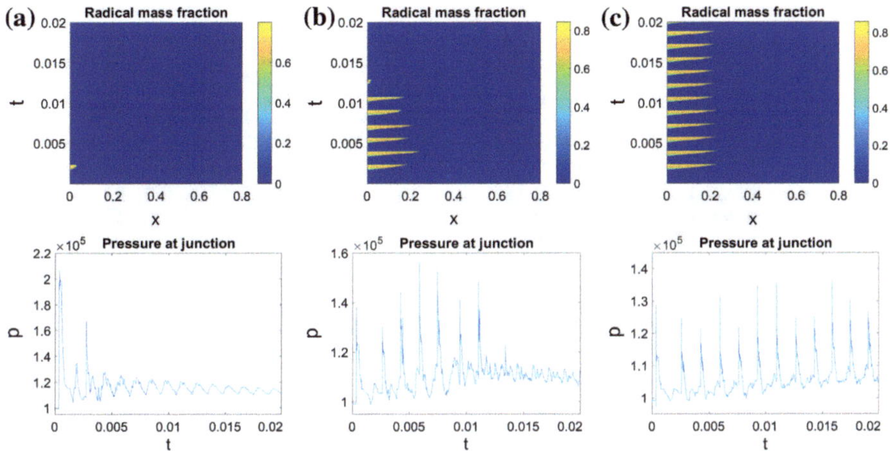

Fig. 4 Influence of plenum radius R_P on the SEC cycle. Radical mass fractions and pressure in Pa at SEC tube–plenum junction are shown versus space and time or only time, respectively, for $R_P = 0.03\,\text{m} = 1.5\,R_T$ (**a**), $R_P = 0.06\,\text{m} = 3\,R_T$ (**b**), and $R_P = 0.08\,\text{m} = 4\,R_T$ (**c**), where R_T is the SEC tube radius

the first recharge the stable (right) and the unstable configurations (middle and left) show markedly different behaviour. The stable cycle takes in a larger total load of fresh gas, and the cycles are repeating robustly. We conclude that a plenum radius of 0.08 m (4 times the SEC tube radius) is sufficient to stabilise the SEC process. This will be the configuration used throughout the following simulations.

3.2 *Length of Plenum*

Here we study the influence of the plenum length, fixing the plenum–to–SEC tube radius ratio to 4. Results from two simulations are represented in Fig. 5 corresponding to plenum lengths of 0.5 m and 4 m, respectively. The SEC tubes are operating nearly independently of this parameter. The most interesting change can be seen in the pressure field of the plenum. Especially when comparing the pressure over space at a fixed point in time in the second row of Fig. 5 one can see the smooth structure in the shorter plenum. A clear wave with three maxima developed. This is due to two effects: The most obvious reason is that the ratio of SEC tube radius to plenum length is smaller with smaller plenum and thus the combustion in the tubes raise the pressure in a broader space interval compared to the plenum length. The more interesting reason is that the configuration shown is close to resonance of plenum and SEC tubes. Therefore, a pressure peak from a combustion in the SEC tube hits the traveling pressure wave in the plenum around its maximum.

A future study will aim at investigating the effects of resonance on the SEC process and how they could be exploited. In the following simulation experiments we fix the

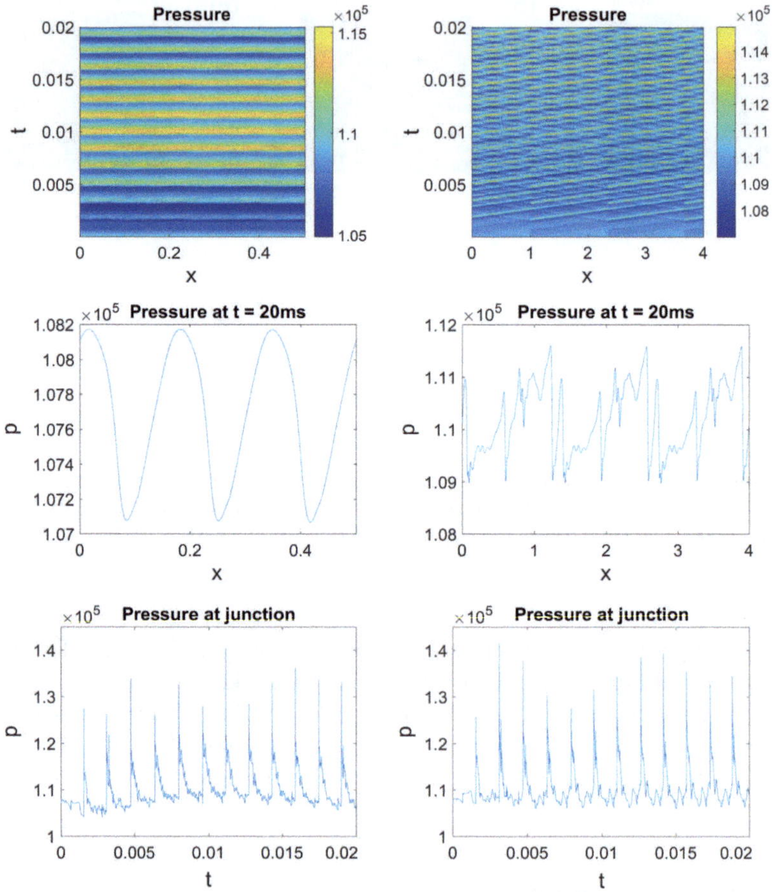

Fig. 5 Influence of plenum length (0.5 m left and 4 m right) on the plenum's pressure field. The figures show the pressure in Pa in the plenum over space and time (first row), at the simulations last time point t = 20 ms over space (second row) and at the junction between SEC tube 1 and the plenum over time (third row)

plenum length to 4 m. For a reference case the very special resonance is not desired as it could shadow the effects of the parameter we wish to study.

3.3 Positioning of SEC Tubes Along the Plenum

For the arrangement of SEC tubes along an annular plenum, an equidistant distribution might seem most natural at a first glance. Nonetheless, especially with the results of Sect. 3.2 in mind, we might expect an asymmetric arrangement to enforce the development of clearer and smoother pressure waves. This supposition seems

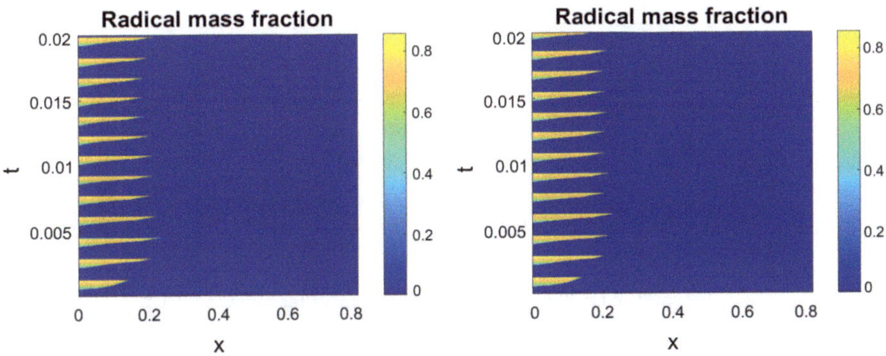

Fig. 6 Influence of SEC tube positioning along the plenum on operation cycle. Left: tube 3 in the bundled case (the one with the highest firing frequency); right: tube 1 in the symmetric case (cycles in the other tubes are equivalent)

to be true as the simulation experiments shown in Figs. 6 and 7 reveal though of course the results differ qualitatively. The current tests represent the extreme cases of equidistantly arranged and very closely bundled SEC tubes with a distance of 0.072 m (~2% of the plenum length) between neighbouring tubes. Figure 6 displays the fuelling cycles mirrored by the radical mass fractions until time $t = 20$ ms in the third bundled SEC tube which is the one with the biggest difference to the reference case and in one of the tubes in the symmetric reference case—the others are equal. Surprisingly, the asymmetric SEC tubes have a slightly higher firing frequency although there are small differences between the bundled tubes. On a longer time scale and with more combustion chambers this could have an important effect on the SEC's efficiency.

The equidistant positioning leads to a pressure field in the plenum (right panel of Fig. 7) with rather fine structures which will result in more homogeneous distributions when turbulent transport is accounted for. In the bundled case we find higher amplitudes intensifying over time and more coarse-grained patterns (left panels of Fig. 7). These could be useful for restarting a shut down SEC tube utilising suction waves in the plenum passing the tube's exit. In-depth investigations of different tube arrangements are to follow in a future study. Nonetheless, within the given time range both configurations work robustly.

3.4 *Interrupted Fuel Pipeline*

The preceding tests were conducted to find a robust configuration to run the SEC as smoothly as possible. In this last investigation we test this robustness. For the time interval of the fourth cycle (5.376–6.99 ms) only compressed "air" without fuel charge is made available for SEC tube 1. The second and third tube keep operating

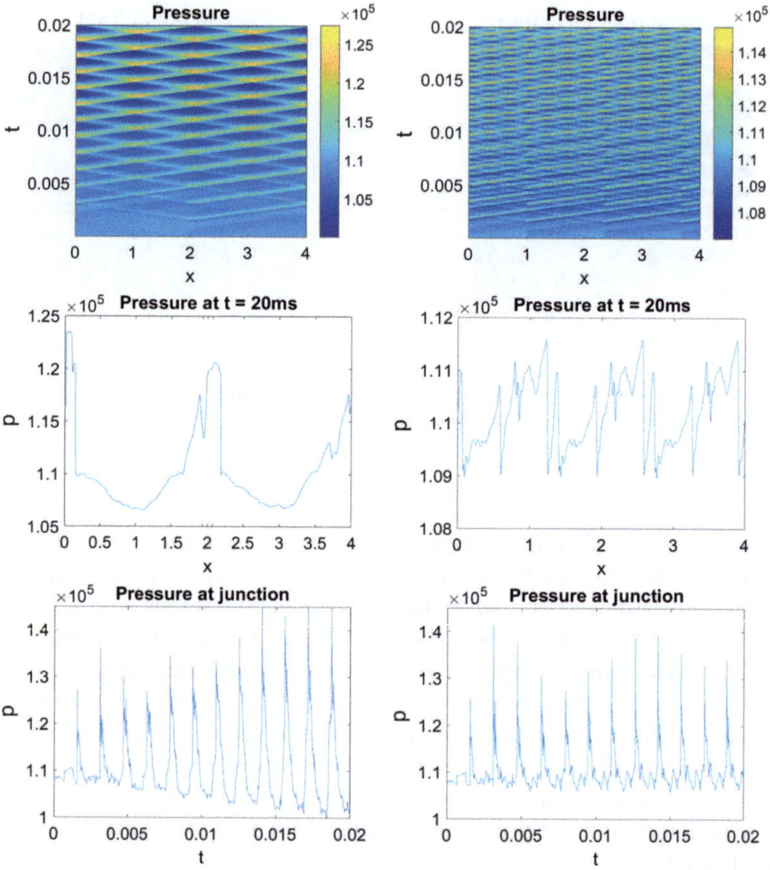

Fig. 7 Influence of SEC tube positioning pressure in the plenum. left: bundled tubes; right: equidistant tubes. The figures show the pressure in Pa in the plenum over space and time (first row), at the simulations last time point t = 20 ms over space (second row) and at the junction between SEC tube 1 and the plenum over time (third row). Please note the different scales of amplitude in the different cases. Equal scales would not reveal patterns in equidistant setting

with minimal disturbances. Hardly visible differences do occur, recognisable when comparing these tube to each other and the undisturbed reference case. This is a consequence of the change in structure in the plenum's pressure field. Unexpectedly, even tube 1 restarts the combustion after the interruption. The cyclic burning is reestablished though unstable. Until now it is unclear whether the combustion will stabilise again over time or die off. This will be the subject of further examinations in the future. Nonetheless, another point can be made for this investigation. In Fig. 9 results from the same test with a slightly different interruption time interval (5.36–6.92 ms) are shown. Here tube 1 restarts its cyclic combustion in a stable way. So obviously, there is a tolerance for interruption of fuel supply of about 1.5 ms probably also depending on the onset of the disturbance. One aim for the future will be to

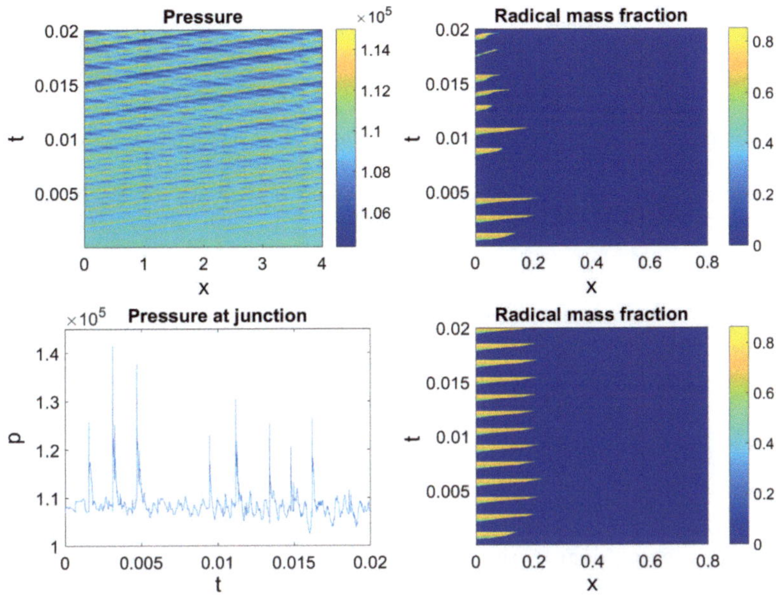

Fig. 8 Interruption of fuel pipeline in tube 1 for time interval 5.376–6.99 ms. The upper figures show the plenum pressure in Pa (left) and the radical mass fraction in tube 1(right). The lower figures display the pressure in Pa at tube 1's junction to the plenum (left) and the radical mass fraction tube 2 (right)

discover the parameter influencing this tolerance interval to find even more robust configurations and to learn how the restart of a failed tube can be positively influenced by suitable controls.

4 Conclusion and Outlook

The previous section has demonstrated the value of the possibility to simulate the highly complex processes going on in SEC tubes coupled to a turbine plenum. We have found interesting hints about what might affect the efficiency and robustness of working cycles and to what extent. Surely, the volume of the plenum must be chosen carefully as we have seen in Sects. 3.1 and 3.2. For the construction of a SEC gas turbine a lower boundary should be found for the studied variables. Although we still need to find a good way to reliably restart a tube after misfiring we can hope for the disturbed tube to reestablish operation and be positive about the others which will keep working nonetheless.

Everything that directly influences the pressure field of the plenum can affect the SEC as can be concluded from the simulation experiments. But not only the plenum configuration and arrangement of the SEC tubes are essential. There are

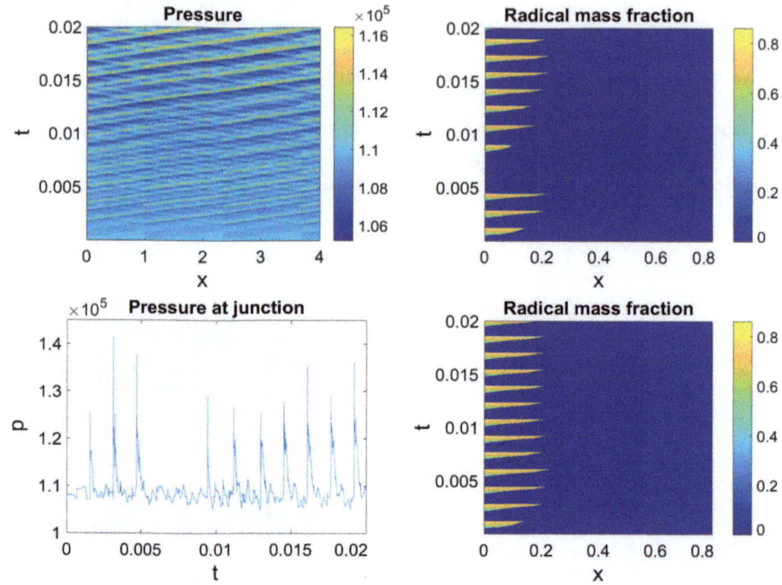

Fig. 9 Interruption of fuel pipeline in tube 1 for 5.36–6.92 ms. The upper figures show the plenum pressure in Pa (left) and the radical mass fraction in tube 1(right). The lower figures display the pressure in Pa at tube 1's junction to the plenum (left) and the radical mass fraction tube 2 (right)

more variables to study such as the firing sequence of the tubes, the amount of fuel burnt in every cycle and of course the fuel itself. Other interesting aspects have not yet been investigated but will be the object of future code development and research, such as more realistic representations of the turbine's characteristics, chemical kinetics, molecular and turbulent transport. Nonetheless, as it is today, the coupled quasi-one-dimensional simulation code can already be used by control engineers for the development and testing of controlling algorithms, and it provides important hints for the design of experimental test rigs in real-world experiments.

Acknowledgements The authors gratefully acknowledge support by Deutsche Forschungsgemeinschaft (DFG) as part of the Collaborative Research Center CRC 1029 "Substantial efficiency increase in gas turbines through direct use of coupled unsteady combustion and flow dynamics", project A03.

References

1. Anand, V. et al.: Dependence of Pressure, Combustion and Frequency Characteristics on Valved Pulsejet Combustor Geometries, Flow Turbulence and Combustion (2017)
2. Berndt, P.: Mathematical modeling of the shockless explosion combustion, PhD thesis, Freie Universität Berlin (2016)
3. Berndt, P., Klein, R.: Modeling the kinetics of the shockless explosion combustion. Combust. Flame **175**, 16–26 (2017)

4. Bobusch, B.C., Berndt, P., Paschereit, C.O., Klein, R.: Shockless explosion combustion: An innovative way of efficient constant volume combustion in gas turbines. Combust. Sci. Technol. **186**(10–11), 1680–1689 (2014)
5. Lee, J.H., Knystautas, R., Yoshikawa, N.: Photochemical initiation of gaseous detonations. Acta Astronaut. **5**(11–12), 971–982 (1978)
6. Shapiro, A.H.: The Dynamics and Thermodynamics of Compressible Fluid Flow. Ronald Press, vol. 1, pp. 1953–54. New York (1953)
7. Zander, L., Tornow, G., Klein, R., Djordjevic, N.: Knock control in shockless explosion combustion by extension of excitation time. AFCC 2018 (same volume) submitted
8. Heiser, W., Pratt, D.T.: Thermodynamic cycle analysis of pulse detonation engines. J. Propuls. Power **18**, 68–76 (2002)

Part VI
Unsteady Cooling

Experimental Study on the Alteration of Cooling Effectivity Through Excitation-Frequency Variation Within an Impingement Jet Array with Side-Wall Induced Crossflow

Arne Berthold and Frank Haucke

Abstract The influence of in-phase variation of the excitation frequency of a 7 by 7 impinging jet array between $f = 0$ and 1000 Hz on the cooling effectivity is investigated experimentally. Liquid crystal thermography is employed to measure a 2-dimensional wall-temperature distribution, which is used to calculate the local Nusselt numbers and evaluate the global and local heat transfer. The cooling effectivity of the dynamic approach is determined by comparison with corresponding steady blowing conditions. The results show that the use of a specific excitation frequency allows a global cooling effectivity increase of more than 50%.

Keywords Heat transfer · Experimental · Internal cooling
Dynamic impingement cooling · Crossflow · Pulsed blowing

1 Introduction

The Collaborative Research Center "SFB 1029" is focused on the overall efficiency enhancement of gas turbines. The classical way to improve the overall gas turbine efficiency is to increase the turbine inlet temperature as well as the turbine pressure ratio. These specific approaches have been implemented over the last decades. Therefore, until today turbine inlet temperature has increased constantly, but also the divergence to the maximum permitted material temperatures. Due to the increase in turbine inlet temperature, convective cooling concepts become more relevant for the design of modern gas turbines or aero engines. Thereby, modern turbine cooling strategies are based on the combination of high-temperature proofed super alloys or

A. Berthold (✉) · F. Haucke
Department of Aeronautics and Astronautics, Chair of Aerodynamics,
Technische Universität Berlin, Marchstr. 12-14, 10587 Berlin, Germany
e-mail: Arne.Berthold@tu-berlin.de

R. King (ed.), *Active Flow and Combustion Control 2018*,
Notes on Numerical Fluid Mechanics and Multidisciplinary Design 141,
https://doi.org/10.1007/978-3-319-98177-2_21

ceramic materials, film cooling for local hot-gas temperatures higher than 1600 K and internal cooling concepts for temperatures between 1300 and 1600 K [1]. Resulting from the steady increase in temperature modern turbines are already operating at the temperature limit of the coating materials. Thus, the logical conclusion is that a further increase in overall turbine efficiency is possible if the efficiency of the cooling concepts is improved as well. A second approach to increase the overall turbine efficiency is to switch the constant-pressure combustion to a constant-volume combustion. Following this approach, the SFB 1029 focuses on classical pulsed detonation [2] and on a new shockless explosion concept [3, 4]. In both cases the combustion process is highly unsteady and induces periodic pressure and temperature changes, which influence the flow characteristics of all gas turbine components. On the one hand, this leads to a higher turbine pressure ratio combined with an increased turbine inlet temperature. On the other hand, turbine inlet conditions can be kept constant, which leads to a reduced number of compressor stages due to the increase in pressure ratio through the combustion process. However, the cooling of the turbine blade is an important limiting parameter and therefore it is necessary to develop improved cooling mechanisms. One starting point for the improvement is the already implemented internal impingement cooling. Steady impinging jets feature high local heat-transfer coefficients compared to standard convective cooling inside turbine blades. After impinging on the hot inner surface, the cooling air mass flow is directed to the trailing edge and is discharged to the main hot gas flow. Thereby, upstream jets generate a cross flow that superimposes downstream impinging jets. The geometrical configuration, including nozzle diameter, nozzle distance, nozzle arrangement and impingement distance as well as Reynolds number of impinging jets, are important influencing parameters, which have been investigated by Florschuetz et al. [5, 6], Weigand and Spring [7] and Xing et al. [8]. To improve this well-established cooling mechanism, one part of the SFB is focused on research and development of dynamically forced impingement jet arrays. Due to the generation of strong vortex structures and their interactions with adjacent ones, the local convective heat transfer on the target surface can be enhanced. Thereby, the efficient exploitation of cooling air mass flow, typically originated from the high-pressure compressor, can be maximized. First experiments with a single forced impingement jet were performed by Liu und Vejrazka [9, 10]. They stated that the forcing of the impingement jets can affect the heat transfer in the wall jet region while the heat transfer within the stagnation area is almost uninfluenced. Additionally, they pointed out that this result is dependent on the nozzle to impingement plate distance. An additional advancement is the mixing effect due to the interaction between jet and environment studied by Hofmann in 2007. The mixing can reduce the jet velocity as well as the heat transfer on the target plate, especially for large impingement distances. The smaller the impingement distances, the less is the mixing effect. The heat transfer can be enhanced if the Strouhal number is of the order of the turbulence magnitude. Thereby, a threshold Strouhal number of $Sr_D = 0.2$ was determined [11]. These results correspond with the findings of Gharib et al. from 1998 [12]. Gharib defines a formation number $t^* = \frac{(u_p \cdot t)}{d}$, which describes the generation of high-energy ring vortices in dependency of the exit velocity of a nozzle u_p, the process time t and the nozzle diameter d. Janetzke [13] interpreted the formation number as the reciprocal of the Strouhal number. Therefore, the limits Gharib introduced to produce ring vortices with maximized size, vorticity and amplitude can be linked to the work of Herwig et al. [14], Middleberg et al. [15] and Janetzke et al. [13]. They describe the production of very strong vortices using square pulses. This enforces periodically strong local and temporal velocity gradients and thus maximizes local convective heat transfer. Influenced by the actuator characteristic there is a dependency between the enhancement of local Nusselt number and the combination of Strouhal number and amplitude. The possible combinations of geometrical and dynamic parameters is very large. The characteristics of an actuation system plays an important role as well. Thereby, the impact of dynamically forced impinging jets on the local heat transfer in the stagnation and wall jet zone needs to be studied in detail.

The present study is focused on the experimental investigation of the local convective heat transfer of an array of 7 by 7 dynamically forced impinging jets with superimposed crossflow. In particular, the study investigates the local convective heat transfer as a function of the excitation frequency and the impingement distance as well as the Reynolds number. The mayor focus thereby is the maximization of the local convective heat transfer, ergo the cooling effectivity inside of the turbine blade.

2 Experimental Setup

The basic experimental setup is schematically displayed in Fig. 1. Except for some minor changes the setup is comparable to previous work [16, 17]. The test rig allows a detailed flow field study under a 7 by 7 impingement jet array on a flat plate with superimposed crossflow. In this setup the crossflow is induced by side walls, which channels the accumulated mass flow from all nozzles towards one exit direction. Thereby, the normalized impingement distance H/D (normalized by nozzle Diameter D) is defined through the height of the crossflow frame. Consequentially, the variation of the impingement distance implies the changing of the crossflow frame.

The nozzles inside of the cooling array are equivalent to the work conducted by Janetzke [13] and consist of a simple drill hole with an exit diameter of $D = 12$ mm. The normalized spacing between two nozzles in every direction is $S/D = 5$. The length over diameter ratio for each nozzle is $L/D = 2.5$. In total the nozzle plate is equipped with 49 individual nozzles, which have an inline arrangement consisting of seven rows in each line.

Seven mass flow control units in cooperation with an in-house compressed air system are providing the required amount of air mass flow with an overall accuracy of 0.1–0.5%. Each of the seven pressure support lines is feeding an individual air divider, which is supporting one row of nozzles transverse to the flow direction. To implement a dynamic forcing, each individual nozzle is equipped with a fast switching solenoid valve. The standard valve parameters are defined as: maximum normalized volume flow rate: $V_N \leq 160\, l_{\mathrm{N}}/\mathrm{min}$ and maximum switching frequency:

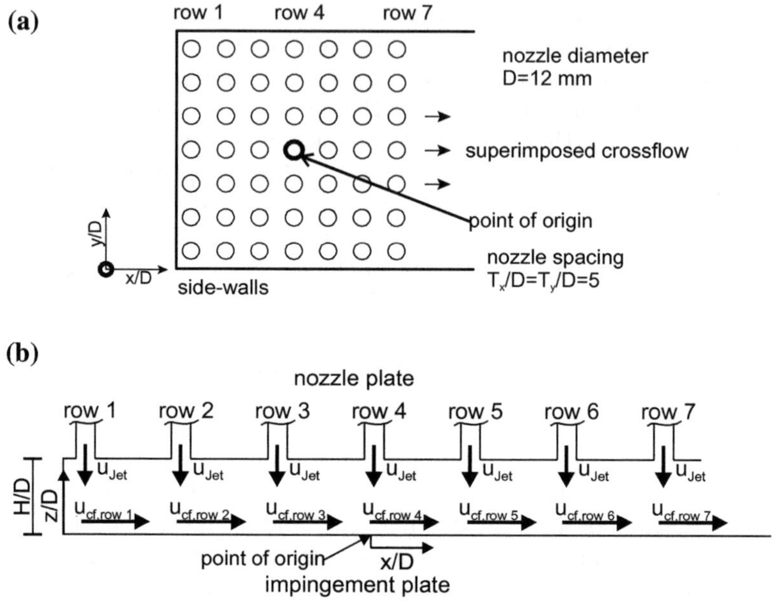

Fig. 1 **a** Schematic experimental setup **b** Schematic origin of u_{cf}/U_{Jet}

$f \leq 1000$ Hz. Since each solenoid valve can be controlled individually, a vast range of possible parameter set-ups can be investigated with this testing rig. The presented data is acquired for the entire frequency range of the solenoid valves. The frequency variation is performed for three impingement distances ($H/D = 2, 3, 5$) and three Reynolds numbers ($Re_D = 3200, 5200, 7200$) at each impingement distance. The superimposed crossflow for all experiments is generated by channelling the entire cooling massflow into one direction. Subsequently, the average crossflow velocity $U_c f$ is increased with every row of impingement nozzles until it reaches its maximum value behind the last row. This concept is comparable to the design of a turbine blade in which the cooling massflow is feeding the superimposed crossflow as well. Figure 1b displays the schematic crossflow velocity increase $U_{cf,row1...7}$ while the exit velocity for each nozzle u_{Jet} is kept constant. Due to the fact that the used cooling massflow, is equivalent to the crossflow massflow the quotient $\frac{U_{cf,rowx}}{u_{Jet}}$ is linear increased with every row of nozzles while it stays constant for different Reynolds numbers at a specific impingement distance H/D. Variation of the impingement distance changes the cross section of the crossflow channel. As a result, the velocity quotient is inverse proportional to the impingement distance if the nozzle Reynolds number is kept constant. Table 1 displays the impingement nozzle position depending quotient for all tested cases.

The implemented measurement method for all the presented data is liquid crystal thermography (LCT). The thermochromic liquid crystal foil (Hallcrest “R35C5W”) has a calibrated measurable temperature range within $T = 35\ \ldots\ 53°$ C. The color

Table 1 Impingement distance dependent increase of crossflow velocity inside of the array

H/D	$\frac{u_{cf,row1}}{U_{Jet}}$ [%]	$\frac{u_{cf,row1}}{U_{Jet}}$ [%]	$\frac{u_{cf,row1}}{U_{Jet}}$ [%]	$\frac{u_{cf,row1}}{U_{Jet}}$ [%]	$\frac{u_{cf,row1}}{U_{Jet}}$ [%]	$\frac{u_{cf,row1}}{U_{Jet}}$ [%]	$\frac{u_{cf,row1}}{U_{Jet}}$ [%]
2	7.9	15.8	23.7	31.6	39.5	47.4	55.3
3	5.3	10.6	15.9	21.2	26.5	31.8	37.1
5	3.2	6.4	9.6	12.8	16	19.2	22.4

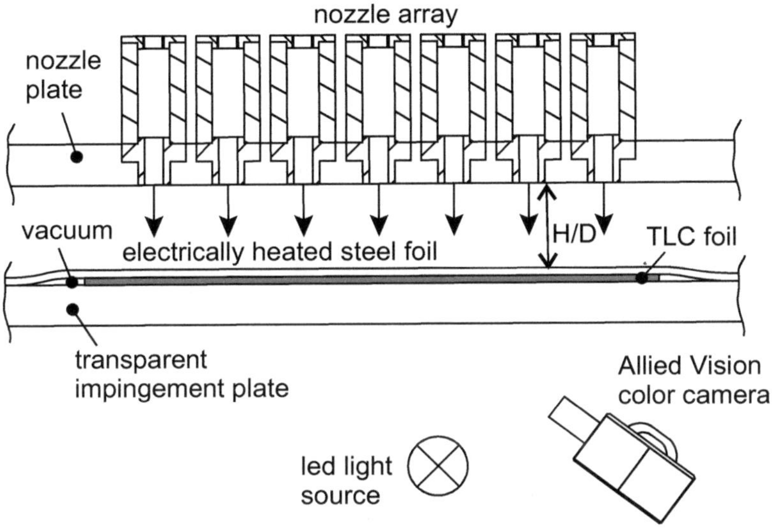

Fig. 2 Schematic setup of impingement plate

range starts with red (cold) and changes over green to blue and violet (hot). If the temperature range is exceeded, the TLC foil appears constantly black and thus temperature information is not analyzable. To minimize the measurement inaccuracy of the TLC foil, a color calibration is performed, which includes the simultaneous temperature depending acquisition of the color parameters hue, saturation and value (HSV). Given that the illumination for all measurements is kept constant, it is possible to determine a temperature band in which every temperature value has a unique combination of the three calibration parameters. Therefore, if the calibration is acquired with the necessary accuracy it is possible to reduce the overall uncertainty for the temperature depending color values to $\Delta T = \pm 0.1$ K.

Figure 2 schematically displays the construction of the impingement plate, which allows the LCT-measurement. The plate is a sandwich construction containing a thin steel foil (600 mm × 600 mm × 0.05 mm), which has the self-adhesive TLC foil attached to its rear side. The two layers are placed on a transparent glass plate (1 m × 1 m × 0.012 m) and the edges are vacuum sealed in order to press the laminate together. The blank side of the steel foil is directed to the impinging jets, while the visible side of the TLC foil is oriented to the transparent impingement plate. The steel foil is connected to a power supply with a maximum electrical output of $P_{max} = 3700$ W. Controlling the electrical current, the steel foil can be heated continuously until a thermal equilibrium for the entire test chamber is obtained. Due to the resulting wall heat flux, the TLC foil is influenced thermally and a time averaged temperature depending color distribution can be measured. The electrical energy is adjusted for each operating point, which is defined by Reynolds number Re_D and temperature range of the TLC foil. Depending on the resulting local wall temperatures, the liquid crystals reflect specific wave lengths of the light source through the glass plate back to a color camera. Through post processing it is possible to convert these RGB-colorspace values into HSV-colorspace values. After dewarping and further post processing of the raw images, a wall temperature distribution is extractable for further processing.

To determine the Nusselt number distribution on the impingement plate, it is assumed that the measured electrical power is equivalent to the emitted heat flux over the given heated area. In this consideration, the heat conduction into the adjacent wall structure as well as radiation effects are neglected. The measurement is performed when the entire system is in a state of thermal equilibrium. Hence, the wall heat flux is transferred completely into the cooling air mass flow of the impinging jets. The local Nusselt numbers can be calculated through the electrical Power P in relation to the heated area A_{heat}. The balance between wall and nozzle temperature $(T_W - T_D)$ (static nozzle temperature), and the thermal conductivity of air λ_{air} as presented in Eq. 1. To acquire the static nozzle temperature, the total nozzle temperature T_0 is measured inside of the nozzle aperture. Due to the low Mach number ($Ma = 0.03$) the relation between the total and the static temperature is around $T_0/T_D \approx 1$. Hence, the measured total temperature value can be estimated as static nozzle temperature T_D.

$$Nu_D = \frac{\dot{q}}{T_W - T_D} \cdot \frac{D}{\lambda_{air}} = \frac{P}{A_{heat} \cdot (T_W - T_D)} \cdot \frac{D}{\lambda_{air}} \tag{1}$$

If all measurement uncertainties are considered, then the overall uncertainty of the Nusselt number can be determined as $\delta Nu_D / Nu_D = 3 - 8\%$. Reproducibility studies on the presented experimental setup showed a maximal random uncertainty of below 3%.

3 Results

Figure 3a presents the calculation basis of the most important quantities, which can be calculated from the 2-dimensional Nusselt number distribution. $\overline{Nu}$ represents the mean global Nusselt number of the entire 2-dimensional field, which is used to categorize the general influence of the variable parameters. The second quantity is the crossflow oriented spatial development of the Nusselt number $\overline{Nu_x}$. This value is the average value normal to the crossflow in y-direction on every position x. For

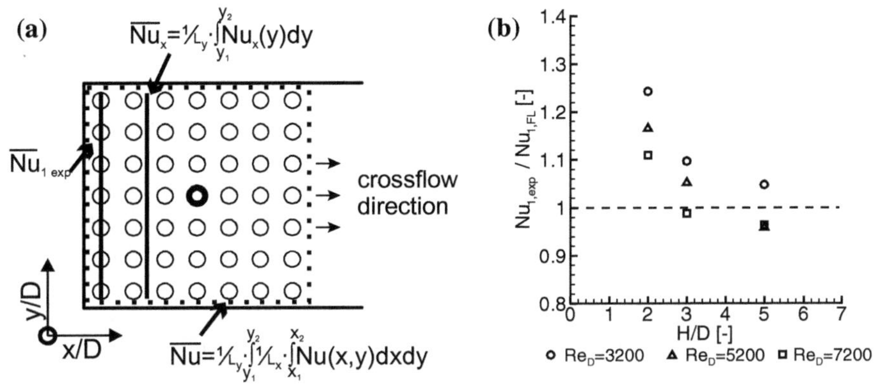

Fig. 3 **a** Schematic display of the calculation base used for the determination of the major quantities **b** Comparison computed data $\overline{Nu}_{1,FL}$ to experimentally acquired data $\overline{Nu}_{1,exp}$

the first focus of the study, the spatial development in the center of the first upstream span wise row of nozzles is calculated and designated $\overline{Nu}_{1,exp}$. The value can be compared to an estimation presented by Florschuetz in 1981 [6]. Equation (2) can be employed to create a theoretical database, which is used to compare the experimental data with.

$$\overline{Nu}_{1,FL} = (x_n/D)^{-0.554} \cdot (y_n/D)^{-0.422} \cdot (z/D)^{0.068} \cdot Re_D^{0.727} \cdot Pr^{1/3} \qquad (2)$$

This equation includes crossflow, which is induced by side walls such as depicted in Fig. 1. Thereby, the crossflow component is a result of the jet nozzle Reynolds number and the geometry formed by the nozzle plate, the impingement surface and the side walls. Additionally, this equation is only valid if a steady blowing impingement jet is assumed. The computed heat transfer coefficients can be compared to the previously introduced value of $\overline{Nu}_{1,exp}$ if all these requirements are implemented into the experimental setup. To allow such a comparison, every data set contains a steady blowing case to check if the acquired data is inside of the expected ratio presented by Florschuetz. Figure 3b shows the ratio between the experimentally acquired data and the calculated $\overline{Nu}_{1,FL}$ value for all tested impingement distances as well as nozzle Reynolds numbers.

For the impingement distances of $H/D = 5$ it is apparent that the maximum deviation, of the quotient from the expected value, is $\Delta\overline{Nu}_{1,exp}/\overline{Nu}_{1,FL} = \pm 4.2\%$. The deviation is slightly decreased as well if the nozzle Reynolds number is increased. A comparable nozzle Reynolds number dependent trend is apparent for an impingement distance of $H/D = 3$, however, both extrema are intensified. Therefore, the maximum deviation is 9.6% and the minimum deviation is −2.2%. This performance is in accordance with Florschuetz's results, who stated an equation accuracy of 11% for 95% of his experimental correlations points. In case of the impingement distance of $H/D = 2$ only the highest nozzle Reynolds number $Re_D = 7200$ is 0.2 %

below the 11% deviation line. Again, a reduction of the nozzle Reynolds number increases the deviation up to $\Delta\overline{Nu}_{1,exp}/\overline{Nu}_{1,FL} = 16\%$ for an nozzle Reynolds number of $Re_D = 5200$ and $\Delta\overline{Nu}_{1,exp}/\overline{Nu}_{1,FL} = 24\%$ for an nozzle Reynolds number of $Re_D = 3200$. This result was to be expected because the used test rig operates at geometrical conditions, which are comparable to a set-up described as critical by Florschuetz. These set-ups tend to create bigger deviations from the estimation function if the impingement distance and the nozzle Reynolds number is decreased. In general, the acquired experimental data is in accordance with the results of Florschuetz.

After the determination and validation of the steady blowing cases, the second focus of the presented study is the determination of the heat transfer coefficients with respect to the excitation frequency, impingement distance and nozzle Reynolds number. Therefore, all impingement jets were phase averaged with an excitation frequency between $f = 0 \ldots 1000$ Hz. The mean Nusselt number for all steady blowing cases $\overline{Nu}_0$ as well as every individual dynamic forcing case $\overline{Nu}$ over the entire flow field is calculated. The nozzle Reynolds number and impingement distance depending quotient $\overline{Nu}/\overline{Nu}_0$ in dependency of the excitation frequency is displayed in Fig. 4. It is apparent that the changeover from steady state blowing to dynamically forced blowing increases the global cooling effectivity $\overline{Nu}/\overline{Nu}_0$ for every test case at any tested frequency. Furthermore, it is evident that the general trend of all investigated nozzle Reynolds numbers for all impingement distances are quite similar. All these cases show an increased Nusselt number quotient in an excitation frequency band between $f = 100$ Hz and $f = 200$ Hz. In case of the impingement distance of $H/D = 2$ and $H/D = 3$ this peak value is followed by a frequency band between $f = 200$ Hz and $f = 500$ Hz, in which the value of the Nusselt number increase is nearly constant. A strong Nusselt number increase becomes observable, if the excitation frequency is increased above $f = 500$ Hz. The increase continues until it reaches the global maximum at a frequency of $f = 700$ Hz. After the maximum the Nusselt number starts to decrease with further increase of the excitation frequency until the minimum is reached at an excitation frequency of $f = 1000$ Hz.

Additional findings need to be discussed, if the impingement distance of $H/D = 5$ is included into the analysis. The first peak in Nusselt number is for a nozzle Reynolds number of $Re_D = 7200$, like all other cases at lower impingement distances, located around $f = 100$ Hz, while both smaller nozzle Reynolds numbers show the first peak at a frequency around $f = 200$ Hz. Additionally, the general trend inside of the frequency band between $f = 300$ Hz and $f = 500$ Hz is changed from a nearly constant development ($H/D = 2$ and $H/D = 3$) to a monotonically increasing one for all nozzle Reynolds numbers. Beginning at a frequency of $f = 500$ Hz the inclination is strongly increased until a global maximum is reached at a frequency of $f = 700$ Hz. Following the global maximum, the development of the Nusselt number is equivalent to previously discussed cases.

After the general determination of the global Nusselt number development, it is interesting to look at the maximum possible gain in global Nusselt number $\overline{Nu}/\overline{Nu}_0$

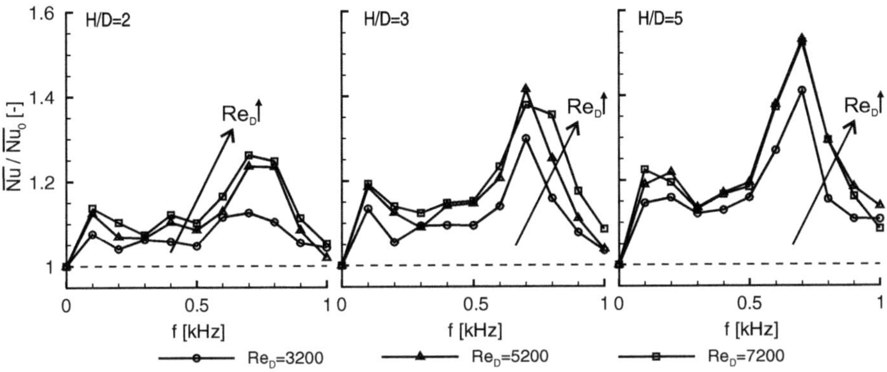

Fig. 4 Global development of the cooling in relation to excitation frequency, nozzle Reynolds number and impingement distance

Table 2 Percental increase in global Nusselt number through dynamic forcing compared to steady blowing

Re_D [−]	$\Delta\overline{Nu}_{max,H/D=2}$ [%]	$\Delta\overline{Nu}_{max,H/D=3}$ [%]	$\Delta\overline{Nu}_{max,H/D=5}$ [%]
3200	12	29	40
5200	23	41	52
7200	26	37	52

in relation to the impingement distance. For an nozzle Reynolds number of $Re_D = 7200$ the global Nusselt number maximum at an impingement distance of $H/D = 2$ is $\overline{Nu}/\overline{Nu}_0 = 1.26$, which equals an increase in cooling effectivity of $\Delta\overline{Nu}_{max} = 26\%$ compared to steady blowing. For an impingement distance of $H/D = 3$ the cooling effectivity at the same frequency is increased by $\Delta\overline{Nu}_{max} = 37\%$ and for an impingement distance of $H/D = 5$ the cooling effectivity is $\Delta\overline{Nu}_{max} = 52\%$ higher than for steady blowing. The same development is evident for the lower nozzle Reynolds numbers as well (see Table 2). Two trends are observable if all values are considered. The first one is that the maximum gain in cooling effectivity is increased with the impingement distance. A physical explanation can be found in the mixing process of the impingement jets. A steady state jet interacts with the surrounding fluid. Therefore, if the impingement distance is increased, the interaction length is increased as well. This leads to an increased mixing of the cooling fluid with the hotter surrounding fluid, which rises the resulting coolant temperature and therefore, reduces the cooling effect.

In case of dynamically forced impingement jets an approximated temporal square wave signal of the jet velocity can be assumed. The duty cycle is a direct driving parameter for estimating the peak velocity. Thereby, it equals the opening time of the fast switching valves during one oscillation period. For the present case, using a duty cycle of $DC = 50\ \%$, the peak nozzle Reynolds number $\hat{Re}_D$ is two times higher than the mean or steady blowing nozzle Reynolds number.

$$\hat{Re}_D = \frac{Re_D}{DC} = 2 \cdot Re_D \tag{3}$$

As a result, the time averaged velocity for both cases is the same if the cooling air mass flow is kept constant. However corresponding to the equation the unsteady velocity distribution shows a doubling of the peak velocity in case of pulsed blowing. The doubling is a consequence of the bisected time in which the entire fluid has to be ejected through the nozzle. In this context strong fluctuating vortex structures can be generated, which are able to perculate the crossflow with a decreased level of interaction [16]. Therefore, the vortex rings can transport more cooling fluid directly to the impingement plate to increase the cooling effectivity. Interesting at this point is that for all investigated impingement distances and nozzle Reynolds numbers the maximum increase in cooling effectivity can be reached at an excitation frequency of $f = 700$ Hz, which is in a frequency band determined by Janetzke [13] to produce mono frequent vortex rings. To determine the exact physical reason for this specific frequency, additional studies with focus on the physical differences inside of the flow field for different excitation frequencies will be conducted.

The second interesting characteristic is the nozzle Reynolds number influence on the global Nusselt number at a fixed impingement distance. In case of an impingement distance $H/D = 2$ the potential gain in global cooling effectivity is steadily increased with the nozzle Reynolds number. If the maximum value at $f = 700$ Hz is used as reference, then the cooling effectivity increase between $Re_D = 3200$ and $Re_D = 5200$ is around 11% while the increase between $Re_D = 5200$ and $Re_D = 7200$ is only around 3%. The behavior of the cooling effectivity deviation between the nozzle Reynolds numbers $Re_D = 5200$ and $Re_D = 7200$ is changed if the impingement distance is increased. While the increase between $Re_D = 3200$ and $Re_D = 5200$ is again around 12% for both greater impingement distances, it appears that the frequency depending global Nusselt number trends start to converge if the nozzle Reynolds number is increased above $Re_D = 5200$. This converging process is also dependent on the impingement distance. Hence at an impingement distance of $H/D = 3$ the trends for both higher nozzle Reynolds numbers are nearly superimposable. Smaller deviations are only noticeable in a frequency band between $f = 700$ Hz and $f = 1000$ Hz. The trends for both nozzle Reynolds numbers stays superimposable up to a frequency around $f = 1000$ Hz if the impingement distance is even further increased up to $H/D = 5$. This result implies that the gain in cooling effectivity through dynamic forcing of the impingement jets is limited to a threshold at an impingement distance dependent nozzle Reynolds number. If this threshold nozzle Reynolds number is surpassed, then there seems to be a plateau in the frequency depending cooling effectivity increase.

In addition to the global development in cooling effectivity, it is commendable to look at the local distribution of the cooling effectivity in crossflow direction, to investigate where the frequency depending increase in cooling effectivity is generated.

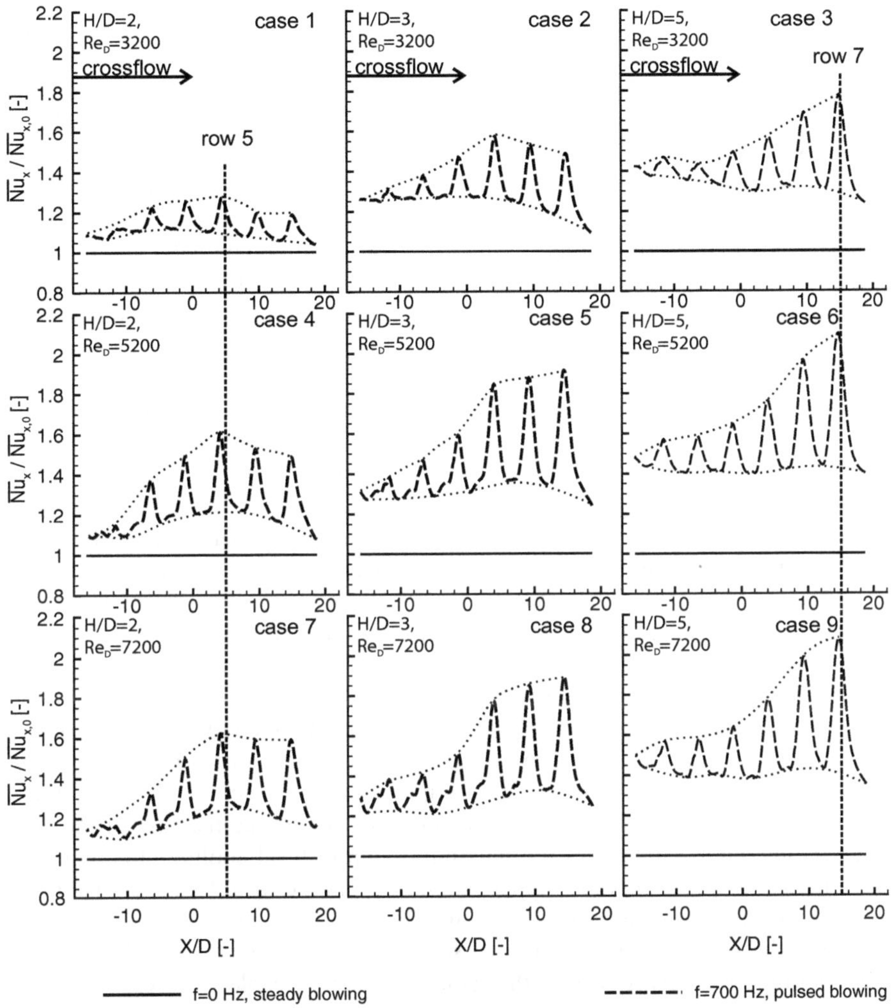

Fig. 5 Spatial distribution of Nusselt number in relation to the impingement distance and nozzle Reynolds number at f = 700 Hz

Therefore, it is necessary to analyze the crossflow oriented spatial development (x/D) of the Nusselt number $(\overline{Nu_x})$. The distribution for steady blowing $(f = 0\,\text{Hz})$ shows a crossflow-oriented reduction of the cooling effectivity, which is increased with the crossflow velocity and, therefore, with the normalized coordinate (x/D). This basic development is comparable to previously published data, which is addressing the specific phenomenon in detail [17]. A way to determine the influence of the excitation frequency on the local cooling effectivity is to normalize the dynamically forced spatial developments with the appropriate spatial development of the steady blowing case $(\overline{Nu_x}/\overline{Nu_{x,0}})$. Figure 5 displays the spatial development of the most effective excitation frequency ($f = 700$ Hz) for all investigated impingement distances and nozzle Reynolds numbers. It is evident that for a fixed nozzle Reynolds number both the lowest spatial cooling effectivity as well as the highest cooling effectivity is constantly raised if the impingement distance is increased. In addition to the raise of the levels, it is noticeable that the range between the low level cooling effectivity and the maximum value of local cooling effectivity is increased with the impingement distance as well. Furthermore, it is noticeable that the location of the maximum increase in cooling effectivity is locally shifted with the impingement distance. For the impingement distance of $H/D = 2$ the global maximum is slightly upstream the fifth row of nozzles (see Fig. 2) while at an impingement distance of $H/D = 5$ the maximum is located slightly upstream the last row of nozzles. A possible reason for this particular behaviour is that with the increase of the impingement distance at a fixed nozzle Reynolds number the crossflow velocity is decreased due to the enlargement of the cross-sectional-area. As a result, the interaction between the generated vortex rings and the resulting crossflow is reduced, which means that more uninfluenced cooling fluid is transported to the impingement plate. In contrast to the similar trends at a fixed nozzle Reynolds number the behaviour of the local Nusselt numbers is quite dissimilar if the impingement distance is kept constant for different nozzle Reynolds numbers. So at a impingement distance of $H/D = 5$ the raise of the nozzle Reynolds number increases the local cooling effectivity (see case 3 and case 9). The general trend of the local cooling effectivity is similar in all three cases, therefore, the global maximum of the cooling effectivity is around the last nozzle and the total value of the gain in cooling effectivity is increased over the downstream rows of nozzles. A slightly different behaviour is noticeable if the nozzle Reynolds number variation at an impingement distance of $H/D = 2$ (case 1, case 4 and case 7) is included into the analysis. Again, the increase of the nozzle Reynolds number increases the maximum gain in local cooling effectivity but now a nozzle Reynolds number dependent change in the local trends is visible. In case 1 the maximum gain in local cooling effectivity is slightly upstream the fifth row of nozzles and the downstream rows show a significant decrease in downstream local cooling effectivity. In case 4 the same trend is noticeable, however the reduction effect at the downstream rows is reduced. This particular trend is continued in case 7. In this particular case the values of the local cooling effectivity maxima are nearly identical to the value of the global maximum. An equal development is also noticeable in the cases at an impingement distance of $H/D = 3$. The local trend of case 2 is thereby qualitatively similar to case 4, while case 5 equals case 7. The most interesting scenario at this impingement distance is case 8. This particular case shows the same qualitative trend of the local cooling effectivity as the previously discussed cases at an impingement distance of $H/D = 5$. The comparative analysis of the nozzle Reynolds number dependent behaviours at fixed impingement distances implies that the increase of the nozzle Reynolds number increases the crossflow component. Given that the increased crossflow is increasing the mixing effects between the impingement jets and the surrounding fluid and, therefore, decreasing the local cooling effectivity. Resulting from that the potential gain in local cooling effectivity through dynamic forcing is increased. The change of the general trends between the impingement distances indicates that the ability of the vortex rings to transport cooling fluid through the crossflow seems to increase stronger than the dampening effects of the crossflow. Additionally, it seems that the nozzle Reynolds number, at which the spatial shift of the global maximum is observable, has a inverse trend to the impingement distance. Hence, to achieve the spatial shift at low impingement distances, the Reynolds number needs to be increased.

4 Conclusion

Experimental research was performed to assess the alteration of the cooling effectivity in dependency of the excitation frequency (0 ... 1000 Hz), Reynolds number ($Re_D = 3200,\ 5200,\ 7200$) and impingement distance ($H/D = 2,\ 3,\ 5$) within a 7 by 7 nozzle impingement jet array with side-wall induced crossflow. Liquid Chrystal Thermography was employed to measure the 2-dimensional wall temperature field, which was used to calculate a Nusselt number distribution. The mean global Nusselt numbers for the steady blowing cases are comparable to well established values from literature and thus the data integrity is confirmed.

The values of global and local cooling effectivity for all combinations of the test parameters were compared to appropriate steady blowing cases to evaluate the impact of dynamically forced impingement jets inside of an array with superimposed crossflow. The analysis of the global cooling effectivity demonstrates an increase for all parameter combinations if any excitation frequency is applied. More detailed results show that the basic influence of the excitation frequency inside of a frequency band between $f = 500$ Hz and $f = 1000$ Hz is identical for all tested cases. The trend is increasing until the maximum Nusselt number ratio $\overline{Nu}/\overline{Nu_0}$ and thus the maximum cooling effectivity is reached at a frequency of $f = 700$ Hz. For the most efficient combination of testing parameters, an increase in cooling effectivity of 52% was discovered. Following the maximum, a continuously decrease in cooling effectivity is observable. For frequencies up to $f = 500$ Hz the impingement distances of $H/D = 2$ and of $H/D = 3$ display an approximately constant Nusselt number ratio, while the ratio for $H/D = 5$ is slightly increasing in the same frequency band. All test cases present a first peak value in cooling effectivity in a frequency band between $f = 100$ Hz and $f = 200$ Hz.

A detailed analysis of the spatial developments in case of the most efficient excitation-frequency revealed a dependency of the location of the maximum gain in cooling effectivity on the impingement distance and the nozzle Reynolds number. Hence, for smaller impingement distances the maximum gain in local cooling effectivity is slightly before the fifth row of nozzles while for higher impingement distances the maximum is located slightly before the last row of nozzle at the same nozzle Reynolds number. It is striking that the potential gain through dynamic forcing is increased if the nozzle Reynolds number is increased at a fixed impingement distance. Additionally, the location of the global cooling effectivity maximum is shifted from slightly upstream the fifth row of nozzles to the last row. The nozzle

Reynolds number that is required for the position shift of the global cooling effectivity maximum is increased if the impingement distance is reduced.

Follow up studies will be performed to understand the physical mechanisms which are responsible for the improvement of the cooling effectivity at an excitation frequency of $f = 700$ Hz as well as to determine the physical behaviour of the dynamic impingement jets and the impingement distance depending alteration of the cooling effectivity.

Acknowledgements The authors gratefully acknowledge support by the Deutsche Forschungsgesellschaft (DFG) as part of the collaborative research centre SFB 1029 "Substantial efficiency increase in gas turbines through direct use of coupled unsteady combustion and flow dynamics" (project B03). Additionally the authors thankfully acknowledge the support of the student research assistants B.Sc. Burcu Ataseven, B.Sc. Lennart Rohlfs and B.Sc. Melik Keller during the measurement process.

References

1. Bräunling, W.J.G.: Flugzeugtriebwerke, vol. 3. Springer, Auflage (2009)
2. Gray, J., Moeck, J., Paschereit, C.: Non-reacting investigations of a pseudo-orifice for the purpose of enhanced deflagration to detonation transition. In: Roy, G.D., Frolov, S.M. (eds.), International Conference on Pulsating and Continuous Detonations, Torus Press (2014)
3. Bobusch, B., Berndt, P., Paschereit, C., Klein, R.: Shockless explosion combustion: an innovative way of efficient constant volume combustion in gas turbines. Combust. Sci. Technol. **186**(10–11):1680Ü1689, 2014, ISSN 0010-2202 (2014)
4. Bobusch, B., Berndt, P., Paschereit, C., Klein, R.: Investigation of fluidic devices for mixing enhancement for the shockless explosion combustion process. Active Flow Combust. Control (2014), S. 281Ü297. Springer, 2015, ISBN 3319119664 (2014)
5. Florschuetz, L.W., Metzger, D.E., Takeuchi, D., Berry, R.: Multziple jet impingement heat transfer characteristic - experimental investigation of in-line and staggered arrays with crossflow. NASA-CR-3217. Arizona State University, Tempe, Department of Mechanicle Engineering (1980)
6. Florschuetz, L.W., Truman, C.R., Metzger, D.E.: Streamwise flow and heat transfer distributions for jet array impingement with crossflow. J. Heat Transf. **103**, 337–342 (1981)
7. Weigand, B., Spring, S.: Multiple jet impingement—a review. Heat Transf. Res. **42**(2), 101–142 (2010)
8. Xing, Y., Spring, S., Weigand, B.: Experimental and numerical investigation of heat transfer characteristics of inline and staggered arrays of impinging jets. J. Heat Transf. **132**, 092201/1–11 (2010)
9. Liu, T., Sullivan, J.P.: Heat transfer and flow structures in an excited circular impingement jet. Int. J. Heat Mass Transf. **39**, 3695–3706 (1996)
10. Vejrazka, J., Tihon, J., Marty, P., Sobolik, V.: Effect of an external excitation on the flow structure in a circular impinging jet. Phys. Fluids **17**, 1051021-01-14 (2005)
11. Hofmann, H.M., Movileanu, D.L., Kind, M., Martin, H.: Influence of a pulsation on heat transfer and flow structure in submerged impinging jets. Int. J. Heat Mass Transf. **50**, 3638–3648 (2007)
12. Gharib, M., Rambod, E., Shariff, K.: A universal time scale for vortex ring formation. J. Fluid Mech. **360**, 121–140 (1998)
13. Janetzke, T.: Experimental investigations of flow field and heat transfer characteristics due to periodically pulsating impinging air jets. Heat Mass Transf. **45**, 193–206 (2008)
14. Herwig, H., Middelberg, G.: The physics of unsteady jet impingement and its heat transfer performance. Acta Mech. **201**, 171–184 (2008)

15. Middelberg, G., Herwig, H.: Convective heat transfer under unsteady impinging jets: the effect of the shape of the unsteadiness. J. Heat Mass Transf. **45**, 1519–1532 (2009)
16. Haucke, F., Nitsche, W., Peitsch, D.: Enhanced convective heat transfer due to dynamically forced impingement jet array. In: Proceedings of ASME Turbo Expo 2016, No. GT2016-57360 (2016)
17. Berthold, A., Haucke, F.: Experimental investigation of dynamically forced impingement cooling. In: Proceedings of ASME Turbo Expo 2017, Vol. 5A: Heat Transfe, ASME Turbo Expo 2017: Turbomachinery Technical Conference and Exposition (2017)

Effects of Wall Curvature on the Dynamics of an Impinging Jet and Resulting Heat Transfer

G.Camerlengo, D. Borello, A. Salvagni and J. Sesterhenn

Abstract The effects of wall curvature on the dynamics of a round subsonic jet impinging on a concave surface are investigated for the first time by direct numerical solution of the compressible Navier-Stokes equations. Impinging jets on curved surfaces are of interest in several applications, such as the impingement cooling of gas turbine blades. The simulation is performed at Reynolds and Mach numbers respectively equal to $3,300$ and 0.8. The impingement wall is kept at a constant temperature, 80 K higher than that of the jet at the inlet. The nozzle-to-plate distance (measured along the jet axis) is set to $5D$, with D the nozzle diameter. In order to highlight the curvature effects, the present results are compared to a previous study of jet impinging on a flat plate. The specific influence of wall curvature is investigated through a frequency analysis based on discrete Fourier transform and dynamic mode decomposition. We found that the peak frequencies of the heat transfer also dominate the dynamics of primary vortices in the free jet region and secondary vortices produced by the interaction of primary vortices and the target plate. These frequencies are approximately 30% lower than those found in the reference study of impinging jet on a flat plate. Imperceptible differences were instead found in the time-averaged integral heat transfer.

Keywords Impinging jet · Curved surface · DNS · Heat transfer · DMD

G. Camerlengo (✉) · J. Sesterhenn
Institut für Strömungmechanik und Technische Akustik,
Technische Universität Berlin, Müller-Breslau-Str. 15, 10623 Berlin, Germany
e-mail: gabriele.camerlengo@tu-berlin.de

D. Borello · A. Salvagni
Dipartimento di Ingegneria Meccanica e Aerospaziale,
Università degli Studi di Roma "La Sapienza", Via Eudossiana 18, 00184 Roma, Italy

R. King (ed.), *Active Flow and Combustion Control 2018*,
Notes on Numerical Fluid Mechanics and Multidisciplinary Design 141,
https://doi.org/10.1007/978-3-319-98177-2_22

1 Introduction

Impinging jets are employed as efficient cooling techniques in several applications. For instance, they are widely used for the cooling of gas turbines components, electronic parts and stock material during material forming processes. Compared to other heat transfer methods (without phase change), the impingement cooling offers efficient use of fluid. For example, in order to produce a given heat transfer coefficient, the flow rate required for the impingement cooling may be two order of magnitude smaller when compared with standard convective confined cooling[1] [1]. In turbine applications, impingement jets are used to cool the combustor case, the combustor can, the turbine casing and the high temperature turbine blades. Specifically for the latter purpose, typical operative temperature differences may lead to required heat fluxes of the order $10^6\,W/m^2$.

The mechanism of heat transfer associated to impinging jets is dominated by turbulence dynamics and the complete comprehension of the whole phenomenology is far to be reached, although, in view of the great interest in their applications, strong research efforts have been made. Moreover, curved cooled surfaced (such as, in turbomachinery applications, turbine vane and blade mid-chord regions) are often approximated as flat surfaces. Many studies are based on this configuration, and regardless of the flow properties (Reynolds number, Prandtl number and Mach number), some general features are shown. With focus on numerics, we recall, among others, the interesting works by [2, 3].

Cornaro et al. [4] analyzed the impingement jet flow on flat, concave and convex surfaces. Several tests were done by changing jet diameter, surface curvature, nozzle-to-plate distance, Reynolds number and turbulence intensity. They observed that low turbulence intensity at the inlet favors the development of well-organized turbulent structures in the free jet that become more and more unstable when the inlet turbulence increases. Such structures generate axial velocity oscillations leading to acceleration-deceleration of the characteristic ring vortices, which have an axial distance that reduces when the Reynolds number increases. The presence of a concave surface made the flow more unstable when compared with convex or flat plate. In fact, the flow leaving the concave surface (here extending for about 210 to 240 degrees) interacts with the primary jet, disturbing the structures here present. As for the influence of the nozzle-to-plate distance, it is noted that, increasing such distance, stronger oscillations of the stagnation point occur, contributing to an earlier breakdown of the vortices reaching the solid surface. Finally, an increase of the relative curvature leads to fewer and less stable vortex structures.

Lee et al. [5] studied the heat transfer and the wall pressure coefficient profiles on a concave surface impacted by a fully developed jet. They found that the heat flux through the plate increases with surface curvature, due to the reduction of the boundary layer thickness and the development of more robust Taylor-Görtler vortices. They also observed a change in the correlation between heat flux and Reynolds

[1]Convective confined cooling occurs when heat is transferred within confined systems such as pipes, closed conduits and heat exchangers.

number when the nozzle-to-plate distance becomes greater than 4, i.e. the impact surface is outside of the potential core of the jet. This is due to the greater turbulence level of the impacting jet. They also showed that the secondary peak of the Nusselt number profile at the wall (see also [6]) cannot be observed with a nozzle-to-plate distance of 10, whereas they are visible when this distance is set to 4. In the latter case, the magnitude of the secondary peak increases with Reynolds number and surface curvature. At a Reynolds of 11,000 and with the minimum considered plate curvature, an inflection point appears in place of the secondary peak.

Choi et al. [7] measured mean velocity and velocity fluctuations for an impinging jet flow over a concave surface. They observed that the increase of the local heat transfer rate and the resulting secondary peak in the Nusselt number profile are related to the existence and strength of velocity fluctuations.

Gilard and Brizzi [8] carried out PIV measurements of the aerodynamic of a slot jet impinging on a concave wall. They focused on the influence of the curvature. They demonstrated that for low curvature, the flow exhibits three different alternating behaviors with large modifications of turbulent variables, indicating the occurrence of strong turbulent instabilities.

Recently, [9] investigated the dominant structures in an impinging jet flow on a concave surface by means of PIV measurements. Although the heat transfer was not directly measured, the authors related such structures to the r.m.s. velocity of the jet in order to extract useful information for the estimation of the position of the secondary peak in the Nusselt number profile at the wall. Results were shown in two perpendicular planes, parallel and perpendicular to the axis of the cylindrical surface. Higher value of the r.m.s. velocity profiles were measured along the curved surface when compared with the flat one, estimating therefore relevant differences in the position of the secondary peak along the considered planes.

Yang et al. [10] analyzed the effect of different nozzle exit on the heat transfer of a slot jet impinging on a concave surface. Although the nozzle-to-plate distance were too large to approximate the impingement cooling of a turbine leading edge, the authors showed that the effect of curvature becomes more prominent as the Reynolds number increases.

Aillaud et al. [11] performed a LES of a round jet impinging on a flat plate, investigating the link between the secondary peak in the Nusselt number distribution and near-wall vortices by means of statistical analysis (PDF, Skewness, and Kurtosis of heat transfer). They found that, where the secondary peak occurs, the wall structures produce a cold fluid flux towards the impingement plate.

As concerns numerical simulations of jets impinging on curved surfaces, very few studies are available in literature. Among these, it is worth mentioning the work by [12]. They carried out a zonal hybrid LES/RANS of flow and heat transfer for a round jet impinging on a concave hemispherical surface, mainly focusing on the assessment of the numerical methods. To the best knowledge of the authors of this paper, no direct numerical simulation (DNS) studies of the configuration under analysis so far exist in literature.

The objective of this DNS study, part of a more extensive research about internal cooling in gas turbines [13–15], is the analysis of the wall curvature effects on a round impinging jet flow, focusing on the modification of heat flux through the impingement plate. To this purpose, numerical results will be compared with a reference case of jet impinging on a flat plate [16–18]. Particularly, techniques devoted to capture the behavior of dynamic systems, such as discrete Fourier transform and dynamic mode decomposition [19], will be used.

2 Computational Details

2.1 Numerical Methods

The governing compressible Navier-Stokes equations are solved in the characteristic pressure-velocity-entropy formulation, as described by [20]. This formulation has particular advantages in the definition of boundary conditions and stability of the numerical solver. Since the smallest scales of turbulence are directly solved, no turbulence model is required. As concerns the space discretization, a 6th order scheme is employed to differentiate the diffusive term, whereas a 5th is applied for the convective term. A 4th order Runge-Kutta scheme is used to advance in time.

2.2 Computational Setup

A direct numerical simulation (DNS) of a subsonic round jet impinging on a semi-cylindrical concave wall is performed using the finite difference code developed in-house at the CFD Group of TU Berlin. The computational domain, sketched in Fig. 1 and consisting of a sector of a cylindrical circular shell, is discretized on a grid with resolution $1024 \times 512 \times 512$ in the azimuthal, radial and longitudinal directions, respectively. Two walls are located at the curved boundaries; the jet issues from an nozzle in the uppermost wall and impinges on the lowermost (target plate), whose relative curvature[2] is 0.125. The grid is refined around the jet axis and in proximity of the wall, leading to a maximum variation of cell spacing less than 1% in all directions and ensuring thereby a maximum value of dimensionless wall distance y^+ at the closest grid points to the walls less than 0.6.

The jet Reynolds number (based on the nozzle diameter D and the inlet bulk velocity U_b) and Prandtl number are set to 3300 and 0.71, respectively. The ratio between the jet pressure at the inlet and the initial ambient pressure is chosen in order to ensure a fully expanded Mach number -equal to 0.8. The initial ambient temperature equals the temperature of the target plate, which is kept uniform and constant; the initial temperature of the jet is also constant and 80 K below the temperature of the plate. A laminar inlet condition is enforced by using a standard hyperbolic tangent profile.

[2]The relative curvature of the target plate is defined as the ratio between the nozzle diameter and the radius of curvature of the plate.

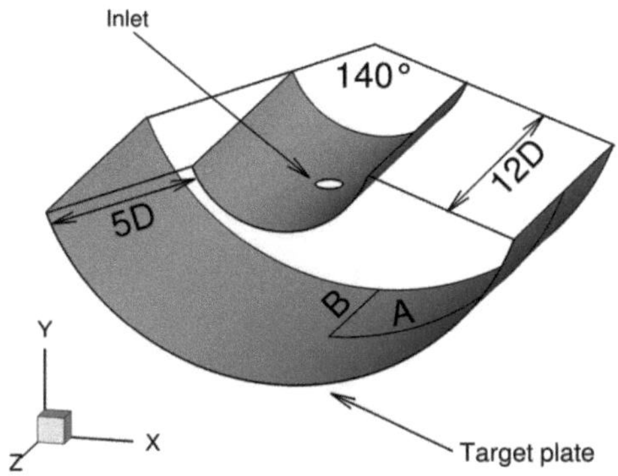

Fig. 1 Sketch of the computational domain. Walls are colored in grey, whereas inlet (orifice in the uppermost wall) and outlet are transparent. Lines A and B will be used to present results on the curved and plane side of the surface, respectively. D indicates the orifice diameter (i.e. jet diameter at the inlet)

The choice of a laminar inlet ensures that spurious frequencies are not artificially inserted into the domain [16]. In the first stage of the computation, a thin annular disturbance is applied to the inlet profiles in order to facilitate the turbulent transition.

Non-slip conditions are enforced at the walls, whereas non-reflecting boundaries are used for the outlet. Furthermore, in order to destroy the vortices leaving the domain, a sponge region is implemented for $r/D > 5$, with r the distance from the jet axis. Namely, forcing terms are added to the right-hand side of the Navier-Stokes equations with magnitude proportional to the difference between the computed and reference quantities, which were evaluated preliminary through a large eddy simulation (LES) performed on a wider domain. It is worth noting that the data computed within the sponge region are disregarded as far as it concerns the discussion of results and the evaluation of statistics.

With the exception of the plate curvature, analogous numerical and physical parameters are employed in the works by Wilke and Sesterhenn [16–18], who showed its validity as a DNS study. Since the grid spacing is, in the present setup, nearly equivalent and the curvature is not deemed to affect noticeably the Kolmogorov microscales (i.e. the smallest scales of turbulence), the validity of the present study can be also ensured.

3 Results and Discussion

In the following, results of the calculation will be shown. Statistics were collected for a time equal to approximately $350\,t_r$, where the reference time t_r is given by D/U_b. This amount of time, corresponding to about 70 times as long as it takes the flow to reach the plate from the inlet, is deemed sufficient for the convergence of first statistical moments, here presented. Since the considered geometry is not axisymmetric, the data cannot be averaged over any statistically homogeneous direction. As shown

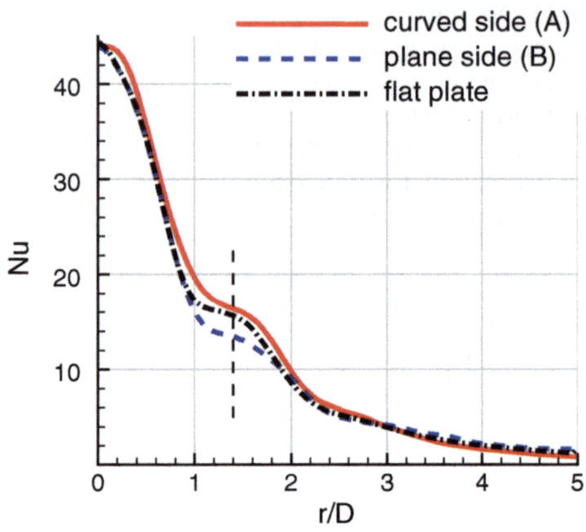

Fig. 2 Averaged local Nusselt number Nu on the target plate as a function of the non-dimensional distance from jet axis r/D. Flat plate data courtesy of Wilke and Sesterhenn [16–18]

in Fig. 1 results will be therefore presented along two[3] characteristic lines, A and B, respectively referred to as *curved side* and *plane side*.

The heat transfer intensity at the target plate may be expressed through the dimensionless Nusselt number, defined as:

$$Nu = \frac{\dot{q}D}{\lambda \Delta T}, \tag{1}$$

being $\dot{q}$ the heat flux, λ the thermal conductivity of the fluid and ΔT the difference between the temperature of the isothermal plate and the bulk temperature of the jet at the inlet. Figure 2 compares the heat transfer calculated in the present case with that presented by Wilke and Sesterhenn [16–18], in the following referred to as *reference* or *flat plate* case. As already mentioned, Wilke and Sesterhenn studied indeed the heat transfer of a jet impinging on a flat plate under analogous conditions (e.g. Reynolds and Mach numbers, nozzle-to-plate distance, velocity and temperature inlet profiles, etc.). It may be observed that Nu on the curved side is everywhere higher than on the plane side and flat plate for dimensionless distances from the jet axis $r/D \lesssim 3$, beyond which it remains below the other curves. Furthermore, the slope of the flat plate curve at the jet axis is the lowest, whereas the plane side and flat plate curves exhibit similar behavior, resulting in a total heat transfer in this region higher than for the reference case. An inflection point, which replaces the characteristic secondary peak appearing for lower nozzle-to-plate distances [3], is also observable in all the curves. It is located on both the curved and plane sides at $r/D \simeq 1.4$, at an advanced location in comparison to the reference case where it was found at $r/D \simeq 1.2$. Elsewhere, the Nusselt number distribution follows more closely the reference case: at $r/D \gtrsim 2.5$ there is no visible difference when compared

[3]On the curved side of the impingement plate, r/D is computed as the length of the arc with origin in the jet axis and running on the surface along line A. Negligible differences appear when computing r/D as the Cartesian distance from the jet axis.

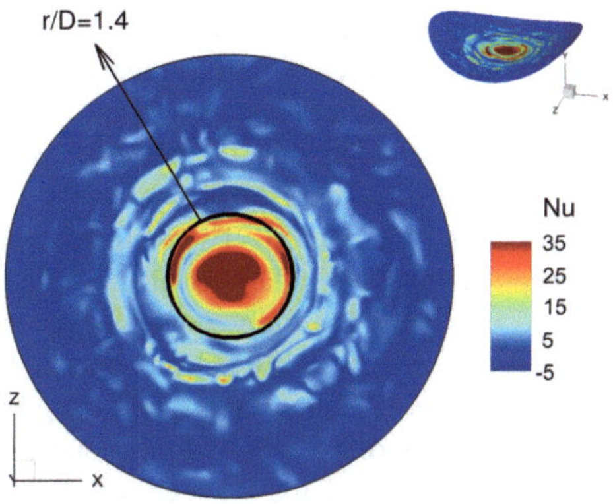

Fig. 3 Instantaneous contours of *Nu* on the target plate. At the top-right corner a 3D rendering, which additionally shows the wall curvature, is plotted

with both the curved and plane sides. The average Nusselt number within a circle of radius $r/D < 5$ differs from the reference case by just 0.03%. Consequently, it may be concluded that the integral heat transfer is not noticeably affected by the plate curvature.

The instantaneous heat transfer follows closely the evolution of turbulent structures at the wall, as shown in Fig. 3 through a snapshot of the instantaneous contours of *Nu* at the target plate. The highest heat flux occurs in a region around the plate center, where the cold jet core impacts the wall; thus, this region fluctuates as the jet core does in proximity of the impingement point. From this region, a series of annuli, characterized by high *Nu*, travels away from the jet axis in a radial direction. At a certain distance from the centre ($r/D \gtrsim 2$), each annulus loses symmetry while decreasing in intensity. The annuli are in fact directly related to the secondary vortices which originates in the wall jet region[4] and are deemed responsible for the inflection point (or secondary peak) in the average *Nu* curve [16–18]. Interestingly, narrow zones of *reverse* heat transfer appear between two following annuli; as a matter of fact, within these regions of negative *Nu* the flow is cooled down by the plate. This phenomenon, counterintuitive at first glance, can be easily seen as an effect of fluid compressibility,[5] friction and injection into the jet of hot fluid from the surrounding environment; indeed, all these physical mechanisms contribute to increase the fluid temperature, to such extent that in some regions, where *reverse* heat transfer occurs, it exceeds the wall temperature. Nevertheless, zones of *reverse* heat transfer are not observable on the average.

Figure 4a shows the time evolution of *Nu* at $r/D = 1.4$ on both the curved and plane sides of the plate, where the inflection point appears. The oscillatory trend of *Nu* confirms the motion of the high heat flux annuli on the wall. Between two successive annuli, zones of low heat flux, characterized at certain times by the aforementioned *reverse* heat transfer, are observable. Within the considered time interval, peaks of

[4]On the other hand, primary vortices appear in the free jet region.

[5]In order to recognize the relevance of fluid compressibility effects to the case in analysis, it is worth recalling that the jet fully expanded Mach number is equal to 0.8.

similar intensity recurs on the curved side, whereas peaks on the plane side exhibit at first lower and then higher intensity.

In order to gain a better insight into the oscillatory nature of the heat transfer, a discrete Fourier transform of the instantaneous Nu signal at specific locations was performed. To this end, 1822 snapshots, covering a total time interval of approximately $66\,t_r$, were used. Two different locations were selected, $r/D = 1.2$ and 1.4, representing the locations of the inflection points in the reference and present case, respectively. By analyzing Fig. 4b, it can be seen that no noticeable difference between the two chosen locations exists. Moreover, two peaks are clearly visible: the first at $St = 0.33$ and the second, of greater intensity, at $St = 0.62$ (the Strouhal number St is the dimensionless frequency, given by $f D/U_b$, with f the frequency). Differently, peaks of the frequency spectrum were found in the reference case at $St = 0.46$ and 0.92 [17], i.e. at frequencies respectively 40% and 50% higher than in the present case. This result indicates that the introduction of a generic curvature in the impingement plate might determine a shift in the peak frequencies of some instantaneous fluid properties.

A Dynamic Mode Decomposition (DMD) of the flow was performed in order to analyze the turbulent structures responsible for the heat transfer at the main frequencies found through the Fourier analysis. The same time-window length and number of samples were used (see above). This method, first introduced by [19], decomposes the flow field as a superposition of modes. It is applicable even when the dynamics of the system is nonlinear and consists in extracting DMD modes and eigenvalues from a time series of collected data. The modes are spatial fields that usually identify coherent structure in the flow, whereas the eigenvalues represent, among other things, the oscillation frequency of each mode. As a result, both modes and

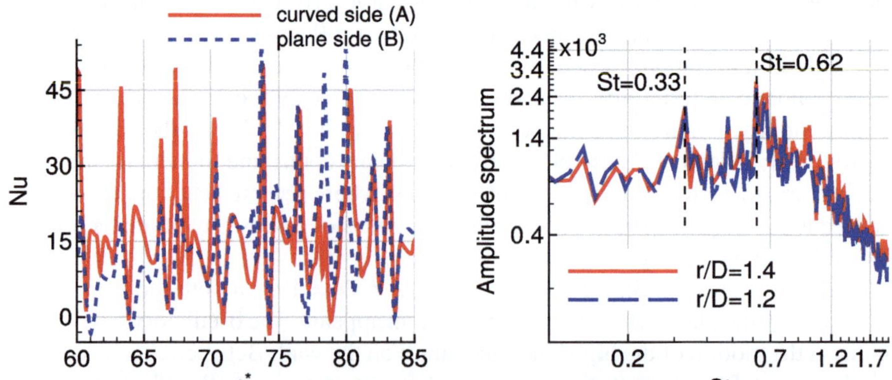

(a) Local instantaneous Nu (b) Fourier transform of Nu

Fig. 4 Local instantaneous Nusselt number at $r/D = 1.4$ versus dimensionless time $t^{\star} = t/t_r$ **a** and amplitude spectrum of its Fourier transform computed at different radii versus Strouhal number **b** Please note that the Fourier spectrum in figure is the average between the spectra computed on both sides (curved and plane) at the indicated r/D

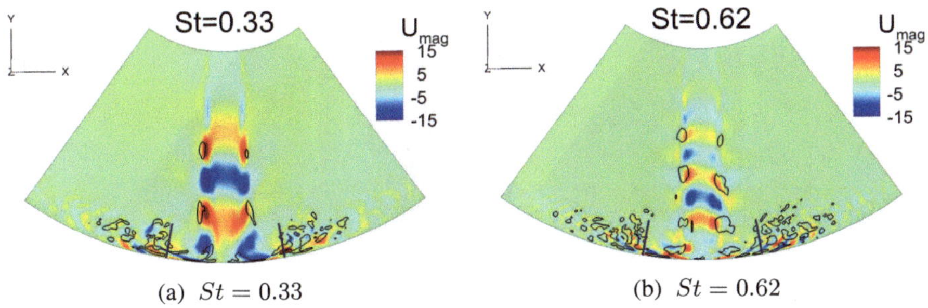

Fig. 5 Snapshots of DMD-reconstructed fields associated with different modes on a plane xy passing through the jet axis, colored by the velocity magnitude and with isolines of Q, i.e. the second invariant of the velocity gradient tensor. The blue lines orthogonal to the wall indicate the position where $r/D = 1.4$

eigenvalues describe the dynamic of an oscillatory flow field [21]. Once the DMD is performed, it is possible to reconstruct the field associated with a specific mode from the frequency back to the time domain. Two snapshots of the reconstructed velocity magnitude[6] associated with the modes oscillating at $St = 0.33$ and 0.62 are shown in Fig. 5a, b respectively. The turbulent structures formed within shear layer of the free jet region are decomposed into two distinct structures; the largest of those are associated with the lowest frequency ($St = 0.33$), whereas the smallest oscillates at the highest frequency ($St = 0.62$). In either case, the structures impinge on the wall and travel outward radially. In this stage, the largest structures lose intensity because of the impact with the target plate, whereas the intensity of the smallest enhances. This causes Nu to oscillate at $St = 0.33$ with an intensity lower than at $St = 0.62$ (see Fig. 4b). This behavior appears likewise in the refrence case, with the only, not negligible, difference in the magnitude of the characteristic frequencies.

Figure 6 shows the contours of the absolute values of the density gradient at four time instants uniformly spaced within a period associated with $St = 0.33$, with the last instant corresponding to the beginning of the following period. The snapshots highlight the life cycle of a typical Kelvin-Helmholtz instability, initially originated within the shear layer of the free jet region at about 1.5 D from the nozzle exit; traveling downwards, the same instability rolls up and grows in size, until when, in the fourth snapshot, it loses symmetry, being stretched in the vertical direction. Within the same period, roughly two of those instabilities are transported towards the wall, in the proximity of which they break down into secondary vortices. It follows that the period corresponding to $St = 0.33$ is needed by single vortices to form, travel towards the wall and break, whereas the frequency corresponding to $St = 0.62$ (roughly double than the first) is associated with the generation of secondary vorticity in the wall jet region.

[6] Note that the reconstructed velocity magnitude can be negative, since it represents the portion of the velocity magnitude that oscillates with the frequency corresponding to the extracted mode.

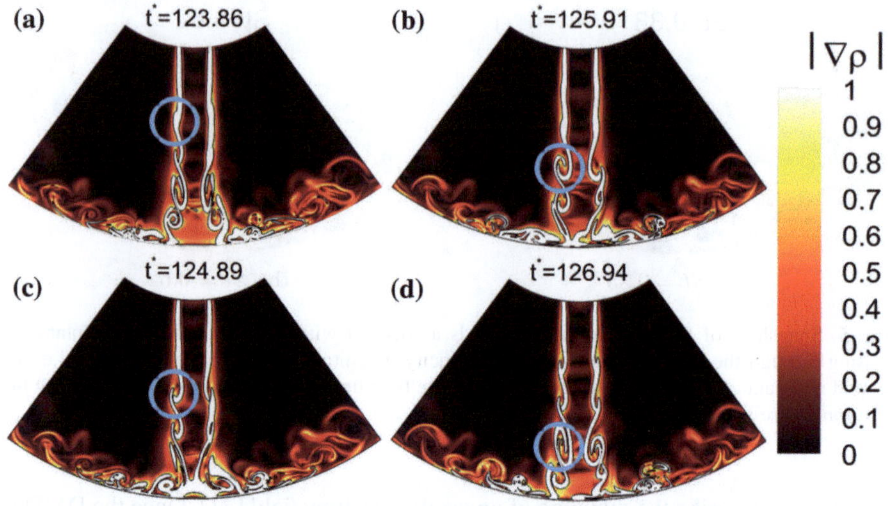

Fig. 6 Absolute value of density gradient on a plane xy passing through the jet axis at four time instants uniformly spaced within the period t_r/St, with $St = 0.33$. The dimensionless time $t^\star$ is given by t/t_r.

4 Conclusions

A round jet impinging on a curved concave surface has been investigated by means of a direct numerical simulation (DNS). The use of this method in a compressible case represents a novelty in the research field, especially in view of the interesting implications on the development of application oriented technologies, such as the internal cooling of gas turbine components. Results have been compared with those presented by Wilke and Sesterhenn [16–18], who studied a jet impinging on a flat surface under otherwise equivalent conditions.

Due to the low Reynolds number (3300), the high plate-to-nozzle distance (5 D) and the laminar inlet condition, the characteristic secondary peak in the average Nusselt number profile at the target plate could not be observed. On the contrary, an inflection point appears. This latter is located at $r/D = 1.4$, whereas Wilke and Sesterhenn found it at $r/D = 1.2$. Despite this difference, the average heat flux integrated over the surface in $r/D < 5$ does not differ in any significant manner.

The Fourier transform and dynamic mode decomposition (DMD) here performed showed, on the other hand, different constituent frequencies in the heat transfer. Indeed, the frequencies governing the generation, transport and breakup of the turbulent structures responsible for the heat transfer were found approximately 30% lower than in the case of jet impinging on a flat plate [17]. Two dominant frequencies have been observed: the lowest ($St = 0.33$) being related with the period needed by a typical Kelvin-Helmholtz instability to be transported in the proximity of the target plate, the highest ($St = 0.62$) governing instead the formation of secondary vorticity in the wall jet region.

This indicates that the dynamic response of the impinging jet flow is affected by the curvature of the target plate, which cannot be therefore disregarded in the implementation of dynamic techniques for heat transfer enhancement, such as pulsating impingement cooling [22]. On the other hand, it is legitimate, for the analyzed geometry, to approximate curved surfaces with flat when time averaged quantities are sought.

It is finally worth noting that, in spite of the interesting results observed, the physics behind the frequency-shift remains to be fully explained. To this end, future work shall address the fluid dynamic stability of the system. This will allow the study of the receptivity of a range of different parameters on the system at acceptable computational cost.

Acknowledgements The authors gratefully acknowledge support by the Deutsche Forschungsgemeinschaft (DFG) as part of collaborative research center SFB 1029 "Substantial efficiency increase in gas turbines through direct use of coupled unsteady combustion and flow dynamics" on project B04.

References

1. Zuckerman, N., Lior, N.: Jet impingement heat transfer: physics, correlations, and numerical modeling. Adv. Heat Transf. **39**, 565–631 (2006)
2. Hadžiabdić, M., Hanjalić, K.: Vortical structures and heat transfer in a round impinging jet. J. Fluid Mech. **596**, 221–260 (2008)
3. Hattori, H., Nagano, Y.: Direct numerical simulation of turbulent heat transfer in plane impinging jet. Int. J. Heat Fluid Flow **25**(5), 749–758 (2004)
4. Cornaro, C., Fleischer, A., Goldstein, R.: Flow visualization of a round jet impinging on cylindrical surfaces. Exp. Therm. Fluid Sci. **20**(2), 66–78 (1999)
5. Lee, D., Chung, Y., Won, S.: Technical note the effect of concave surface curvature on heat transfer from a fully developed round impinging jet. Int. J. Heat Mass Transf. **42**(13), 2489–2497 (1999)
6. Viskanta, R.: Heat transfer to impinging isothermal gas and flame jets. Exp. Therm. Fluid Sci. **6**(2), 111–134 (1993)
7. Choi, M., Yoo, H.S., Yang, G., Lee, J.S., Sohn, D.K.: Measurements of impinging jet flow and heat transfer on a semi-circular concave surface. Int. J. Heat Mass Transf. **43**(10), 1811–1822 (2000)
8. Gilard, V., Brizzi, L.E.: Slot jet impinging on a concave curved wall. J. Fluids Eng. **127**(3), 595–603 (2005)
9. Hashiehbaf, A., Baramade, A., Agrawal, A., Romano, G.: Experimental investigation on an axisymmetric turbulent jet impinging on a concave surface. Int. J. Heat Fluid Flow **53**, 167–182 (2015)
10. Yang, G., Choi, M., Lee, J.S.: An experimental study of slot jet impingement cooling on concave surface: effects of nozzle configuration and curvature. Int. J. Heat Mass Transf. **42**(12), 2199–2209 (1999)
11. Aillaud, P., Duchaine, F., Gicquel, L.: LES of a round impinging jet: investigation of the link between Nusselt secondary peak and near-wall vorical structures. In: ASME Turbo Expo 2016: Turbomachinery Technical Conference and Exposition, vol. 5B: Heat Transfer, p. V05BT11A002. American Society of Mechanical Engineers (2016)
12. Jefferson-Loveday, R., Tucker, P.: LES of impingement heat transfer on a concave surface. Numer. Heat Transf. Part A Appl. **58**(4), 247–271 (2010)

13. Borello, D., Salvagni, A., Hanjalić, K.: Effects of rotation on flow in an asymmetric rib-roughened duct: LES study. Int. J. Heat Fluid Flow **55**, 104–119 (2015a)
14. Borello, D., Salvagni, A., Rispoli, F., Hanjalic, K.: LES of the flow in a rotating rib-roughened duct. In: Direct and Large-Eddy Simulation IX, pp. 283–288. Springer (2015b)
15. Borello, D., Rispoli, F., Properzi, E., Salvagni, A.: LES-based assessment of rotation-sensitized turbulence models for prediction of heat transfer in internal cooling channels of turbine blades. In: ASME Turbo Expo 2016: Turbomachinery Technical Conference and Exposition, vol. 5B: Heat Transfer, p. V05BT11A012. American Society of Mechanical Engineers (2016)
16. Wilke, R., Sesterhenn, J.: Numerical simulation of impinging jets. In: High Performance Computing in Science and Engineering vol. 14, pp. 275–287. Springer (2015)
17. Wilke, R., Sesterhenn, J.: Numerical simulation of subsonic and supersonic impinging jets. In: High Performance Computing in Science and Engineering vol. 15, pp. 349–369. Springer (2016)
18. Wilke, R., Sesterhenn, J.: Statistics of fully turbulent impinging jets. J. Fluid Mech. **825**, 795–824 (2017)
19. Schmid, P.J.: Dynamic mode decomposition of numerical and experimental data. J. Fluid Mech. **656**, 5–28 (2010)
20. Sesterhenn, J.: A characteristic-type formulation of the Navier-Stokes equations for high order upwind schemes. Comput. Fluids **30**(1), 37–67 (2000)
21. Tu, J.H., Rowley, C.W., Luchtenburg, D.M., Brunton, S.L., Kutz, J.N.: On dynamic mode decomposition: theory and applications (2013). arXiv preprint arXiv:1312.0041
22. Janetzke, T., Nitsche, W.: Time resolved investigations on flow field and quasi wall shear stress of an impingement configuration with pulsating jets by means of high speed PIV and a surface hot wire array. Int. J. Heat Fluid Flow **30**(5), 877–885 (2009)

Map Estimation for Impingement Cooling with a Fast Extremum Seeking Algorithm

Benjamin Fietzke, Matthias Kiesner, Arne Berthold, Frank Haucke and Rudibert King

Abstract In many actively controlled processes, such as active flow or combustion control, a set of actuation parameters has to be specified, ranging from actuation frequency to pulse width to geometrical parameters such as actuator spacing. As a specific example, impingement cooling is considered here. Finding the optimal parameters for impingement cooling with steady-state measurements is a time consuming process because of the necessary time to reach thermal equilibrium. This work presents an algorithm for fast extremum seeking to reduce the amount of time needed. It is inspired by an Extremum Seeking Controller, which is a simple but powerful feedback control technique. The first results using this concept are promising, as the magnitude of the optimal pulse frequency for the cooling efficiency of pulsed impingement jets could be found with sufficient precision in a short period of time. The main advantages of this concept are the simple execution on a test rig, its versatility, and the fact that almost no information about the investigated system is necessary.

Keywords Impingement cooling · Extremum seeking · Map estimation

---

B. Fietzke (✉) · M. Kiesner · R. King
Institute of Process and Plant Technology, Technische Universität Berlin,
Chair of Measurement and Control, Hardenbergstr. 36a, 10623 Berlin, Germany
e-mail: b.fietzke@tu-berlin.de

R. King
e-mail: rudibert.king@tu-berlin.de

A. Berthold · F. Haucke
Institute of Aeronautics and Astronautics, Technische Universität Berlin,
Chair of Aerodynamics Marchstr. 12–14, 10587 Berlin, Germany
e-mail: arne.berthold@tu-berlin.de

F. Haucke
e-mail: frank.haucke.1@tu-berlin.de

R. King (ed.), *Active Flow and Combustion Control 2018*,
Notes on Numerical Fluid Mechanics and Multidisciplinary Design 141,
https://doi.org/10.1007/978-3-319-98177-2_23

1 Introduction

The efficiency of gas turbines is steadily being improved by increasing their pressure ratio and the turbine inlet temperature. An increased inlet temperature is acceptable because cooling mechanisms are improved and more resistent blade materials are used, such as superalloys and ceramic thermal barrier coatings [4]. However, despite significant achievements in the field of materials research, additional blade cooling concepts, such as film cooling, are needed for temperatures higher than 1300 °C. In addition to film cooling, impingement cooling inside the turbine blades leads to higher local heat transfer rates compared to conventional convective cooling.

By dynamically forcing the impingement jets, it is possible to further improve the cooling effect via the generation of strong vortex structures that possess increased convective heat transfer capability, which results in higher local convective heat transfer coefficients [7–11, 14].

In this paper, an impingement cooling test rig [3, 5] with a 7×7 impingement array is considered (see Fig. 1). One challenge for an encompassing study is the large number of possible actuation parameters. This results in a time-consuming optimization procedure to find the best combination of parameters. Traditionally, the influence of individual actuation parameters is revealed by steady-state measurements for which it is mandatory to wait until the test rig is in a state of thermal equillibrium. For the specific task considered here, this leads to a measurement time of up to 30 min for each possible combination of actuation parameters. Therefore, an alternative method

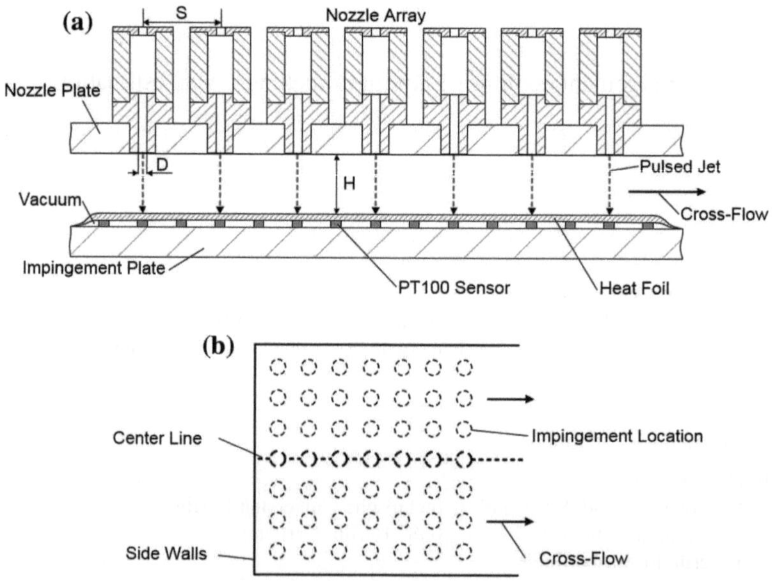

Fig. 1 **a** Schematic of the experimental setup. **b** Impingement locations with side walls and induced cross-flow from the top. Reproduced from [3]

is proposed that makes it possible to locate extrema in a shorter period of time. The presented concept is inspired by a well-known closed-loop controller, namely the Extremum Seeking Controller (ESC). Generally, an ESC drives a system to an optimal operating point without much knowledge about the system. More specifically, no mathematical model of the process is required but just the knowledge that an extremum exists. However, some estimate of its response time must be known [2]. A closed-loop ESC will find the next local optimum in a steady-state map of a system and will remain in its vicinity. While this suffices for many applications, the goal of the present contribution is to produce an estimate of the complete map, possibly comprising several local extrema. Fewer precise steady-state measurements near the located optima can then be performed to obtain detailed information about the best heat transfer conditions.

In this work, this idea will be exploited to find a relationship between the actuation parameters of a pulsed jet and the realized heat transfer coefficients. More specifically, the optimal actuation frequency f of the pulsed jets and its impact on the cooling effect is studied.

The paper is organized as follows. Section 2 introduces the experimental setup. The concept of using an ESC-like algorithm in fast map estimation is detailed in Sect. 3. Results are presented in Sect. 4, and conclusions are drawn in Sect. 5.

2 Experimental Setup

The experimental setup for this work is based on previous investigations by Haucke and Berthold [3, 5]. A schematic of the setup is given in Fig. 1a. As shown, the nozzles are arranged in a 7×7 inline array on a nozzle plate. The nozzles have an exit diameter of $D = 12\,\text{mm}$ and a distance of $S = 60\,\text{mm}$ to the next nozzles. The impingement distance between the nozzles and the target plate is $H = 36\,\text{mm}$. The target plate consists of a thin steel foil ($600 \times 600 \times 0.05\,\text{mm}$) and a wooden impingement plate ($1\,\text{m}\times\ 1\,\text{m} \times 12\,\text{mm}$). Three sides of the foil are enclosed by side walls that induce a cross-flow, which superimposes the pulsed jets and therefore reduces the efficiency of the impingement cooling.

The edges of the foil are sealed, and a vacuum is applied to press the heat foil to the impingement plate. The steel foil itself is connected to a constant current power source. Because the foil being a resistor, it serves as a heat source for the setup.

Under the center line of the impingement jets (Fig. 1b) and between the heat foil and the impingement plate, 15 Pt100 temperature sensors are integrated to measure the temperature distribution in a straight line. For each jet, a Pt100 sensor is placed under the center of the impingement location. Additional sensors are mounted in the middle between two adjacent jets.[1]

The precision of the class A Pt100 sensors is based on DIN EN 60751. Therefore, the measurement error of the sensors is lower than 0.25°C in the temperature range $T_i = \{20°\text{C} \ldots 40°\text{C}\}$ considered here. Additionally, a Pt100 sensor is placed in

[1]Thermocouples instead of Pt100 sensors would allow for an even higher bandwidth if necessary.

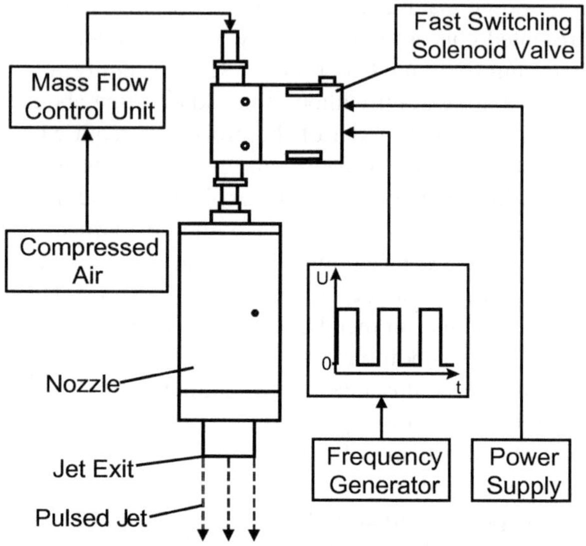

Fig. 2 Setup of a valve-nozzle combination in the 7 × 7 nozzle array [3]

the center nozzle to measure the temperature of the cooling air, $T_{\text{jet}}(t)$. In [3, 5], a thermography liquid crystal (TLC) foil is used for temperature measurements on the plate. Because measuring with a TLC foil is based on the evaluation of image analytical data, it is more difficult to obtain information about transient behavior.

To generate pulsed jets, every nozzle is equipped with a fast switching solenoid valve (*FESTO MHJ9-QS-4-MF*), as shown in Fig. 2. The permissible maximum normalized volume flow of the valves is $V_{\text{N}} = 160\,\text{l}_{\text{N}}/\text{min}$, and the maximum state switching frequency is $f_{\text{st,max}} = 1000\,\text{Hz}$. The air mass flow is supplied by an in-house compressor. A mass flow control unit is plugged in between each of the seven lines of valves and the compressor. The control unit adjusts the desired amount of cooling air and directs it to an air divider. The latter splits the flow equally into seven smaller flows for the corresponding valve line. In this setup, the operating point for the mass flow control unit is set so that the resulting averaged Reynolds number for each nozzle is $\overline{Re} \approx 7200$, calculated with respect to the mean jet velocity. The frequency of the pulsed jets is determined by a square wave signal of an Field Programmable Gate Array (FPGA) frequency generator, making it possible to precisely change the frequency with a high resolution whereby the valves open and close dynamically with a duty cycle of 50%.

3 Concept

As mentioned, the applied concept for the fast map estimation is inspired by the basic ESC algorithm, which has been successfully implemented in various systems [6, 13] and which will be described in some detail below. The main advantage of the ESC is the possibility of driving a dynamic system to an optimal operating point despite the fact that no further information is needed besides a rough estimate of the dominant time constant of the system. This will now be exploited in an open-loop manner to estimate the gradient of an unknown map. Thus, local extrema can be found. To keep the contribution compact, the explanation will be built on the task of increasing cooling efficiency, which is investigated in this paper, instead of separate treatments of the general and specific cases. Cooling efficiency depends on the actuation frequency f of the pulsed jets, that is, an optimal value for this actuation frequency is sought.

As an output of the considered system, an average temperature difference $\varDelta T(t) = \overline{T}(t) - T_{\mathrm{jet}}(t)$ is introduced. Here, $\overline{T}(t)$ is the arithmetic mean of the Pt100 measurements (see Sect. 2), while $T_{\mathrm{jet}}(t)$ is the temperature of the cooling jet. For a steady state, this difference is denoted by $\varDelta T_{\mathrm{s}}$.

The central idea of ESC is explained in Fig. 3. Assume that the system input $f(t)$ is on the left of the map's minimum and $f(t)$ is perturbed by $a_{\mathrm{f}}\,\sin(\omega t)$ with the perturbation amplitude a_{f} and the perturbation frequency ω. The output $\varDelta T_{\mathrm{s}}$ then has a shape similar to a negative sine wave, which means a phase shift of 180° in comparison to the perturbation $a_{\mathrm{f}}\,\sin(\omega t)$. The amplitude of the output is roughly the input amplitude a_{f} multiplied by the unknown local gradient of the map, $\mathrm{d}(\varDelta T_{\mathrm{s}})/\mathrm{d}f$. However, if an input is applied to the right of the minimum, the same input signal would cause an output signal similar to a positive sine wave, that is, with a zero phase shift. Therefore, due to a harmonic input perturbation, an output phase shift yields information whether a minimum exists towards lower or larger input values.

In a classical (closed-loop) ESC, the output signal would be further processed to obtain an estimate of the size of the slope of the map to then apply a gradient-based optimization that drives the input to the optimal value. For this, the excitation frequency has to be small enough to obtain a steady-state input–output relationship

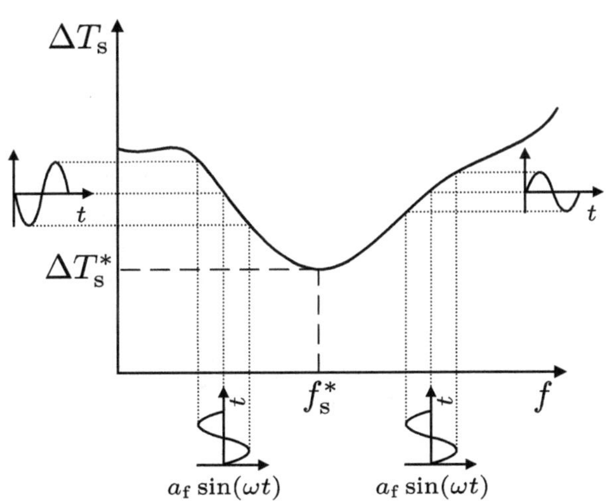

Fig. 3 Phase switch of the output $\varDelta T_{\mathrm{s}}$ in a steady-state map when moving across an extremum

[2]. In the case considered here, the loop will not be closed, but the input f will be continiously increased in addition to the harmonic perturbation to scan the complete input range. Moreover, to speed up evaluation, this scan will not be performed in a quasi-steady manner so that dynamic effects will appear. That is, additional, unwanted phase shifts between the input and output will be observed.

The concept for fast map estimation is sketched in Fig. 4. The process itself can either be described by its steady-state map or by a block in the corresponding block diagram representing the dynamic behavior of the system (see Fig. 4). Note that both blocks are unknown. In this paper, the steady-state map characterizes the gain from the input signal, which is the pulse frequency $f(t)$ of the jets, to the average temperature difference $\Delta T(t)$. When the corresponding thermal equillibrium $\overline{T}_s$ is reached, it results in a steady temperature difference ΔT_s.

To generate the input, the on–off solenoid valves are operated periodically with a frequency $f(t)$ that is ramped up $(m_f t)$ and perturbed harmonically, that is, the excitation frequency reads as follows:

$$f(t) = a_f \sin(\omega t) + m_f t. \tag{1}$$

Now, as a first-order approximation, the output of the process $\Delta T(t)$ can be described as:

$$\Delta T(t) \approx \Delta T_s + a_f \frac{d(\Delta T_s)}{df} \sin(\omega t). \tag{2}$$

If the map was linear and the excitation of the system quasi-steady, equality would result in Eq. (2). The measured temperature difference $\Delta T(t)$ is then bandpass (BP)-filtered with zero phase shift to cut off the steady part ΔT_s. For that, a fourth-order

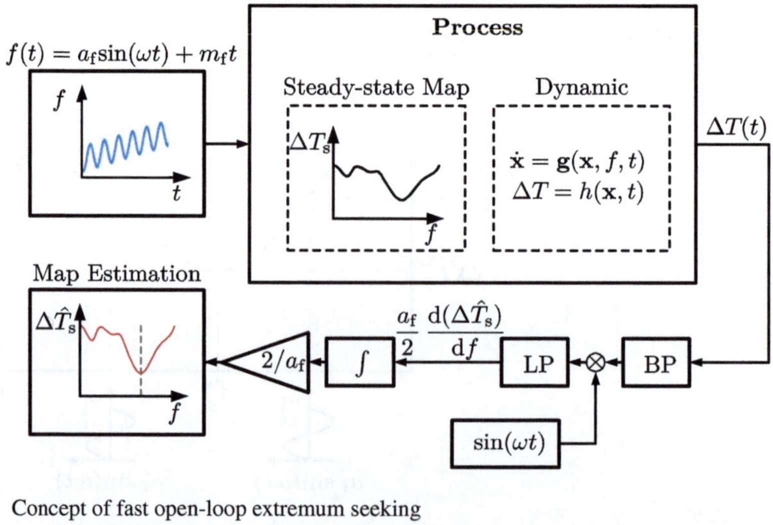

Fig. 4 Concept of fast open-loop extremum seeking

BP filter with identical upper and lower corner frequencies equal to the perturbation frequency ω is applied, as in:

$$G_{\mathrm{BP}}(s) = \left(\frac{2\omega s}{(s+\omega)^2}\right)^2, \tag{3}$$

where s is the Laplace variable. A fourth-order BP is chosen to obtain a larger roll-off of its gain for other frequencies. For zero initial conditions, the BP-filtered data can be described as follows:

$$\Delta T_{\mathrm{BP}}(t) \approx a_{\mathrm{f}} \frac{\mathrm{d}(\Delta T_{\mathrm{s}})}{\mathrm{d}f} \sin(\omega t). \tag{4}$$

After this step, $\Delta T_{\mathrm{BP}}(t)$ is demodulated with the perturbation signal. The demodulated signal $\Delta T_{\mathrm{Dem}}(t)$ can be described with the following expression, where the first term on the right-hand side will vary with time when the actuation frequency is ramped up:

$$\Delta T_{\mathrm{Dem}}(t) = \sin(\omega t)\, \Delta T_{\mathrm{BP}}(t) \approx \underbrace{\frac{a_{\mathrm{f}}}{2} \frac{\mathrm{d}(\Delta T_{\mathrm{s}})}{\mathrm{d}f}}_{\text{non-periodic}} - \underbrace{\frac{a_{\mathrm{f}}}{2} \frac{\mathrm{d}(\Delta T_{\mathrm{s}})}{\mathrm{d}f} \cos(2\omega t)}_{\text{periodic}}. \tag{5}$$

After the demodulation, the signal has a non-periodic and a periodic part with doubled frequency. The latter can be filtered out with a lowpass (LP) filter, such as a Butterworth filter. The outcome of filtering can be improved by using an acausal filter, as all calculations are performed after the experiment is finished. As a result, the LP output $\Delta T_{\mathrm{LP}}(t)$ mainly consists of the non-periodic part

$$\Delta T_{\mathrm{LP}}(t) \approx \frac{a_{\mathrm{f}}}{2} \frac{\mathrm{d}(\Delta T_{\mathrm{s}})}{\mathrm{d}f}. \tag{6}$$

The factor $a_{\mathrm{f}}/2$ can be removed easily by proper scaling. To finally obtain an estimate $\Delta\hat{T}_{\mathrm{s}} = \Delta\hat{T}_{\mathrm{s}}(f)$, the gradient $\mathrm{d}(\Delta T_{\mathrm{s}})/\mathrm{d}f$ must be integrated with respect to the frequency f. In the experiment, however, only an integration over time is possible. To that end,

$$\Delta\hat{T}_{\mathrm{s}}(f) = \frac{2}{a_{\mathrm{f}}} \int \Delta T_{\mathrm{LP}}\, \mathrm{d}f \approx \int \frac{\mathrm{d}(\Delta T_{\mathrm{s}})}{\mathrm{d}f}\, \mathrm{d}f \tag{7}$$

is reformulated as

$$\Delta\hat{T}_{\mathrm{s}}(f) \approx \int \frac{\mathrm{d}(\Delta T_{\mathrm{s}})}{\mathrm{d}f} \frac{\mathrm{d}f}{\mathrm{d}t}\, \mathrm{d}t. \tag{8}$$

The time derivative of f contains a constant and a fast harmonic part (see Eq. (1)). The latter was only introduced to obtain fast local gradient information and leads to high-frequency oscillation of the integrated result. Therefore, only the slope of the ramp m_f is used for the time derivative. Hence,

$$\Delta \hat{T}_\mathrm{s}(f) \approx m_\mathrm{f} \int \frac{\mathrm{d}(\Delta T_\mathrm{s})}{\mathrm{d}f}\, \mathrm{d}t. \tag{9}$$

is obtained. Low values of the estimated temperature difference $\Delta \hat{T}_\mathrm{s}(f)$ between the cooling jets and the heated foil indicate actuation frequencies resulting in a large heat transfer.

4 Results

At the start of every measurement series, the frequency of the pulsed jets is set to a constant value of 200 Hz to reach steady-state. The ramp then starts and continues until $f = 1000\,\mathrm{Hz}$ is reached. In Fig. 5, an example with $a_\mathrm{f} = 20\,\mathrm{Hz}$, $\omega = 0.2\,\mathrm{rad/s}$, and a ramp duration of 30 min, that is, $m_\mathrm{f} = 0.\bar{4}\,\mathrm{Hz/s}$, is displayed. The measured unsteady temperature of the cooling air $T_\mathrm{jet}(t)$ is shown in panel b). To decrease the influence of this unwanted dynamic, $T_\mathrm{jet}(t)$ is LP-filtered, resulting in smoother progress. In addition, the plate's mean temperature $\overline{T}(t)$ is displayed. To evaluate the cooling efficiency, the temperature difference between $\overline{T}(t)$ and the filtered data $T_\mathrm{jet,LP}(t)$ is used:

$$\Delta T(t) = \overline{T}(t) - T_\mathrm{jet,LP}(t). \tag{10}$$

This is shown at the bottom of Fig. 5. The lowest difference is seen for $t \approx 1550\,\mathrm{s}$ when $f \approx 770\,\mathrm{Hz}$. However, this is not the optimal frequency, as will be shown below.

As mentioned in the previous section and as a result of the non-quasi-steady excitation, the output signal $\Delta T(t)$ lags with respect to the input signal $f(t)$ due to the dynamic behavior of the process. Additionally, this lag depends on whether the process is left or right of an extremum (see discussion of Fig. 3). This can be illustrated by comparing the perturbation $\sin(\omega t)$ with $\Delta T(t)$ or, as the BP features a zero phase shift, with the normalized output signal of the BP $\Delta \tilde{T}_\mathrm{BP}(t) = \Delta T_\mathrm{BP}(t)/\max(\Delta T_\mathrm{BP}(t))$, as done at the top of Fig. 6. Note that in this Figure, t=0 marks the start of the sine ramp, so the first ~200 s to reach steady-state are cut off.

Assuming the process is left of a minimum of the unknown map, then, according to Fig. 3, the temporal minimum of the output signal should coincide with a temporal maximum of the input. As this is not exactly the case in the first part of Fig. 6 (up to

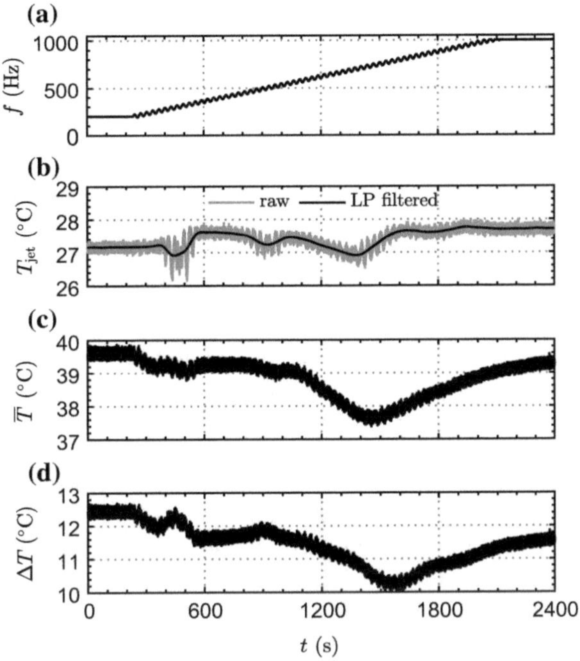

Fig. 5 **a** Ramp of sinusoidal frequency with $a_f = 20\,\text{Hz}$ and $\omega = 0.2\,\text{rad/s}$ starting at 200 Hz and reaching 1000 Hz in 30 min. Waiting times at both the beginning and the end are introduced as well. **b** Temperature of the cooling air, original and LP-filtered. **c** Averaged temperature of the heat foil's center line. **d** Difference between averaged temperature $\overline{T}(t)$ and LP-filtered jet temperature $T_{\text{jet,LP}}(t)$

90 s), the difference can be attributed to the unknown phase shift caused by the slow response of $\Delta T(t)$.

However, for the experiment shown in Fig. 6, it is obvious that the process moves to the right of the minimum somewhere in the time span from 90 to 100 s, as, following the arguments for Fig. 3, a maximal output value is now near a maximal input value. As a result, estimating the phase shift would require the information from the map, which is not available at this point. To resolve the problem, the phase shift is determined from squared signals:

$$\alpha = \sin(\omega t)^2 \tag{11}$$

$$\beta = \Delta T_{\text{BP}}^2(t), \tag{12}$$

as now both the maxima and minima of the signal will result in a maximum. The phase shift can now easily be estimated for those time frames where the process crosses an extremum on the map. To that end, the cross-correlation $R_{\alpha,\beta}(\tau)$ between α and β is calculated, where $R_{\alpha,\beta}(\tau)$ is a function of the time shift variable τ. The maximum value of $R_{\alpha,\beta}(\tau)$ determines the average time shift τ_{sh} of the process. Since α is periodic, only a frame of the period length of α is considered for this determination:

$$\tilde{R}_{\alpha,\beta}(\tau) = \frac{R_{\alpha,\beta}(\tau)}{R_{\alpha,\beta}(0)}. \tag{13}$$

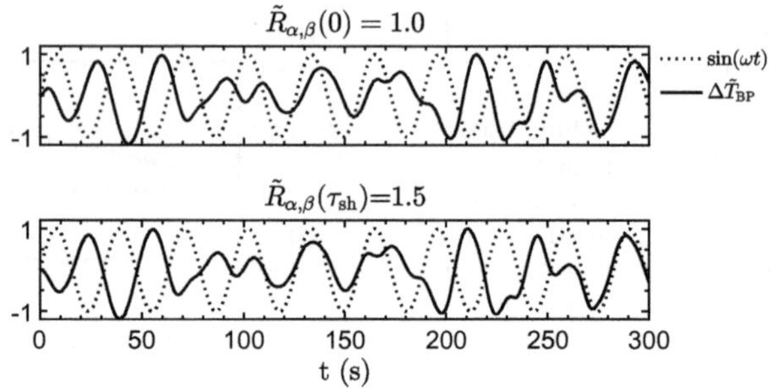

Fig. 6 Top: Phase comparison between the perturbation and the normalized BP output. Bottom: Phase comparison after phase correction. $t = 0$ describes the start of the sine ramp

At the bottom of Fig. 6, the phase-corrected signal, obtained with this procedure, is displayed. The extrema of the phase-corrected signal are better aligned with the sine extrema. $\tilde{R}_{\alpha,\beta}(\tau_{sh})$ is around 50% higher than before phase correction. Comparing minimum–minimum and minimum–maximum pairs in the time traces, the process seems to be left of a map's local minimum for $0 < t < 90$ s and $190 < t < 250$ s but right of a map's local minimum for $100 < t < 160$ s and $260 < t < 300$ s.

The results of the map estimation exploring the ESC algorithm are displayed in the upper panel of Fig. 7. Estimated maps are shown with three different perturbation frequencies ω, a perturbation amplitude $a_f = 20$ Hz, and a sine ramp duration of 30 min. Note that the map is normalized:

$$\Delta\hat{\tilde{T}}_s = \frac{\Delta\hat{T}_s}{\Delta\hat{T}_s(f = 200\,\text{Hz})}. \tag{14}$$

For all three settings, the location of the estimated optimum is near 700 Hz, which is significantly different from the apparent optimum in Fig. 5; this is likely due to dynamic effects. It must be pointed out that, taking the waiting time before and after the ramp into account, each of these experiments lasted around 40 min. For comparison, at the bottom of Fig. 7, the averaged and normalized Nusselt number $\tilde{Nu}_{av}$[2] of the impingement test rig is given. In general, the Nusselt number characterizes the relationship between convective and conductive heat transfer. The data for the Nusselt number determination is obtained with the same impingement parameters but using a TLC foil and a camera system for temperature measurement ([3] for more details). The determination of each individual data point lasted 30 min in the steady-state approach, that is, on the coarse measurement grid, the expenditure of time increased by a factor of almost 7.

Compared to the estimated optimal frequency obtained through ESC map estimation, the optimal frequency in the Nusselt number map is nearly the same. The minimum of the temperature difference $\Delta T_s(t)$ between the steel foil and the cooling air should always go hand in hand with the maximum of the Nusselt number.

[2]To obtain a normalization, the calculated Nusselt number is divided by the Nusselt number determined from an experiment with steady, non-pulsed impingement jets. In both the steady and the non-steady cases, identical average massflow rates are used.

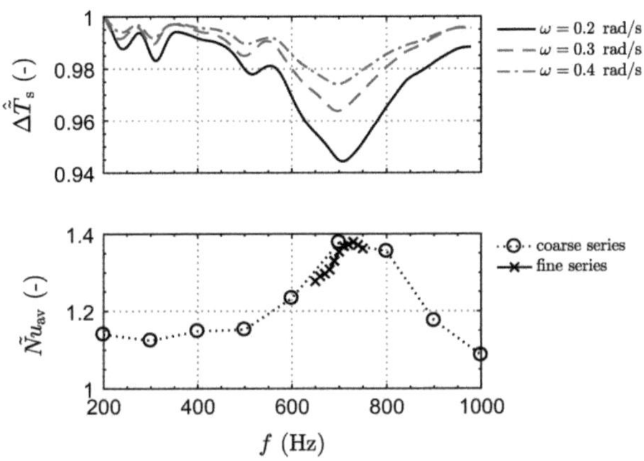

Fig. 7 Top: Estimated steady-state map for different perturbation frequencies. Bottom: Nusselt number of the heat transfer for impingement cooling calculated based on steady-state measurements

As crossing any extremum in the map results in a phase shift of 180°, such extrema with respect to the actuation frequency are likely to have rather high precision. However, $\Delta\hat{T}_s$ has to be obtained from the integration of a noisy gradient. Therefore, this estimate is deemed rather uncertain.

5 Conclusion

Adapting the ESC algorithm for open-loop fast map estimation results in a concept that delivers promising estimates of the locations of extrema on a steady-state map. The big advantage of this concept is the simple execution on the test rig and a robustness to a variation of the algorithm's parameters. Moreover, there are few requirements with respect to the knowledge about the system under study.

Three different perturbation frequencies were tested. In all cases, optima were found near an optimum that was confirmed by time-consuming steady-state measurements. An increase in the sine ramp duration from 30 to 45 min resulted in no significant differences in the found locations. In contrast, there is no indication that a sine ramp with a shorter duration than 30 min would result in significantly different locations. Hence, a ramp with shorter duration could further increase the time-saving potential. This will be part of future studies.

Even with a 30-min ramp, the saving of time is crucial compared to steady-state measurements using a TLC foil. While in steady-state measurements a measurement point needs around 30 min, it is possible to scan the whole frequency range with one sine ramp at the same time. As the ESC approach only yields an estimate, its results can be used in a second step to precisely characterize an optimum with the classical approach, as was done here with a fine steady-state measurement grid (see Fig. 7). Another promising idea is to extend the concept for a multidimensional parameter variation, as has already been done in terms of closed-loop ESC [1, 2, 12].

Acknowledgements The authors gratefully acknowledge the support of the Deutsche Forschungsgemeinschaft (DFG) as part of the Collaborative Research Center CRC 1029 "Substantial efficiency increase in gas turbines through direct use of coupled unsteady combustion and flow dynamics" in projects B03 and B06. Special thanks go to Joachim Kraatz and Klaus Noack who helped with the electrical setup for the measurements.

References

1. Ariyur, K.B., Krstic, M.: Analysis and design of multivariable extremum seeking. Am. Control Conf. (2002). https://doi.org/10.1109/ACC.2002.1025231
2. Ariyur, K.B., Krstic, M.: Real-Time Optimization by Extremum-Seeking Control. Wiley (2003)
3. Berthold, A., Haucke, F.: Experimental investigation of dynamically forced impingement cooling. ASME Turbo Expo (2017). https://doi.org/10.1115/GT2017-63140
4. Clarke, D.R., Oechsner, M., Padture, N.P.: Thermal-barrier coatings fore more efficient gas-turbine engines. MRS Bull. **37**(10), 891–898 (2012)
5. Haucke, F., Nitsche, W., Peitsch, D.: Enhanced convective heat transfer due to dynamically forced impingement jet array. ASME Turbo Expo (2016). https://doi.org/10.1115/GT2016-57360
6. Henning, L., Becker, R., Feuerbach, G., Muminovic, R., King, R., Brunn, A., Nitsche, W.: Extensions of adaptive slope-seeking for active flow control. Proc. Inst. Mech. Eng. Part I: J. Syst. Control Eng. **222**(5), 309–322 (2008)
7. Herwig, H., Middelberg, G.: The physics of unsteady jet impingement and its heat transfer performance. Acta Mech. **201**, 171–184 (2008)
8. Hofmann, H.M., Movileanu, D.L., Kind, M., Martin, H.: Influence of a pulsation on heat transfer and flow structure in submerged impinging jets. Int. J. Heat Mass Transf. **50**, 3638–3648 (2007)
9. Janetzke, T.: Experimental investigations of flow field and heat transfer characteristics due to periodically pulsating impinging air jets. J. Heat Mass Transf. **45**, 193–206 (1998)
10. Liu, T., Sullivan, J.P.: Heat transfer and flow structures in an excited circular impingement jet. J. Heat Mass Transf. **39**, 3695–3706 (1996)
11. Middelberg, G., Herwig, H.: Convective heat transfer under unsteady impinging jets: The effect of the shape of the unsteadiness. J. Heat Mass Transf. **45**, 1519–1532 (2009)
12. Rotea, M.A.: Analysis of multivariable extremum seeking algorithms. Am. Control Conf. (2000). https://doi.org/10.1109/ACC.2000.878937
13. Tan, Y., Moase, W.H., Manzie, C., Nešić, D., Mareels, I.M.Y.: Extremum seeking from 1922 to 2010. In: Chinese Control Conference (CCC), pp. 14–26 (2010)
14. Vejrazka, J., Tihon, J., Marty, Ph., Sobolik, V.: Effect of an external excitation on the flow structure in a circular impinging jet. Phys. Fluids **17**, 105102-01-14 (2005)

Author Index

A
Abel, Dirk, 167
Albin, Thivaharan, 167
Alvi, Farrukh S., 33
Anand, Vijay, 197
Andert, Jakob, 167
Antoulas, Athanasios C., 255
An, Xuanhong, 19
Arnold, Florian, 135

B
Bellenoue, Marc, 215
Berthold, Arne, 339, 367
Bettrich, Valentin, 53
Bilbow, William M., 33
Bitter, Martin, 53
Borello, D., 353
Boust, Bastien, 215

C
Camerlengo, G., 353

D
Djordjevic, Neda, 151
Dubljevic, Stevan S., 75

E
Edgington-Mitchell, D., 237
Engels, Thomas, 305

F
Fernandez, Erik, 33
Fietzke, Benjamin, 367

G
Gavin, Jennifer, 33
Gosea, Ion Victor, 255
Gray, Joshua A. T., 185, 237
Greenblatt, David, 105
Gutmark, Ephraim, 197

H
Haghdoost, M. Rezay, 237
Hanraths, Niclas, 185
Hasin, David, 105
Haucke, Frank, 339, 367
Heinkenschloss, Matthias, 255
He, Xiaowei, 19

I
Izadi, Mojtaba, 75

K
Keisar, David, 105
Kiesner, Matthias, 367
King, Rudibert, 135, 367
Klein, Rupert, 151, 237, 321
Koch, Charles R., 75
Krah, Philipp, 305

L
Le Provost, Mathieu, 19

R. King (ed.), *Active Flow and Combustion Control 2018*,
Notes on Numerical Fluid Mechanics and Multidisciplinary Design 141,
https://doi.org/10.1007/978-3-319-98177-2

M
Mehrmann, Volker, 271
Michael, Louisa, 289
Michalski, Quentin, 215
Mihalyovics, Jan, 91
Moeck, Jonas P., 185
Mutzel, Sophie, 305

N
Nadolski, M., 237
Niehuis, Reinhard, 53
Nikiforakis, Nikolaos, 289
Nuss, Eugen, 167

O
Oberleithner, K., 237

P
Paschereit, Christian O., 121, 185
Peitsch, Dieter, 91

R
Radespiel, Rolf, 3
Reiss, Julius, 271, 305
Ritter, Dennis, 167

S
Salvagni, A., 353
Schneider, Kai, 305
Schulze, Philipp, 271
Sellappan, Prabu, 33
Semaan, Richard, 3
Sesterhenn, J., 353
Sroka, Mario, 305
Staats, Marcel, 91

T
Tornow, Giordana, 135, 151, 321

V
Völzke, Fabian E. , 121, 185

W
Wick, Maximilian, 167
Williams, David R., 19

X
Xiang, Sun Lin, 33

Y
Yosef El Sayed, M., 3
Yücel, Fatma C., 121, 185

Z
Zander, Lisa, 151

MIX
Papier aus verantwortungsvollen Quellen
Paper from responsible sources
FSC® C105338
www.fsc.org

If you have any concerns about our products,
you can contact us on
ProductSafety@springernature.com

In case Publisher is established outside the EU,
the EU authorized representative is:
Springer Nature Customer Service Center GmbH
Europaplatz 3, 69115 Heidelberg, Germany

Printed by Libri Plureos GmbH
in Hamburg, Germany